STUDENT SOLUTIONS MANUAL

to the third editions of

CHEMISTRY

Bailar, Moeller, Kleinberg, Guss, Castellion, Metz

and

CHEMISTRY
with Inorganic Qualitative Analysis

Moeller, Bailar, Kleinberg, Guss, Castellion, Metz

John Williams
Miami University at Hamilton

Clyde Metz
College of Charleston

Harcourt Brace Jovanovich, Publishers
and its subsidiary, Academic Press

San Diego New York Chicago Austin Washington, D.C.
London Sydney Tokyo Toronto

Front Cover: Paul Silverman, Test Tube Abstract. © Fundamental Photographs.

ISBN: 0-15-506462-2

Printed in the United States of America

TABLE OF CONTENTS

CHAPTER 2

UNITS, MEASUREMENTS, AND NUMBERS

Solutions to Exercises

2.1 (a) 4 significant figures

1 2 3 4
234.7

(b) 3 significant figures

1 2 3
0.0300

(c) 2 (or more?) significant figures

1 2 ?
630

(d) 3 significant figures

1 2 3
700.

(e) 4 significant figures

1 2 3 4
1036

2.2 (a)

```
  16.3|
 -10.0|2
   6.2|8
  6.3
```

(b)

```
  6000.|
 -  32 |
  5968 |
```

(c) $\dfrac{16.3}{10.02} = [1.6267465] = 1.63$

(d) (6000.)(32) = [192,000] = 190,000

(e) $\dfrac{(0.0327)}{(0.0147)} = [2.2244898] = 2.22$

2.3 $x = \dfrac{(44.008)}{(0.0820568)(273.15)(1.976757)} - \dfrac{1}{1.3}$

$= [0.9932591 - 0.7692308] = 0.99326 - 0.77 = 0.22$

2.4 (a) $0.0203 = 2.03 \times 10^{-2}$ (→→)

(b) $5{,}260{,}000 = 5.26 \times 10^{6}$ (←←← ←←←)

(c) $0.0010 = 1.0 \times 10^{-3}$ (→→→)

(d) $6.3 \times 10^{-6} = 0.0000063$ (←←←←←←)

(e) $1.01 \times 10^{4} = 10{,}100$ (→ →→→)

(f) $3.0 \times 10^{-2} = 0.030$ (←←)

2.5 (a)

```
3.62 x 10^-3 = 36.2|  x 10^-4
2.68 x 10^-4 =  2.6|8 x 10^-4
               38.8|8 x 10^-4
               3.89 x 10^-3
```

(b) $(3.14 \times 10^{-4})(5.1 \times 10^{-5})$

$= [16.014 \times 10^{-9}]$

$= 1.6 \times 10^{-8}$

(c) $\dfrac{(4.58 \times 10^{4})}{(5.44 \times 10^{-14})} = [0.8419118 \times 10^{18}]$

$= 8.42 \times 10^{17}$

(d)

$$\begin{array}{lrr} 7.39 \times 10^{5} & = & 739,|000 \\ \underline{7.39 \times 10^{-5}} & = & \underline{0.0000739} \\ & & 739,|000.0000739 \\ & & 7.39 \times 10^{5} \end{array}$$

2.6 $T(°C) = (1\ C°/1.8\ F°)[T(°F) - 32]$

$= (1\ C°/1.8\ F°)(-15 - 32) = -26\ °C$

$T(K) = T(C°) + 273.15 = (-26\ °C) + 273.15 = 247\ K$

2.7 (a) $(265\ \mu L)\left[\dfrac{1\ L}{10^{6}\ \mu L}\right] = 2.65 \times 10^{-4}\ L$

(b) $(2.65 \times 10^{-4}\ L)\left[\dfrac{10^{3}\ mL}{1\ L}\right] = 0.265\ mL$

2.8 $(362\ g)\left[\dfrac{1\ oz}{28.35\ g}\right] = 12.8\ oz$

2.9 $(1\ h)\left[\dfrac{60\ min}{1\ h}\right]\left[\dfrac{60\ s}{1\ min}\right] = 3600\ s$

2.10 $(6.35\ g)\left[\dfrac{1\ cm^{3}}{11.3\ g}\right] = 0.562\ cm^{3}$

2.11 $(16.3\ mL)\left[\dfrac{86.2\ g}{100.\ mL}\right] = 14.1\ g$

2.12 $(\$8300)\left[\dfrac{1\ kg}{\$1300}\right]\left[\dfrac{1\ day}{5.4\ kg}\right] = 1.2\ days$

2.13 $(1\ light\ yr)\left[\dfrac{365\ days}{1\ yr}\right]\left[\dfrac{24\ h}{1\ day}\right]\left[\dfrac{60\ min}{1\ h}\right]\left[\dfrac{60\ s}{1\ min}\right]\left[\dfrac{3.00 \times 10^{8}\ m}{1\ s}\right] = 9.46 \times 10^{15}\ m$

Solutions to Odd-Numbered Questions and Problems

2.1 (a) ±1 (b) ±0.0001 (c) probably ±10

(d) ±0.001 (e) probably ±1000 (f) ±0.0001

(g) ±0.01 (h) probably ±10

2.3 (a) 3 significant figures

1 2 3
454

(b) 2 significant figures

1 2
2.2

(c) 4 significant figures

1 2 3 4
2.205

(d) 3 significant figures

1 2 3
0.0353

(e) 5 significant figures

1 2 3 4 5
1.0080

(f) 4 significant figures

1 2 3 4
14.00

(g) 3 (or more) significant figures

$$\overset{123?}{1030}$$

2.5 (a)
$$\begin{array}{r|l} 4 & 23.1 \\ & 0.256 \\ 1 & 00 \\ \hline 5 & 23.356 \end{array}$$
500 (or 520 or 523)

(b)
$$\begin{array}{r|l} 52.98 & 7 \\ 9.35 & 45 \\ 6.12 & \\ \hline 68.46 & 15 \end{array}$$
68.46

(c)
$$\begin{array}{r|l} 14.3 & 920 \\ -4.4 & \\ \hline 9.9 & 920 \end{array}$$
10.0

(d) (5183)(2.2) = [11402.6] = 11,000

(e) $\dfrac{14.000}{6.1}$ = [2.295082] = 2.3

(f) (6.11)(π) = [19.195131] = 19.2

(g) (14.3)(60) = [858] = 900 (or 860)

(h) $\dfrac{1020}{1.2}$ = 850

(i) $\dfrac{(3.2)(454)}{(8.6214)}$ = [168.51091] = 170

(j) $(4/3)(\pi)(2.16)^3$ = [42.213354] = 42.2 (Both 4 and 3 are exact.)

(k) (6.0 + 9.57 + 0.61)(1.113) = [18.00834] = 18.0

The number of significant figures in the final answer is limited by the addition step, which should be rounded to the tenths place giving 3 significant figures.

$$\begin{array}{r|l} 6.0 & \\ 9.5 & 7 \\ 0.6 & 1 \\ \hline 16.1 & 8 \end{array}$$

(l) (2.93)(14.7) + (1203)(0.0296) + (9.38)(5.2) = [127.4558] = 127

The final answer should not be expressed beyond the units place:

$$\begin{array}{r|l} 43. & 1 \leftarrow \text{3 significant figures} \\ 35. & 6 \leftarrow \text{3 significant figures} \\ 49 & \leftarrow \text{2 significant figures} \\ \hline 127. & 7 \leftarrow \text{round to units place} \end{array}$$

2.7 (a) $$\text{average} = \frac{(2.50 + 2.42 + 2.43 + 2.40 + 2.41)\ \text{cm}}{5} = 2.432\ \text{cm}$$

(b) 2.432 ± 0.07 cm includes all of the data

(c) $$\%\ \text{error} = \frac{2.432\ \text{cm} - 2.44\ \text{cm}}{2.44\ \text{cm}} \times 100 = -0.3\ \%$$

(The negative sign indicates that the experimental value is lower than the accepted true value.)

2.9 (a) $6500 = 6.5 \times 10^3$ (←←←)

(b) $0.0041 = 4.1 \times 10^{-3}$ (→→→)

(c) $0.003050 = 3.050 \times 10^{-3}$ (→→→)

(d) $810. = 8.10 \times 10^2$ (←←)

(e) $0.0000003 = 3 \times 10^{-7}$ (f) $9{,}352{,}000 = 9.352 \times 10^{6}$

(g) $42 \times 10^{3} = 4.2 \times 10^{4}$

2.11 (a) $5.26 \times 10^{3} = 5260$ (b) $4.10 \times 10^{-6} = 0.00000410$

(c) $5 \times 10^{5} = 500{,}000$ (d) $0.3 \times 10^{4} = 3000$

(e) $16.2 \times 10^{-3} = 0.0162$ (f) $9.346 \times 10^{3} = 9346$

2.13 (a)

$$\begin{array}{rcr} 5.29 \times 10^{3} & = & 5.29 \quad\; \times 10^{3} \\ -1.609 \times 10^{2} & = & -0.16|09 \times 10^{3} \\ \hline & & 5.12|91 \times 10^{3} \\ & & 5.13 \times 10^{3} \end{array}$$

(b) $(2.547 \times 10^{2})(3.2 \times 10^{-1}) = [8.1504 \times 10^{1}] = 8.2 \times 10^{1}$

(c) $(6.1 \times 10^{-2})(5.800 \times 10^{-6}) = [3.538 \times 10^{-7}] = 3.5 \times 10^{-7}$

(d) $$\frac{(3.261 \times 10^{-3}) + (2.58 \times 10^{4})}{1.2 \times 10^{-7}} = \frac{(0.0000003261 \times 10^{4}) + (2.58 \times 10^{4})}{1.2 \times 10^{-7}}$$

$$= \frac{2.58 \times 10^{4}}{1.2 \times 10^{-7}} = [2.15 \times 10^{11}] = 2.2 \times 10^{11}$$

2.15

$$d = \frac{\left[\begin{array}{l}(0.99983952) + (1.6945176 \times 10^{-2})(25.00) - (7.9870401 \times 10^{-6})(25.00)^{2} \\ \quad - (4.6170461 \times 10^{-8})(25.00)^{3} + (1.0556032 \times 10^{-10})(25.00)^{4} \\ \quad - (2.8054253 \times 10^{-13})(25.00)^{5}\end{array}\right]}{1.0000000000 + (1.6879850 \times 10^{-2})(25.00)}$$

$$= \frac{\left[\begin{array}{l}(0.99983952) + (0.4236) - (4.992 \times 10^{-3}) - (7.214 \times 10^{-4}) \\ \quad + (4.124 \times 10^{-5}) - (2.740 \times 10^{-6})\end{array}\right]}{1.0000000000 + 0.4220}$$

$$= \frac{1.4178}{1.4220} = 0.99704 \text{ g/cm}^{3}$$

2.17 The seven base physical quantities specified by the SI system are length, mass, time, electric current, temperature, luminous intensity, and amount of substance. All other physical quantities are derived from the seven base quantities.

2.19 (a) Mass is expressed in units of (ix) g, (xvi) kg, and (xvii) mg.

(b) Energy is expressed in units of (i) erg, (vi) J, and (xiii) cal.

(c) Length is expressed in terms of (ii) cm, (v) Å, (vii) km, (xi) nm, and (xiv) m.

(d) Volume is expressed in terms of (iii) cm^3, (viii) mL, (xii) L, and (xv) dm^3.

(e) Temperature is expressed in terms of (iv) K and (x) °C.

2.21 (a) cm = 10^{-2} m
pm = 10^{-12} m
km = 10^{3} m
largest is km

(b) MV = 10^{6} V
mV = 10^{-3} V
nV = 10^{-9} V
largest is MV

(c) fm = 10^{-15} m
mm = 10^{-3} m
pm = 10^{-12} m
nm = 10^{-9} m
largest is mm

(d) TJ = 10^{12} J
kJ = 10^{3} J
μJ = 10^{-6} J
largest is TJ

(e) cm^3 = 10^{-6} m^3
dm^3 = 10^{-3} m^3
km^3 = 10^{9} m^3
largest is km^3

2.23 T(°C) = (1 C°/1.8 F°)[T(°F) - 32]

(a) T(°C) = (1 C°/1.8 F°)(98.6 - 32) = 37.0 °C
(b) T(°C) = (1 C°/1.8 F°)(-10. - 32) = -23 °C
(c) T(°C) = (1 C°/1.8 F°)(78 - 32) = 26 °C
(d) T(°C) = (1 C°/1.8 F°)(250 - 32) = 120 °C
(e) T(°C) = (1 C°/1.8 F°)(32.00 - 32) = 0.00 °C

2.25 T(K) = T(°C) + 273.15

(a)
37.0|
273.1|5
310.1|5
310.2 K

(b)
-23 |
273.|15
250.|15
250. K

(c)
26|
273|.15
299|.15
299 K

(d)
12|0
27|3.15
39|3.15
390 K

(e)
0.00|
273.15|
273.15|
273.15 K

2.27 T(°F) = (1.8 F°/1 C°)T(°C) + 32

(a) T(°F) = (1.8 F°/1 C°)(100.00 °C) + 32 = 212.00 °F
(b) T(°F) = (1.8 F°/1 C°)(-151.8 °C) + 32 = -241.2 °F
(c) T(°F) = (1.8 F°/1 C°)(444.7 °C) + 32 = 832.5 °F
(d) T(°F) = (1.8 F°/1 C°)(2750 °C) + 32 = 4980 °F
(e) T(°F) = (1.8 F°/1 C°)(338 °C) + 32 = 640. °F

2.29 The two conversion factors are (1 atm/760 Torr) and (760 Torr/1 atm).

2.31 $$\left[\frac{6\ \text{ft}}{1\ \text{fathom}}\right]\left[\frac{12\ \text{in}}{1\ \text{ft}}\right]\left[\frac{2.54\ \text{cm}}{1\ \text{in}}\right]\left[\frac{1\ \text{m}}{100\ \text{cm}}\right] = \frac{1.83\ \text{m}}{1\ \text{fathom}}$$

(a) $$(2.00\ \text{fathom})\left[\frac{1.83\ \text{m}}{1\ \text{fathom}}\right] = 3.66\ \text{m}$$

(b) $(20{,}000\ \text{league})\left[\frac{3040\ \text{fathom}}{1\ \text{league}}\right]\left[\frac{1.83\ \text{m}}{1\ \text{fathom}}\right] = 1 \times 10^{8}\ \text{m}$

2.33 (a) $(10.3\ \text{Å})\left[\frac{1\ \text{nm}}{10\ \text{Å}}\right] = 1.03\ \text{nm}$

(b) $(635\ \text{cal})\left[\frac{4.184\ \text{J}}{1\ \text{cal}}\right] = 2660\ \text{J}$

(c) $(14.6\ \text{L})\left[\frac{1\ \text{dm}^3}{1\ \text{L}}\right] = 14.6\ \text{dm}^3$

(d) $(14.6\ \text{kg})\left[\frac{10^3\ \text{g}}{1\ \text{kg}}\right] = 1.46 \times 10^{4}\ \text{g}$

(e) $(14.6\ \text{atm})\left[\frac{101{,}325\ \text{Pa}}{1\ \text{atm}}\right] = 1.48 \times 10^{6}\ \text{Pa}$

(f) $(1.2\ \text{eV})\left[\frac{1.602 \times 10^{-19}\ \text{J}}{1\ \text{eV}}\right] = 1.9 \times 10^{-19}\ \text{J}$

(g) $(735.2\ \text{Torr})\left[\frac{1\ \text{atm}}{760\ \text{Torr}}\right] = 0.9674\ \text{atm}$

(h) $(21.65\ \text{mL})\left[\frac{1\ \text{cm}^3}{1\ \text{mL}}\right] = 21.65\ \text{cm}^3$

2.35 (a) $(25\ \text{m})\left[\frac{100\ \text{cm}}{1\ \text{m}}\right]\left[\frac{1\ \text{in}}{2.54\ \text{cm}}\right]\left[\frac{1\ \text{yd}}{36\ \text{in}}\right] = 27\ \text{yd}$

(b) $\left[\frac{13.7\ \text{g}}{1\ \text{mL}}\right]\left[\frac{1\ \text{kg}}{1000\ \text{g}}\right]\left[\frac{1\ \text{mL}}{1\ \text{cm}^3}\right]\left[\frac{100\ \text{cm}}{1\ \text{m}}\right]^3 = 13{,}700\ \text{kg/m}^3$

(c) $(3.2\ \text{Torr})\left[\frac{1\ \text{atm}}{760\ \text{Torr}}\right]\left[\frac{101{,}325\ \text{Pa}}{1\ \text{atm}}\right] = 430\ \text{Pa}$

(d) $(4.6\ \text{atm})\left[\frac{101{,}325\ \text{Pa}}{1\ \text{atm}}\right]\left[\frac{1\ \text{bar}}{10^5\ \text{Pa}}\right] = 4.7\ \text{bar}$

(e) $(14.6\ \text{lb})\left[\frac{454\ \text{g}}{1\ \text{lb}}\right]\left[\frac{1\ \text{kg}}{1000\ \text{g}}\right] = 6.63\ \text{kg}$

2.37 (a) $\left[\frac{8.314\ \text{J}}{\text{K mol}}\right]\left[\frac{1\ \text{erg}}{10^{-7}\ \text{J}}\right] = 8.314 \times 10^{7}\ \text{erg/K mol}$

(b) $\left[\frac{8.314\ \text{J}}{\text{K mol}}\right]\left[\frac{1\ \text{cal}}{4.184\ \text{J}}\right] = 1.987\ \text{cal/K mol}$

(c) $\left[\frac{8.314\ \text{J}}{\text{K mol}}\right]\left[\frac{1\ \text{L atm}}{101.325\ \text{J}}\right] = 0.08205\ \text{L atm/K mol}$

2.39 $(5 \times 10^{21}\ \text{molecules})\left[\frac{12\ \text{carbon atoms}}{1\ \text{molecule}}\right] = 6 \times 10^{22}\ \text{carbon atoms}$

2.41 $(1\text{ week})\left[\frac{7\text{ days}}{1\text{ week}}\right]\left[\frac{5.0\text{ ton gas}}{1\text{ day}}\right]\left[\frac{5\text{ ton sulfur dioxide}}{100\text{ ton gas}}\right] = 2\text{ ton sulfur dioxide}$

2.43 $\left[\frac{0.58\text{ Å}}{1.5\times 10^{8}\text{ km}}\right]\left[\frac{1\text{ nm}}{10\text{ Å}}\right]\left[\frac{1\text{ m}}{10^{9}\text{ nm}}\right]\left[\frac{1\text{ km}}{10^{3}\text{ m}}\right] = 3.9\times 10^{-22}$

2.45 (a) 16 | 273 | .15

3 significant figures needed

(b) 0.094 | 273.150 |

6 significant figures needed

(c) 103.7 | 273.1 | 5

4 significant figures needed

(d) 729.65 | −273.15 |

5 significant figures needed

(e) (2.303)(273)(5.26)

3 significant figures needed

(f) $\frac{1}{305} - \frac{1}{273}$

3 significant figures needed

2.47 $(1.0000\text{ L})\left[\frac{10^{3}\text{ mL}}{1\text{ L}}\right]\left[\frac{1\text{ cm}^{3}}{1\text{ mL}}\right]\left[\frac{1.0056\text{ g}}{1\text{ cm}}\right] = 1.0056\times 10^{3}\text{ g}$

2.49 volume = (1.11 mm)(1.09 mm)(1.12 mm) = 1.36 mm^3

$$\text{density} = \left[\frac{2.236\text{ mg}}{1.36\text{ mm}^{3}}\right]\left[\frac{1\text{ g}}{10^{3}\text{ mg}}\right]\left[\frac{10\text{ mm}}{1\text{ cm}}\right]^{3} = 1.64\text{ g/cm}^{3}$$

2.51 mass of water = 93.34 g − 68.31 g = 25.03 g

$$\text{volume of water} = (25.03\text{ g})\left[\frac{1\text{ cm}^{3}}{1.0000\text{ g}}\right] = 25.03\text{ cm}^{3} = \text{volume of the container}$$

mass of unknown liquid = 88.42 g − 68.31 g = 20.11 g

$$\text{density of unknown liquid} = \frac{20.11\text{ g}}{25.03\text{ cm}^{3}} = 0.8034\text{ g/cm}^{3}$$

2.53 1. Study the problem and be sure you understand it.

(a) What is unknown?

The number of diskettes that is equivalent to a fixed disk.

(b) What is known?

The number of tracks, the number of sectors per track, and the the number of bytes per sector for the diskette and the fixed disk.

2. Decide how to solve the problem.

(a) What is the connection between the known and the unknown?

The connection is provided by conversion factors based on the known relationship between bytes, sectors, and tracks for the diskette

(1 diskette = 80 tracks, 1 track = 9 sectors, 1 sector = 512 bytes) and for the fixed disk (1 fixed disk = 1227 tracks, 1 track = 17 sectors, 1 sector = 512 bytes).

(b) What is necessary to make the connection?

Set up a decimal calculation that includes conversion factors that allow conversion from fixed disk to diskettes.

3. Set up the problem and solve it.

$$(1 \text{ fixed disk})\left[\frac{1227 \text{ track}}{1 \text{ fixed disk}}\right]\left[\frac{17 \text{ sector}}{1 \text{ track}}\right]\left[\frac{512 \text{ byte}}{1 \text{ sector}}\right]\left[\frac{1 \text{ sector}}{512 \text{ byte}}\right] \times \left[\frac{1 \text{ track}}{9 \text{ sector}}\right]\left[\frac{1 \text{ diskette}}{80 \text{ track}}\right]$$

$$= [28.9708333\ldots] \text{ diskettes}$$

$$= \text{nearly } 29 \text{ diskettes}$$

4. Check the result.

(a) Are significant figures and the location of the decimal point correct?

Yes. Note: all numbers are exact, but it is not necessary to include all significant figures because the question asked "approximately how many".

(b) Did the answer come out in the correct units?

Yes.

(c) Is the answer reasonable?

Yes. The answer is reasonable, a fixed disk is designed to hold the information stored on many diskettes.

CHAPTER 3

CHEMISTRY: THE SCIENCE OF MATTER

Solutions to Exercises

3.1 $Z = 11$; thus, 11 protons

$N = A - Z = 23 - 11 = 12$; thus 12 neutrons

This composition of the nucleus is no different from that of the ion.

3.2 $$(3.3198 \times 10^{-23}\ \text{g})\left[\frac{1\ \text{u}}{1.66057 \times 10^{-24}\ \text{g}}\right] = 19.992\ \text{u}$$

3.3 atomic mass = (0.9092)(19.992 u) + (0.00257)(20.994 u) + (0.0882)(21.991 u)

= 18.18 u + 0.0540 u + 1.94 u

= 20.17 u

Solutions to Odd-Numbered Questions and Problems

3.1 The respective symbols and names of the elements are:

(a) Be beryllium (b) B boron (c) V vanadium

(d) As arsenic (e) Ba barium

3.3 The respective names and symbols of the elements are:

(a) xenon Xe (b) nickel Ni (c) magnesium Mg

(d) cobalt Co (e) silicon Si (f) lead Pb

(g) potassium K (h) silver Ag (i) fluorine F

(j) radon Rn

3.5 (a) A drop of mercury is one phase and is further classified as matter, pure substance, element.

(b) A mound of sugar crystals is one phase and is further classified as matter, pure substance, compound.

(c) An ice cube is one phase and is further classified as matter, pure substance, compound.

(d) A melting ice cube consists of two phases and is further classified as matter, heterogeneous mixture.

(e) A puddle of water is one phase and is further classified as matter, pure substance, compound.

(f) A scoop of "smooth" peanut butter is one phase and is further classified as matter, homogeneous mixture.

(g) A scoop of "crunchy" peanut butter consists of two phases and is further classified as matter, heterogeneous mixture.

3.7 The properties that are chemical properties include (i) noncombustible, (j) forms $[I_3]^-$ ions in aqueous solutions, and (k) poisonous.

3.9 See Figure 3.5 of the text. No, cathode rays are always the same whatever the identity of the gas in the tube. Cathode rays must be fundamental to all matter.

3.11 The charge-to-mass ratio of an electron was determined by simultaneously imposing perpendicular magnetic and electric fields upon a beam of cathode rays in an arrangement such that the beam continued in the same direction as the initial direction. The *e/m* value was calculated from the field strengths. The mass value proved that the electron was a subatomic particle.

3.13 There would be little or no deflection of any of the α-particles in the beam. The α-particle scattering experiments demonstrated that each atom has a small, dense central core in which virtually all the mass and positive charge of the atom are concentrated.

3.15 $V_{atom} = (4/3)(\pi)(5.29 \times 10^{-11}\ m)^3 = 6.20 \times 10^{-31}\ m^3$

$V_{proton} = (4/3)(\pi)(1.5 \times 10^{-15}\ m)^3 = 1.4 \times 10^{-44}\ m^3$

The fraction of space occupied by the nucleus is

$$\text{fraction occupied} = \frac{V_{proton}}{V_{atom}} = \frac{1.4 \times 10^{-44}\ m^3}{6.20 \times 10^{-31}\ m^3} = 2.3 \times 10^{-14}$$

3.17 (a) The denominator is larger giving a smaller value for the force.

(b) The numerator is larger giving a larger value for the force.

(c) The numerator is larger giving a larger value for the force.

(d) The denominator is 4 times larger and the numerator is 4 times larger giving no net change in the strength of the force.

3.19 An isotope of an element has a certain number of protons and neutrons. Different isotopes of an element have the same number of protons, but different numbers of neutrons. The various hydrogen isotopes are named protium ("regular" hydrogen with no neutrons), deuterium (with 1 neutron), and tritium (with 2 neutrons).

3.21 The number of protons in an atom is equal to the atomic number of that atom. The number of neutrons in an atom is found by subtracting the atomic number from the mass number ($N = A - Z$).

(a) ^{100}Rh has 45 protons and 100 - 45 = 55 neutrons.
(b) ^{146}Nd has 60 protons and 146 - 60 = 86 neutrons.
(c) ^{79}Br has 35 protons and 79 - 35 = 44 neutrons.
(d) ^{7}Li has 3 protons and 7 - 3 = 4 neutrons.
(e) ^{159}Tb has 65 protons and 159 - 65 = 94 neutrons.

3.23

Symbol	Z	N	A	Number of Electrons	Electrical Charge
$^{29}_{14}Si$	14	15	29	14	0
$^{34}_{16}S^{2-}$	16	18	34	18	-2
$^{56}_{26}Fe^{2+}$	26	30	56	24	+2
$^{188}_{79}Au^{3+}$	79	109	188	76	+3

3.25 (a) Isotopes of the same element (N): (i) ^{12}N, (iii) ^{13}N, (v) ^{14}N, (vi) ^{15}N, (vii) ^{16}N, (ix) ^{17}N.

(b) Same number of neutrons (8): (ii) ^{13}B, (iv) ^{14}C, (vi) ^{15}N, (viii) ^{16}O, (x) ^{17}F, (xi) ^{18}Ne.

(c) Same mass number (13): (ii) ^{13}B, (iii) ^{13}N
(14): (iv) ^{14}C, (v) ^{14}N
(16): (vii) ^{16}N, (viii) ^{16}O
(17): (ix) ^{17}N, (x) ^{17}F

3.27 (a) $$(77.9204\ u)\left[\frac{1.6605655 \times 10^{-24}\ g}{1\ u}\right] = 1.29392 \times 10^{-22}\ g$$

(b) $$(79.9164\ u)\left[\frac{1.6605655 \times 10^{-24}\ g}{1\ u}\right] = 1.32706 \times 10^{-22}\ g$$

(c) $(81.9135\ u)\left[\frac{1.6605655 \times 10^{-24}\ g}{1\ u}\right] = 1.36023 \times 10^{-22}\ g$

3.29 (a) $(2.506119 \times 10^{-22}\ g)\left[\frac{1\ u}{1.6605655 \times 10^{-24}\ g}\right] = 150.9196\ u$

(b) $(2.539352 \times 10^{-22}\ g)\left[\frac{1\ u}{1.6605655 \times 10^{-24}\ g}\right] = 152.9209\ u$

3.31 (a) $(55.93904\ u)\left[\frac{1.6605655 \times 10^{-24}\ g}{1\ u}\right] = 9.289044 \times 10^{-23}\ g$

(b) $(55.9349\ u)\left[\frac{1.6605655 \times 10^{-24}\ g}{1\ u}\right] = 9.28836 \times 10^{-23}\ g$

(c) $(55.94002\ u)\left[\frac{1.6605655 \times 10^{-24}\ g}{1\ u}\right] = 9.289207 \times 10^{-23}\ g$

The masses differ because each isotope contains a different number of protons and neutrons--the mass of a proton is slightly different than the mass of a neutron. Each isotope also has a different binding energy, so the amounts of mass converted to energy differ.

3.33 atomic mass = (0.7899)(23.98504 u) + (0.1000)(24.98584 u) + (0.1101)(25.98259 u)

= 24.31 u

3.35 Let x = fraction of ^{63}Cu and $1 - x$ = fraction of ^{65}Cu.

$$(x)(62.9298\ u) + (1 - x)(64.9278\ u) = (63.546\ u)$$

$$(62.9298)x - (64.9278)x = (63.546) - (64.9278)$$

$$-(1.998)x = -1.382$$

$$x = 0.6917$$

$$1 - x = 0.3083$$

69.17% ^{63}Cu and 30.83% ^{65}Cu

3.37 (a) The four molecules would $^{1}H^{35}Cl$, $^{1}H^{37}Cl$, $^{2}H^{35}Cl$, and $^{2}H^{37}Cl$. (b) The respective masses are 36 u, 38 u, 37 u, and 39 u. (c) $^{1}H^{35}Cl > {}^{1}H^{37}Cl > {}^{2}H^{35}Cl > {}^{2}H^{37}Cl$.

CHAPTER 4

ATOMS, MOLECULES, AND IONS

Solutions to Exercises

4.1 The formulas of the six compounds are NH_4NO_3, $(NH_4)_2SO_4$, $Ca(NO_3)_2$, $CaSO_4$, $Fe(NO_3)_3$, and $Fe_2(SO_4)_3$.

4.2

	number of atoms		atomic mass (u/atom)		mass (u)
K	1	x	39.10	=	39.10
B	1	x	10.81	=	10.81
H	4	x	1.01	=	4.04

molecular mass of KBH_4 = 53.95 u

4.3 (a) The name of the Cr^{3+} cation is chromium(III) ion and for the Sc^{3+} cation the name is scandium ion [or scandium(III) ion].

(b) The symbol for the cesium ion is Cs^+ and for the lead ion the symbol is Pb^{2+}.

4.4 (a) The name of the F^- anion is the fluoride ion.

(b) The symbol for the iodide ion is I^- and the symbol for the sulfide ion is S^{2-}.

4.5 (a) The formula for potassium nitride is K_3N and the formula for strontium carbonate is $SrCO_3$.

(b) The formula $Na(CH_3COO)$ represents sodium acetate and $NiCl_2$ represents nickel(II) chloride.

4.6 (a) The formula for dialuminum hexachloride is Al_2Cl_6 and the formula for iodine trifluoride is IF_3.

(b) The name of IF_7 is iodine heptafluoride and the name of P_4O_{10} is tetraphosphorus decoxide.

4.7 $PBr_5(s) + 4H_2O(l) \rightarrow H_3PO_4(aq) + 5HBr(aq)$

4.8 $(7.8 \times 10^{24} \text{ formula units CsBr})\left[\dfrac{1 \text{ mol CsBr}}{6.022 \times 10^{23} \text{ formula units CsBr}}\right]$

$= 13 \text{ mol CsBr}$

4.9

	moles of atoms		molar mass (g/mol)		mass (g)
C	7	x	12.01	=	84.07
H	6	x	1.01	=	6.06
O	2	x	16.00	=	32.00

molar mass benzoic acid = 122.13 g

4.10 $(34 \text{ kg sucrose})\left[\dfrac{1000 \text{ g sucrose}}{1 \text{ kg sucrose}}\right]\left[\dfrac{1 \text{ mol sucrose}}{342.34 \text{ g sucrose}}\right] = 99 \text{ mol } C_{12}H_{22}O_{11}$

4.12 $(2.1 \text{ g FeO})\left[\dfrac{1 \text{ mol FeO}}{71.85 \text{ g FeO}}\right] = 0.029 \text{ mol FeO}$

$(2.9 \text{ g } Fe_2O_3)\left[\dfrac{1 \text{ mol } Fe_2O_3}{159.70 \text{ g } Fe_2O_3}\right] = 0.018 \text{ mol } Fe_2O_3$

The number of moles of FeO is greater than the number of moles of Fe_2O_3.

4.13 $(14.6 \text{ g } CaCO_3)\left[\dfrac{1 \text{ mol } CaCO_3}{100.09 \text{ g } CaCO_3}\right]\left[\dfrac{1 \text{ mol C}}{1 \text{ mol } CaCO_3}\right]\left[\dfrac{6.022 \times 10^{23} \text{ C atoms}}{1 \text{ mol C}}\right]$

$= 8.78 \times 10^{22} \text{ C atoms}$

4.14 $(9.68 \text{ g } Ca(NO_3)_2)\left[\dfrac{1 \text{ mol } Ca(NO_3)_2}{164.10 \text{ g } Ca(NO_3)_2}\right] = 0.0590 \text{ mol } Ca(NO_3)_2$

$\dfrac{(0.0590 \text{ mol } Ca(NO_3)_2)}{(250.0 \text{ mL})(1 \text{ L}/1000 \text{ mL})} = 0.236 \text{ mol } Ca(NO_3)_2/\text{L} = 0.236 \text{ M}$

4.15 $(10.5 \text{ mL})\left[\dfrac{1 \text{ L}}{1000 \text{ mL}}\right]\left[\dfrac{6.0 \text{ mol NaOH}}{1 \text{ L}}\right] = 0.063 \text{ mol NaOH}$

4.16 $(3.0 \text{ mmol HCl})\left[\dfrac{1 \text{ mol}}{1000 \text{ mmol}}\right]\left[\dfrac{1 \text{ L}}{6.0 \text{ mol HCl}}\right]\left[\dfrac{1000 \text{ mL}}{1 \text{ L}}\right] = 0.50 \text{ mL}$

4.17

	moles of atoms		molar mass (g/mol)		mass (g)
Cl	5	x	35.45	=	177.3
C	14	x	12.01	=	168.1
H	9	x	1.01	=	9.09

molar mass of DDT = 354.5

$\% \text{ Cl} = \dfrac{177.3 \text{ g}}{354.5 \text{ g}} \times 100 = 50.01 \%$

$\% \text{ C} = \dfrac{168.1 \text{ g}}{354.5 \text{ g}} \times 100 = 47.42 \%$

$\% \text{ H} = \dfrac{9.09 \text{ g}}{354.5 \text{ g}} \times 100 = 2.56 \%$

4.18 mass of Cl = 13.73 g - 3.10 g = 10.63 g Cl

$$\% \text{ P} = \frac{3.10 \text{ g}}{13.73 \text{ g}} \times 100 = 22.6 \%$$

$$\% \text{ Cl} = \frac{10.63 \text{ g}}{13.73 \text{ g}} \times 100 = 77.42 \%$$

4.19

$$(3.10 \text{ g P})\left[\frac{1 \text{ mol P}}{30.97 \text{ g P}}\right] = 0.100 \text{ mol P}$$

$$(10.64 \text{ g Cl})\left[\frac{1 \text{ mol Cl}}{35.45 \text{ g Cl}}\right] = 0.3001 \text{ mol Cl}$$

$$\frac{n_P}{n_P} = \frac{0.100}{0.100} = 1.00 \qquad \frac{n_{Cl}}{n_P} = \frac{0.3001}{0.100} = 3.00$$

Empirical formula is PCl_3.

4.20 Assume exactly 100 g sodium thiosulfate.

	Na	S	O
mass	29.1 g	40.6 g	30.4 g
molar mass	22.99 g/mol	32.06 g/mol	16.00 g/mol
no. of moles	$(29.1 \text{ g})\left[\frac{1 \text{ mol}}{22.99 \text{ g}}\right]$ = 1.27 mol	$(40.6 \text{ g})\left[\frac{1 \text{ mol}}{32.06 \text{ g}}\right]$ = 1.27 mol	$(30.4 \text{ g})\left[\frac{1 \text{ mol}}{16.00 \text{ g}}\right]$ = 1.90 mol
mole ratio n/n_{Na}	$\frac{1.27}{1.27} = 1.00$	$\frac{1.27}{1.27} = 1.00$	$\frac{1.90}{1.27} = 1.50$
rel. moles of atoms	2	2	3

Empirical formula is $Na_2S_2O_3$.

4.21 empirical formula molar mass = (3)(10.81 g) + (5)(1.01 g) = 37.48 g

$$\frac{115 \text{ g/mol}}{37.48 \text{ g/mol}} = 3.07$$

Molecular formula is B_9H_{15}.

Solutions to Odd-Numbered Questions and Problems

4.1 (a) iii, (b) ii, (c) iii, (d) ii, (e) iii, (f) i

4.3 (a) ii, (b) iii, (c) i, (d) ii, (e) iii, (f) iii

4.5 (a) i, (b) ii, (c) i, (d) ii, (e) i, (f) i

4.7 (a) $Ca(ClO_3)_2$, (b) $Al_2(SO_4)_3$, (c) $AuBr_3$, (d) NH_4CN, (e) K_3PO_4, (f) $Na_2Cr_2O_7$, (g) $Ca(MnO_4)_2$

4.9 Columns: number of atoms, atomic mass (u/atom), mass (u)

(a) Cl_2

	number of atoms		atomic mass (u/atom)		mass (u)
Cl	2	x	35.45	=	70.90
			molecular mass Cl_2	=	70.90 u

(b) Fe

	number of atoms		atomic mass (u/atom)		mass (u)
Fe	1	x	55.85	=	55.85
			atomic mass Fe	=	55.85 u

(c) Fe^{3+}

	number of atoms		atomic mass (u/atom)		mass (u)
Fe^{3+}	1	x	55.85	=	55.85
			atomic mass of Fe^{3+}	=	55.85 u

(d) $C_{12}H_{22}O_{11}$

	number of atoms		atomic mass (u/atom)		mass (u)
C	12	x	12.01	=	144.1
H	22	x	1.01	=	22.2
O	11	x	16.00	=	176.0
			molecular mass $C_{12}H_{22}O_{11}$	=	342.3 u

(e) $KClO_3$

	number of atoms		atomic mass (u/atom)		mass (u)
K	1	x	39.10	=	39.10
Cl	1	x	35.45	=	35.45
O	3	x	16.00	=	48.00
			molecular mass $KClO_3$	=	122.55 u

(f) $CoWO_4$

	number of atoms		atomic mass (u/atom)		mass (u)
Co	1	x	58.93	=	58.93
W	1	x	183.85	=	183.85
O	4	x	16.00	=	64.00
			molecular mass $CoWO_4$	=	306.78 u

(g) $Pt_2(CO)_3Cl_4$

	number of atoms		atomic mass (u/atom)		mass (u)
Pt	2	x	195.09	=	390.18
C	3	x	12.01	=	36.03
O	3	x	16.00	=	48.00
Cl	4	x	35.45	=	141.8
			molecular mass $Pt_2(CO)_3Cl_4$	=	616.0 u

(h) $Cu_2(CO_3)(OH)_2$

	number of atoms		atomic mass (u/atom)		mass (u)
Cu	2	x	63.55	=	127.1
C	1	x	12.01	=	12.01
O	5	x	16.00	=	80.00
H	2	x	1.01	=	2.02
			molecular mass $Cu_2(CO_3)(OH)_2$	=	221.1 u

(i) H_3PO_4

	number of atoms		atomic mass (u/atom)		mass (u)
H	3	x	1.01	=	3.03
P	1	x	30.97	=	30.97
O	4	x	16.00	=	64.00
			molecular mass H_3PO_4	=	98.00 u

(j) $(NH_4)_3AsO_4$

	number of atoms		atomic mass (u/atom)		mass (u)
N	3	x	14.01	=	42.03
H	12	x	1.01	=	12.1
As	1	x	74.92	=	74.92
O	4	x	16.00	=	64.00
			molecular mass $(NH_4)_3AsO_4$	=	193.1 u

4.11 The respective symbols and names are

(a) Li^+, lithium ion
(b) Cd^{2+}, cadmium ion,
(c) Fe^{2+}, iron(II) ion
(d) Mn^{2+}, manganese(II) ion,
(e) Al^{3+}, aluminum ion.

4.13 The respective names and symbols are

(a) sodium ion, Na^+
(b) zinc ion, Zn^{2+}
(c) silver ion, Ag^+
(d) mercury(II) ion, Hg^{2+}
(e) iron(III) ion, Fe^{3+}

4.15 The respective symbols and names are

(a) N^{3-}, nitride ion
(b) O^{2-}, oxide ion
(c) Se^{2-}, selenide ion
(d) F^-, fluoride ion
(e) Br^-, bromide ion

4.17 The respective formulas and names are

(a) LiS, lithium sulfide
(b) SnO_2, tin(IV) oxide
(c) RbI, rubidium iodide
(d) Li_2O, lithium oxide
(e) UO_2, uranium(IV) oxide
(f) Ba_3N_2, barium nitride
(g) NaF, sodium fluoride

4.19 The respective names and formulas are

(a) sodium fluoride, NaF
(b) zinc oxide, ZnO
(c) barium peroxide, BaO_2
(d) magnesium bromide, $MgBr_2$
(e) hydrogen iodide, HI
(f) copper(I) chloride, CuCl
(g) potassium iodide, KI

4.21 The respective formulas and names are

(a) $(NH_4)_2SO_4$, ammonium sulfate
(b) $K_2Cr_2O_7$, potassium dichromate
(c) $Fe(ClO_4)_2$, iron(II) perchlorate
(d) $CaCO_3$, calcium carbonate
(e) $NaNO_2$, sodium nitrite
(f) K_2CrO_4, potassium chromate
(g) Na_2SO_3, sodium sulfite

4.23 The respective names and formulas are

(a) potassium sulfite, K_2SO_3
(b) calcium permanganate, $Ca(MnO_4)_2$
(c) barium phosphate, $Ba_3(PO_4)_2$
(d) copper(I) sulfate, Cu_2SO_4
(e) ammonium acetate, $NH_4(CH_3COO)$
(f) silver nitrate, $AgNO_3$
(g) uranium(IV) sulfate, $U(SO_4)_2$

4.25 The respective formulas and names are

(a) HCl, hydrochloric acid
(b) H_3PO_4, phosphoric acid
(c) $HClO_4$, perchloric acid
(d) HNO_3, nitric acid
(e) H_2SO_3, sulfurous acid
(f) H_3PO_3, phosphorous acid

4.27 The name of the acid with the formula H_2CO_3 is carbonic acid. The two anions derived from it are HCO_3^-, the hydrogen carbonate ion, and CO_3^{2-}, the carbonate ion.

4.29 The respective formulas and names are

(a) CO, carbon monoxide
(b) CO_2, carbon dioxide

(c) SF_6, sulfur hexafluoride (d) $SiCl_4$, silicon tetrachloride
(e) IF, iodine monofluoride

4.31 The respective names and formulas are
(a) diboron trioxide, B_2O_3 (b) silicon dioxide, SiO_2
(c) phophorus trichloride, PCl_3 (d) sulfur tetrachloride, SCl_4
(e) bromine trifluoride, BrF_3 (f) hydrogen telluride, H_2Te
(g) diphosphorus trioxide, P_2O_3

4.33 (a) The reactant is N_2O.
(b) The products are N_2 and O_2.
(c) The physical state of each substance is a gas.
(d) The capital Greek letter delta, Δ, represents that heat is necessary to make the reaction take place.

4.35 (a) Gaseous nitrogen reacts with gaseous hydrogen at 400 °C and 250 atm pressure in the presence of FeO as a catalyst to produce gaseous ammonia.
(b) When gaseous carbon monoxide and gaseous oxygen are heated, gaseous carbon dioxide is formed.
(c) Heating solid silicon dioxide and carbon at 3000 °C gives liquid silicon and gaseous carbon monoxide.

4.37 (a) $Cl_2O_7(g) + H_2O(l) \rightarrow 2HClO_4(aq)$
(b) $Br_2(l) + H_2O(l) \rightarrow HBr(aq) + HBrO(aq)$
(c) $Ca_3(PO_4)_2(s) + 3H_2SO_4(aq) \rightarrow 3CaSO_4(s) + 2H_3PO_4(aq)$
(d) $V_2O_4(s) + 4HClO_4(aq) \rightarrow 2VO(ClO_4)_2(aq) + 2H_2O(l)$
(e) $Fe_2O_3(s) + 3H_2(g) \rightarrow 2Fe(s) + 3H_2O(l)$
(f) $2K(s) + 2H_2O(l) \rightarrow 2KOH(aq) + H_2(g)$
(g) $MgCO_3(s) \xrightarrow{\Delta} MgO(s) + CO_2(g)$
(h) $Al_2S_3(s) + 6H_2O(l) \rightarrow 2Al(OH)_3(s) + 3H_2S(g)$
(i) $BaO_2(s) + H_2SO_4(aq) \rightarrow H_2O_2(aq) + BaSO_4(s)$

4.39 (a) $$(9.5 \times 10^{21} \text{ Cs atoms})\left[\frac{1 \text{ mol Cs}}{6.022 \times 10^{23} \text{ Cs atoms}}\right] = 1.6 \times 10^{-2} \text{ mol Cs}$$

(b) $$(4.7 \times 10^{27} \text{ molecules } CO_2)\left[\frac{1 \text{ mol } CO_2}{6.022 \times 10^{23} \text{ molecules } CO_2}\right]$$

$$= 7.8 \times 10^{3} \text{ mol } CO_2$$

(c) $(1.63 \times 10^{23} \text{ formula units } BaCl_2)\left[\frac{1 \text{ mol } BaCl_2}{6.022 \times 10^{23} \text{ formula units } BaCl_2}\right]$

$= 0.271 \text{ mol } BaCl_2$

(d) $(1.2 \times 10^{22} \text{ Cu atoms})\left[\frac{1 \text{ mol Cu}}{6.022 \times 10^{23} \text{ Cu atoms}}\right] = 0.020 \text{ mol Cu}$

4.41 (a) $(13.4 \text{ mol Ga})\left[\frac{6.022 \times 10^{23} \text{ Ga atoms}}{1 \text{ mol Ga}}\right] = 8.07 \times 10^{24} \text{ Ga atoms}$

(b) $(13.4 \text{ mol } C_6H_5CH_3)\left[\frac{6.022 \times 10^{23}\ C_6H_5CH_3 \text{ molecules}}{1 \text{ mol } C_6H_5CH_3}\right]$

$= 8.07 \times 10^{24}\ C_6H_5CH_3 \text{ molecules}$

(c) $(13.4 \text{ mol } AgNO_3)\left[\frac{6.022 \times 10^{23}\ AgNO_3 \text{ formula units}}{1 \text{ mol } AgNO_3}\right]$

$= 8.07 \times 10^{24}\ AgNO_3 \text{ formula units}$

4.43

	moles of atoms		molar mass (g/mol)	mass (g)
(a) NO_2				
N	1	x	14.01	= 14.01
O	2	x	16.00	= 32.00
			molar mass NO_2	= 46.01 g
(c) XeF_6				
Xe	1	x	131.30	= 131.30
F	6	x	19.00	= 114.0
			molar mass XeF_6	= 245.3 g
(e) $H_2PO_4^-$				
H	2	x	1.01	= 2.02
P	1	x	30.97	= 30.97
O	4	x	16.00	= 64.00
			molar mass $H_2PO_4^-$	= 96.99 g
(g) $KAuI_4$				
K	1	x	39.10	= 39.10
Au	1	x	196.97	= 196.97
I	4	x	126.90	= 507.60
			molar mass $KAuI_4$	= 743.67 g

	moles of atoms		molar mass (g/mol)	mass (g)
(b) $Ba(OH)_2$				
Ba	1	x	137.33	= 137.33
O	2	x	16.00	= 32.00
H	2	x	1.01	= 2.02
			molar mass $Ba(OH)_2$	= 171.35 g
(d) Mn^{2+}				
Mn^{2+}	1	x	54.94	= 54.94
			molar mass Mn^{2+}	= 54.94 g
(f) N_2				
N	2	x	14.01	= 28.02
			molar mass N_2	= 28.02 g
(h) $C_6H_5N_3O_4$				
C	6	x	12.01	= 72.06
H	5	x	1.01	= 5.05
N	3	x	14.01	= 42.03
O	4	x	16.00	= 64.00
			molar mass $C_6H_5N_3O_4$	= 183.14 g

(i) $Cu(IO_3)_2$

Cu	1	x	63.55	=	63.55
I	2	x	126.90	=	253.80
O	6	x	16.00	=	96.00

molar mass $Cu(IO_3)_2$ = 413.35 g

4.45 (a) $(5.3 \text{ mol C})\left[\frac{12.01 \text{ g C}}{1 \text{ mol C}}\right] = 64 \text{ g C}$

(b) $(0.1273 \text{ mol } N_2O_5)\left[\frac{108.02 \text{ g } N_2O_5}{1 \text{ mol } N_2O_5}\right] = 13.75 \text{ g } N_2O_5$

(c) $(1.3 \text{ µmol AmBr}_3)\left[\frac{1 \text{ mol}}{10^6 \text{ µmol}}\right]\left[\frac{483 \text{ g AmBr}_3}{1 \text{ mol AmBr}_3}\right] = 6.3 \times 10^{-4} \text{ g AmBr}_3$

(d) $(1 \times 10^{-10} \text{ mol HCl})\left[\frac{36.46 \text{ g HCl}}{1 \text{ mol HCl}}\right] = 4 \times 10^{-9} \text{ g HCl}$

4.47 (a) $(7.9 \text{ mg Tc})\left[\frac{1 \text{ g}}{1000 \text{ mg}}\right]\left[\frac{1 \text{ mol Tc}}{98.91 \text{ g Tc}}\right] = 8.0 \times 10^{-5} \text{ mol Tc}$

(b) $(16.8 \text{ g NH}_3)\left[\frac{1 \text{ mol NH}_3}{17.04 \text{ g NH}_3}\right] = 0.986 \text{ mol NH}_3$

(c) $(3.25 \text{ kg NH}_4\text{Br})\left[\frac{1000 \text{ g}}{1 \text{ kg}}\right]\left[\frac{1 \text{ mol NH}_4\text{Br}}{97.95 \text{ g NH}_4\text{Br}}\right] = 33.2 \text{ mol NH}_4\text{Br}$

(d) $(5.6 \text{ g PCl}_5)\left[\frac{1 \text{ mol PCl}_5}{208.22 \text{ g PCl}_5}\right] = 0.027 \text{ mol PCl}_5$

4.49 $(48.00 \text{ g O}_3)\left[\frac{1 \text{ mol O}_3}{48.00 \text{ g O}_3}\right]\left[\frac{6.022 \times 10^{23} \text{ O}_3 \text{ molecules}}{1 \text{ mol O}_3}\right] = 6.022 \times 10^{23} \text{ O}_3 \text{ molecules}$

$(6.022 \times 10^{23} \text{ O}_3 \text{ molecules})\left[\frac{3 \text{ O atoms}}{1 \text{ O}_3 \text{ molecule}}\right] = 1.807 \times 10^{24} \text{ O atoms}$

4.51 $(222.99 \text{ g AuCN})\left[\frac{1 \text{ mol AuCN}}{222.99 \text{ g AuCN}}\right]\left[\frac{6.022045 \times 10^{23} \text{ formula units}}{1 \text{ mol AuCN}}\right]$

$= 6.0220 \times 10^{23} \text{ formula units}$

$(6.0220 \times 10^{23} \text{ formula units AuCN})\left[\frac{1 \text{ Au}^+ \text{ ion}}{1 \text{ formula unit AuCN}}\right]$

$= 6.0220 \times 10^{23} \text{ Au}^+ \text{ ions}$

$(6.0220 \times 10^{23} \text{ formula units AuCN})\left[\frac{1 \text{ CN}^- \text{ ion}}{1 \text{ formula unit AuCN}}\right]$

$= 6.0220 \times 10^{23} \text{ CN}^- \text{ ions}$

$$(6.0220 \times 10^{23} \text{ formula units AuCN})\left[\frac{3 \text{ atoms}}{1 \text{ formula unit AuCN}}\right]$$

$$= 1.8066 \times 10^{24} \text{ atoms}$$

4.53 (a) $$(10.0 \text{ g } Na_2SO_4)\left[\frac{1 \text{ mol } Na_2SO_4}{142.04 \text{ g } Na_2SO_4}\right] = 0.0704 \text{ mol } Na_2SO_4$$

$$\frac{0.0704 \text{ mol } Na_2SO_4}{1.00 \text{ L}} = 0.0704 \text{ mol } Na_2SO_4/\text{L} = 0.0704 \text{ M}$$

(b) $$(56 \text{ g } CaCl_2)\left[\frac{1 \text{ mol } CaCl_2}{110.98 \text{ g } CaCl_2}\right] = 0.50 \text{ mol } CaCl_2$$

$$\frac{0.50 \text{ mol } CaCl_2}{1.00 \text{ L}} = 0.50 \text{ mol } CaCl_2/\text{L} = 0.50 \text{ M}$$

(c) $$(42.6 \text{ g } Al(NO_3)_3)\left[\frac{1 \text{ mol } Al(NO_3)_3}{213.01 \text{ g } Al(NO_3)_3}\right] = 0.200 \text{ mol } Al(NO_3)_3$$

$$\frac{0.200 \text{ mol } Al(NO_3)_3}{1.00 \text{ L}} = 0.200 \text{ mol } Al(NO_3)_3/\text{L} = 0.200 \text{ M}$$

4.55 $$(250.0 \text{ mL})\left[\frac{1 \text{ L}}{1000 \text{ mL}}\right]\left[\frac{0.100 \text{ mol NaOH}}{1 \text{ L}}\right]\left[\frac{40.00 \text{ g NaOH}}{1 \text{ mol NaOH}}\right] = 1.00 \text{ g NaOH}$$

4.57 $$(108 \text{ mL})\left[\frac{1 \text{ L}}{1000 \text{ mL}}\right]\left[\frac{0.62 \text{ mol acid}}{1 \text{ L}}\right] = 0.067 \text{ mol acid}$$

Dilution to 0.300 L does not change the number of moles of acid; thus, still 0.067 mol.

$$\frac{0.067 \text{ mol acid}}{0.300 \text{ L}} = 0.22 \text{ mol acid/L} = 0.22 \text{ M}$$

4.59 mass of Cl = 10.39 g - 3.56 g Fe = 6.83 g Cl

$$\% \text{ Fe} = \frac{3.56 \text{ g}}{10.39 \text{ g}} \times 100 = 34.3 \%$$

$$\% \text{ Cl} = \frac{6.83 \text{ g}}{10.39 \text{ g}} \times 100 = 65.7 \%$$

4.61 (a) KClO

$$\% \text{ K} = \frac{39.10 \text{ g K}}{90.55 \text{ g KClO}} \times 100 = 43.18 \%$$

$$\% \text{ Cl} = \frac{35.45 \text{ g Cl}}{90.55 \text{ g KClO}} \times 100 = 39.15 \%$$

$$\% \text{ O} = \frac{16.00 \text{ g O}}{90.55 \text{ g KClO}} \times 100 = 17.67 \%$$

(b) $KClO_2$

$$\% K = \frac{39.10 \text{ g K}}{106.55 \text{ g KClO}_2} \times 100 = 36.70 \%$$

$$\% Cl = \frac{35.45 \text{ g Cl}}{106.55 \text{ g KClO}_2} \times 100 = 33.27 \%$$

$$\% O = \frac{32.00 \text{ g O}}{106.55 \text{ g KClO}_2} \times 100 = 30.03 \%$$

(c) $KClO_3$

$$\% K = \frac{39.10 \text{ g K}}{122.55 \text{ g KClO}_3} \times 100 = 31.91 \%$$

$$\% Cl = \frac{35.45 \text{ g Cl}}{122.55 \text{ g KClO}_3} \times 100 = 28.93 \%$$

$$\% O = \frac{48.00 \text{ g O}}{122.55 \text{ g KClO}_3} \times 100 = 39.17 \%$$

(d) $KClO_4$

$$\% K = \frac{39.10 \text{ g K}}{138.55 \text{ g KClO}_4} \times 100 = 28.22 \%$$

$$\% Cl = \frac{35.45 \text{ g Cl}}{138.55 \text{ g KClO}_4} \times 100 = 25.59 \%$$

$$\% O = \frac{64.00 \text{ g O}}{138.55 \text{ g KClO}_4} \times 100 = 46.19 \%$$

4.63 $$(5.5 \text{ g KClO}_3)\left[\frac{1 \text{ mol KClO}_3}{122.55 \text{ g KClO}_3}\right]\left[\frac{3 \text{ mol O}}{1 \text{ mol KClO}_3}\right]\left[\frac{16.00 \text{ g O}}{1 \text{ mol O}}\right] = 2.2 \text{ g O}$$

4.65 azurite $$\% Cu = \frac{190.65 \text{ g Cu}}{344.69 \text{ g azurite}} \times 100 = 55.311 \%$$

chalcocite $$\% Cu = \frac{127.10 \text{ g Cu}}{159.16 \text{ g chalcocite}} \times 100 = 79.857 \%$$

chalcopyrite $$\% Cu = \frac{63.55 \text{ g Cu}}{183.52 \text{ g chalcopyrite}} \times 100 = 34.63 \%$$

covellite $$\% Cu = \frac{63.55 \text{ g Cu}}{95.61 \text{ g covellite}} \times 100 = 66.47 \%$$

cuprite $$\% Cu = \frac{127.10 \text{ g Cu}}{143.10 \text{ g cuprite}} \times 100 = 88.819 \%$$

malachite $$\% Cu = \frac{127.10 \text{ g Cu}}{221.13 \text{ g malachite}} \times 100 = 57.478 \%$$

Cuprite has the highest copper content on a mass percentage basis.

4.67 Assume exactly 100 g epinephrine.

	C	H	O	N
mass	56.8 g	6.56 g	28.4 g	8.28 g
molar mass	12.01 g/mol	1.01 g/mol	16.00 g/mol	14.01 g/mol
no. of moles	$(56.8\text{ g})\left[\frac{1\text{ mol}}{12.01\text{ g}}\right]$ = 4.73 mol	$(6.56\text{ g})\left[\frac{1\text{ mol}}{1.01\text{ g}}\right]$ = 6.50 mol	$(28.4\text{ g})\left[\frac{1\text{ mol}}{16.00\text{ g}}\right]$ = 1.78 mol	$(8.28\text{ g})\left[\frac{1\text{ mol}}{14.01\text{ g}}\right]$ = 0.591 mol
mole ratio n/n_N	$\frac{4.73}{0.591} = 8.00$	$\frac{6.50}{0.591} = 11.0$	$\frac{1.78}{0.591} = 3.01$	$\frac{0.591}{0.591} = 1.00$
rel. moles of atoms	8	11	3	1

Empirical formula is $C_8H_{11}O_3N$.

4.69 (a) Assume exactly 100 g compound.

	C	H	O
mass	52.2 g	13.0 g	34.8 g
molar mass	12.01 g/mol	1.01 g/mol	16.00 g/mol
no. of moles	$(52.2\text{ g})\left[\frac{1\text{ mol}}{12.01\text{ g}}\right]$ = 4.35 mol	$(13.0\text{ g})\left[\frac{1\text{ mol}}{1.01\text{ g}}\right]$ = 12.9 mol	$(34.8\text{ g})\left[\frac{1\text{ mol}}{16.00\text{ g}}\right]$ = 2.18 mol
mole ratio n/n_O	$\frac{4.35}{2.18} = 2.00$	$\frac{12.9}{2.18} = 5.92$	$\frac{2.18}{2.18} = 1.00$
rel. moles of atoms	2	6	1

Empirical formula is C_2H_6O.

(b) empirical formula molar mass

$$= (2 \times 12.01\text{ g}) + (6 \times 1.01\text{ g}) + (16.00\text{ g}) = 46.08\text{ g}$$

$$\frac{91.6\text{ g/mol}}{46.08\text{ g/mol}} = 1.99$$

Molecular formula is $C_4H_{12}O_2$.

(c) exact molar mass = (2)(46.08 g) = 92.16 g

4.71 $(4.86\text{ g }CO_2)\left[\frac{1\text{ mol }CO_2}{44.01\text{ g }CO_2}\right]\left[\frac{1\text{ mol C}}{1\text{ mol }CO_2}\right]\left[\frac{12.01\text{ g C}}{1\text{ mol C}}\right] = 1.33\text{ g C}$

$(2.03\text{ g }H_2O)\left[\frac{1\text{ mol }H_2O}{18.02\text{ g }H_2O}\right]\left[\frac{2\text{ mol H}}{1\text{ mol }H_2O}\right]\left[\frac{1.01\text{ g H}}{1\text{ mol H}}\right] = 0.228\text{ g H}$

mass of O = 2.00 g - 1.33 g C - 0.228 g H = 0.44 g O

	C	H	O
mass	1.33 g	0.228 g	0.44 g
molar mass	12.01 g/mol	1.01 g/mol	16.00 g/mol
no. of moles	$(1.33\ g)\left[\frac{1\ mol}{12.01\ g}\right]$ = 0.111 mol	$(0.228\ g)\left[\frac{1\ mol}{1.01\ g}\right]$ = 0.226 mol	$(0.44\ g)\left[\frac{1\ mol}{16.00\ g}\right]$ = 0.028 mol
mole ratio n/n_O	$\frac{0.111}{0.028} = 4.0$	$\frac{0.226}{0.028} = 8.1$	$\frac{0.028}{0.028} = 1.0$
rel. moles of atoms	4	8	1

Empirical formula is C_4H_8O.

4.73 (a) The largest common divisor of 8 and 18 is 2 giving C_4H_9. Likewise, dividing (b) 8, 10, 4, and 2 by 2 gives $C_4H_5N_2O$; (c) 6, 12, and 6 by 6 gives CH_2O; (d) 6, 8, and 6 by 2 gives $C_3H_4O_3$.

4.75 $$\left[\frac{65{,}000\ u\ hemoglobin}{1\ molecule\ hemoglobin}\right]\left[\frac{0.0035\ u\ Fe}{1\ u\ hemoglobin}\right]\left[\frac{1\ atom\ Fe}{55.85\ u\ Fe}\right]$$

$$= 4.1\ Fe\ atoms/1\ molecule\ hemoglobin$$

There are 4 iron atoms in one molecule of hemoglobin.

4.77 (a) $$(1.00\ g\ CaO)\left[\frac{1\ mol\ CaO}{56.08\ g\ CaO}\right] = 0.0178\ mol\ CaO$$

$$(0.0178\ mol\ CaO)\left[\frac{6.022 \times 10^{23}\ formula\ units\ CaO}{1\ mol\ CaO}\right]$$

$$\times \left[\frac{2\ atoms}{1\ formula\ unit\ CaO}\right] = 2.14 \times 10^{22}\ atoms$$

(b) $$(1.00\ g\ Ca(OH)_2)\left[\frac{1\ mol\ Ca(OH)_2}{74.10\ g\ Ca(OH)_2}\right] = 0.0135\ mol\ Ca(OH)_2$$

$$(0.0135\ mol\ Ca(OH)_2)\left[\frac{6.022 \times 10^{23}\ formula\ units\ Ca(OH)_2}{1\ mol\ Ca(OH)_2}\right]$$

$$\times \left[\frac{5\ atoms}{1\ formula\ unit\ Ca(OH)_2}\right] = 4.06 \times 10^{22}\ atoms$$

(c) $$(1.00\ g\ Mg)\left[\frac{1\ mol\ Mg}{24.31\ g\ Mg}\right] = 0.0411\ mol\ Mg$$

$$(0.0411\ mol\ Mg)\left[\frac{6.022 \times 10^{23}\ Mg\ atoms}{1\ mol\ Mg}\right] = 2.48 \times 10^{22}\ Mg\ atoms$$

(d) $(1.00\text{ g HOCH}_2\text{CH}_2\text{OH})\left[\frac{1\text{ mol HOCH}_2\text{CH}_2\text{OH}}{62.08\text{ g HOCH}_2\text{CH}_2\text{OH}}\right] = 0.0161\text{ mol HOCH}_2\text{CH}_2\text{OH}$

$$(0.0161\text{ mol HOCH}_2\text{CH}_2\text{OH})\left[\frac{6.022 \times 10^{23}\text{ molecules HOCH}_2\text{CH}_2\text{OH}}{1\text{ mol HOCH}_2\text{CH}_2\text{OH}}\right]$$

$$\times \left[\frac{10\text{ atoms}}{1\text{ molecule HOCH}_2\text{CH}_2\text{OH}}\right] = 9.70 \times 10^{22}\text{ atoms}$$

(e) $(1.00\text{ g H}_2\text{O})\left[\frac{1\text{ mol H}_2\text{O}}{18.02\text{ g H}_2\text{O}}\right] = 0.0555\text{ mol H}_2\text{O}$

$$(0.0555\text{ mol H}_2\text{O})\left[\frac{6.022 \times 10^{23}\text{ molecules H}_2\text{O}}{1\text{ mol H}_2\text{O}}\right]\left[\frac{3\text{ atoms}}{1\text{ molecule H}_2\text{O}}\right]$$

$$= 1.00 \times 10^{23}\text{ atoms}$$

The sample of water contains the largest number of moles and the largest number of atoms.

4.79 Assume exactly 100 g oxide.

mass of O = 100 g - 68.4 g = 31.6 g O

$$(31.6\text{ g O})\left[\frac{1\text{ mol O}}{16.00\text{ g O}}\right]\left[\frac{2\text{ mol M}}{3\text{ mol O}}\right] = 1.32\text{ mol M}$$

$$\frac{68.4\text{ g M}}{1.32\text{ mol M}} = 51.8\text{ g M/mol M}$$

Therefore, atomic mass is 51.8 u and the identity of the metal is Cr.

CHAPTER 5

CHEMICAL REACTIONS AND STOICHIOMETRY

Solutions to Exercises

5.1 $Ni(s) + H_2O(g\ or\ l) \rightarrow$ no reaction

$Ni(s) + 2HCl(aq) \rightarrow NiCl_2(aq) + H_2(g)$

$NiO(s) + H_2(g) \rightarrow Ni(s) + H_2O(g)$

5.2 (a) Reaction between aluminum and copper(II) ion will occur because aluminum is above copper in the activity series.

(b) Reaction between tin and iron(II) ion will not occur because tin is below iron in the activity series.

(c) Platinum will not react with HCl because Pt is below hydrogen in the activity series.

5.3 (a) The compound reacts to form two other compounds, a decomposition reaction.

(b) Two compounds combine to form a single compound, a combination reaction.

(c) The two ionic reactants that are both dissolved in water yield two products that are solids, a partner-exchange reaction.

5.4 (a) KNO_2 is soluble in water; the formula is written as $K^+ + NO_2^-$.

(b) $BaCO_3$ is nearly insoluble in water; the formula is written as $BaCO_3(s)$.

(c) Hg_2I_2 is nearly insoluble in water; the formula is written as $Hg_2I_2(s)$.

5.5 The equation rewritten using ionic forms is

$$2Na^+ + CO_3^{2-} + Ca^{2+} + 2OH^- \rightarrow 2Na^+ + 2OH^- + CaCO_3(s)$$

Cancelling the sodium and hydroxide spectator ions gives the net equation as

$$Ca^{2+} + CO_3^{2-} \rightarrow CaCO_3(s)$$

5.6 (a) Reaction between magnesium nitrate and sodium hydroxide will occur by partner exchange because because magnesium hydroxide is insoluble will precipitate.

$$Mg(NO_3)_2(aq) + 2NaOH(aq) \rightarrow Mg(OH)_2(s) + 2NaNO_3(aq)$$

$$Mg^{2+} + 2OH^- \rightarrow Mg(OH)_2(s)$$

(b) Reaction between potassium chloride and sodium nitrate will not occur because both possible products, potassium nitrate and sodium chloride, are soluble salts.

5.7 (a) $\frac{2\text{ mol } H_3AsO_3}{3\text{ mol } H_2S}$ (b) $\frac{1\text{ mol } As_2S_3}{2\text{ mol } H_3AsO_3}$ (c) $\frac{1\text{ mol } As_2S_3}{3\text{ mol } H_2S}$

5.8 $(15.0\text{ mol } CH_4)\left[\frac{1\text{ mol } CO_2}{1\text{ mol } CH_4}\right] = 15.0\text{ mol } CO_2$

$(15.0\text{ mol } CH_4)\left[\frac{2\text{ mol } H_2O}{1\text{ mol } CH_4}\right] = 30.0\text{ mol } H_2O$

5.10 $(120\text{ g NaCl})\left[\frac{1\text{ mol NaCl}}{58.44\text{ g NaCl}}\right] = 2.1\text{ mol NaCl}$

$(2.1\text{ mol NaCl})\left[\frac{1\text{ mol HCl}}{1\text{ mol NaCl}}\right] = 2.1\text{ mol HCl}$

$(2.1\text{ mol HCl})\left[\frac{36.46\text{ g HCl}}{1\text{ mol HCl}}\right] = 77\text{ g HCl}$

5.11 $(1.5\text{ ton-mol S})\left[\frac{1\text{ ton-mol } SO_2}{1\text{ ton-mol S}}\right]\left[\frac{1\text{ ton-mol } CaCO_3}{1\text{ ton-mol } SO_2}\right]\left[\frac{100.09\text{ ton } CaCO_3}{1\text{ ton-mol } CaCO_3}\right] = 150\text{ ton } CaCO_3$

5.12 The overall equation for the process is

$$2Sb_2S_3 + 9O_2 \rightarrow Sb_4O_6 + 6SO_2$$
$$Sb_4O_6 + 6C \rightarrow 4Sb + 6CO$$
$$\overline{2Sb_2S_3 + 9O_2 + 6C \rightarrow 4Sb + 6CO + 6SO_2}$$

$$(1.00\text{ kg ore})\left[\frac{23.2\text{ kg } Sb_2S_3}{100.0\text{ kg ore}}\right]\left[\frac{1000\text{ g}}{1\text{ kg}}\right]\left[\frac{1\text{ mol } Sb_2S_3}{339.68\text{ g } Sb_2S_3}\right] \times \left[\frac{4\text{ mol Sb}}{2\text{ mol } Sb_2S_3}\right]\left[\frac{121.75\text{ g Sb}}{1\text{ mol Sb}}\right] = 166\text{ g Sb}$$

5.13 Let x = mass Cu_2O and $1.351 - x$ = mass CuO.

$$(x\text{ g } Cu_2O)\left[\frac{1\text{ mol } Cu_2O}{143.10\text{ g } Cu_2O}\right]\left[\frac{2\text{ mol Cu}}{1\text{ mol } Cu_2O}\right]\left[\frac{63.55\text{ g Cu}}{1\text{ mol Cu}}\right] = (0.8882)x\text{ g Cu}$$

$$[(1.351 - x)\text{ g CuO}]\left[\frac{1\text{ mol CuO}}{79.55\text{ g CuO}}\right]\left[\frac{1\text{ mol Cu}}{1\text{ mol CuO}}\right]\left[\frac{63.55\text{ g Cu}}{1\text{ mol Cu}}\right] = [1.079 - (0.7989)x]\text{ g Cu}$$

$$(0.8882)x\text{ g Cu} + [1.079 - (0.7989)x]\text{ g Cu} = 1.152\text{ g Cu}$$
$$(0.0893)x = 0.073$$
$$x = 0.82\text{ g } Cu_2O$$

$$\frac{0.82\ g\ Cu_2O}{1.351\ g\ mixture} \times 100 = 61\ \%\ Cu_2O\ (by\ mass)$$

$$100\ \% - 61\ \% = 39\ \%\ CuO\ (by\ mass)$$

5.14 $$(25.00\ mL)\left[\frac{1\ L}{1000\ mL}\right]\left[\frac{0.112\ mol\ NaOH}{1\ L}\right] = 0.00280\ mol\ NaOH$$

$$(0.00280\ mol\ NaOH)\left[\frac{1\ mol\ HCl}{1\ mol\ NaOH}\right] = 0.00280\ mol\ HCl$$

$$(0.00280\ mol\ HCl)\left[\frac{1\ L}{0.103\ mol\ HCl}\right]\left[\frac{1000\ mL}{1\ L}\right] = 27.2\ mL$$

5.15 $$(22.7\ mL)\left[\frac{1\ L}{1000\ mL}\right]\left[\frac{0.106\ mol\ NaOH}{1\ L}\right] = 2.41 \times 10^{-3}\ mol\ NaOH$$

$$(2.41 \times 10^{-3}\ mol\ NaOH)\left[\frac{1\ mol\ KHC_8H_4O_4}{1\ mol\ NaOH}\right] = 2.41 \times 10^{-3}\ KHC_8H_4O_4$$

$$(2.41 \times 10^{-3}\ mol\ KHC_8H_4O_4)\left[\frac{204.2\ g\ KHC_8H_4O_4}{1\ mol\ KHC_8H_4O_4}\right] = 0.492\ g\ KHC_8H_4O_4$$

$$\frac{0.492\ g}{0.863\ g} \times 100 = 57.0\ mass\ \%\ KHC_8H_4O_4$$

5.16 $2H_2(g) + O_2(g) \rightarrow 2H_2O(l)$

$$(10.0\ g\ H_2)\left[\frac{1\ mol\ H_2}{2.02\ g\ H_2}\right] = 4.95\ mol\ H_2$$

$$(10.0\ g\ O_2)\left[\frac{1\ mol\ O_2}{32.00\ g\ O_2}\right] = 0.313\ mol\ O_2$$

$$(0.313\ mol\ O_2)\left[\frac{2\ mol\ H_2}{1\ mol\ O_2}\right] = 0.626\ mol\ H_2$$

Therefore, O_2 is limiting reagent.

$$(0.313\ mol\ O_2)\left[\frac{2\ mol\ H_2O}{1\ mol\ O_2}\right]\left[\frac{18.02\ g\ H_2O}{1\ mol\ H_2O}\right] = 11.3\ g\ H_2O$$

5.17 $$(1.00\ g\ NH_4Cl)\left[\frac{1\ mol\ NH_4Cl}{53.50\ g\ NH_4Cl}\right] = 0.0187\ mol\ NH_4Cl$$

$$(0.0187\ mol\ NH_4Cl)\left[\frac{1\ mol\ NH_3}{1\ mol\ NH_4Cl}\right] = 0.0187\ NH_3$$

$$(0.0187\ mol\ NH_3)\left[\frac{17.04\ g\ NH_3}{1\ mol\ NH_3}\right] = 0.319\ g\ NH_3$$

$$(0.319\ g\ NH_3)(0.78) = 0.25\ g\ NH_3$$

Solutions to Odd-Numbered Questions and Problems

5.1 (i) The combination reaction (A + B → C) is reaction a.

(ii) The decomposition reactions (C → A + B) are reactions e and g.

(iii) The displacement reactions (A + BC → AC + B) are reactions d, f, and h.

(iv) The partner-exchange reactions (AC + BD → AD + BC) are reactions b and c.

5.3 (a) Reaction between iron and magnesium ion will not occur because iron is below magnesium in the activity series.

(b) Reaction between nickel and copper(II) ion will occur because nickel is above copper in the activity series.

(c) Reaction between copper and hydrogen ion will not occur because copper is below hydrogen in the activity series.

(d) The reaction between magnesium and steam will occur because magnesium is high enough in the activity series to displace hydrogen from steam.

5.5 (a) $PbSO_4$ is listed as an exception to the generally soluble sulfates, $PbSO_4(s)$.

(b) Na^+ compounds and acetates are generally soluble, $Na^+ + CH_3COO^-$.

(c) NH_4^+ compounds are generally soluble and $(NH_4^+)_2CO_3$ is listed as an exception to the generally insoluble carbonates, $2NH_4^+ + CO_3^{2-}$.

(d) Sulfides are generally insoluble, $MnS(s)$.

(e) Chlorides are generally soluble, $Ba^{2+} + 2Cl^-$.

5.7 (a) $2Fe^{3+} + 2I^- \rightarrow 2Fe^{2+} + I_2(s)$

(b) $2IO_3^- + 6HSO_3^- \rightarrow 2I^- + 6SO_4^{2-} + 6H^+$

(c) $BrCl(g) + H_2O(l) \rightarrow H^+ + Cl^- + HBrO(aq)$

(d) $PCl_5(g) + 4H_2O(l) \rightarrow H_3PO_4(aq) + 5H^+ + 5Cl^-$

(e) $2S_2O_3^{2-} + I_2(s) \rightarrow S_4O_6^{2-} + 2I^-$

5.9 (a) Reaction between mercury nitrate and sodium sulfide will occur by partner exchange because mercury sulfide is insoluble and would precipitate. $Hg^{2+} + S^{2-} \rightarrow HgS(s)$

(b) Reaction between aluminum nitrate and lithium hydroxide will occur by partner exchange because aluminum hydroxide is insoluble and would precipitate. $Al^{3+} + 3OH^- \rightarrow Al(OH)_3(s)$

(c) Reaction between lithium sulfite and sodium chloride will not occur because both possible products, lithium chloride and sodium sulfite, are soluble salts.

(d) Reaction between iron(III) hydroxide and potassium nitrate will not occur because both possible products, iron(III) nitrate and potassium hydroxide, are water soluble.

5.11 (a) One mol C_7H_{16} reacts with 11 mol O_2 to give 7 mol CO_2 and 8 mol H_2O.

(b) One molecule of C_7H_{16} reacts with 11 molecules of O_2 to give 7 molecules of CO_2 and 8 molecules of H_2O.

(c) A 100.23 g sample of C_7H_{16} reacts with 352.00 g of O_2 to give 308.07 g of CO_2 and 144.16 g of H_2O.

5.13 (a) The mole ratio of O_2 to ZnS is (3 mol O_2)/(2 mol ZnS). Likewise, the mole ratio of (b) ZnO to ZnS is (2 mol ZnO)/(2 mol ZnS) = (1 mol ZnO)/(1 mol ZnS), and (c) SO_2 to ZnS is (2 mol SO_2)/(2 mol ZnS) = (1 mol SO_2)/(1 mol ZnS).

5.15 (a) $(1.00 \text{ mol } KClO_3)\left[\frac{3 \text{ mol } O_2}{2 \text{ mol } KClO_3}\right] = 1.50 \text{ mol } O_2$

(b) $(1.00 \text{ mol } H_2O_2)\left[\frac{1 \text{ mol } O_2}{2 \text{ mol } H_2O_2}\right] = 0.500 \text{ mol } O_2$

(c) $(1.00 \text{ mol } HgO)\left[\frac{1 \text{ mol } O_2}{2 \text{ mol } HgO}\right] = 0.500 \text{ mol } O_2$

(d) $(1.00 \text{ mol } NaNO_3)\left[\frac{1 \text{ mol } O_2}{2 \text{ mol } NaNO_3}\right] = 0.500 \text{ mol } O_2$

(e) $(1.00 \text{ mol } KClO_4)\left[\frac{2 \text{ mol } O_2}{1 \text{ mol } KClO_4}\right] = 2.00 \text{ mol } O_2$

5.17 $4NH_3(g) + 5O_2(g) \rightarrow 4NO(g) + 6H_2O(l)$

(a) $(1.50 \text{ mol } NH_3)\left[\frac{5 \text{ mol } O_2}{4 \text{ mol } NH_3}\right] = 1.88 \text{ mol } O_2$

(b) $(1.50 \text{ mol } NH_3)\left[\frac{4 \text{ mol } NO}{4 \text{ mol } NO_2}\right] = 1.50 \text{ mol } NO$

(c) $(1.50 \text{ mol } NH_3)\left[\frac{6 \text{ mol } H_2O}{4 \text{ mol } NH_3}\right] = 2.25 \text{ mol } H_2O$

5.19 (a) $(1.00 \text{ mol } NH_3)\left[\frac{3 \text{ mol } O_2}{4 \text{ mol } NH_3}\right] = 0.750 \text{ mol } O_2$

(b) $(1.00 \text{ mol } NH_3)\left[\frac{2 \text{ mol } N_2}{4 \text{ mol } NH_3}\right] = 0.500 \text{ mol } N_2$

(c) $(1.00 \text{ mol } NH_3)\left[\frac{6 \text{ mol } H_2O}{4 \text{ mol } NH_3}\right] = 1.50 \text{ mol } H_2O$

5.21 $(80.0 \text{ g NaOH})\left[\frac{1 \text{ mol NaOH}}{40.00 \text{ g NaOH}}\right] = 2.00 \text{ mol NaOH}$

$(2.00 \text{ mol NaOH})\left[\frac{2 \text{ mol Na}}{2 \text{ mol NaOH}}\right] = 2.00 \text{ mol Na}$

$(2.00 \text{ mol Na})\left[\frac{22.99 \text{ g Na}}{1 \text{ mol Na}}\right] = 46.0 \text{ g Na}$

5.23 $(1.50 \text{ g } CaCl_2)\left[\frac{1 \text{ mol } CaCl_2}{110.98 \text{ g } CaCl_2}\right] = 0.0135 \text{ mol } CaCl_2$

$(0.0135 \text{ mol } CaCl_2)\left[\frac{2 \text{ mol AgCl}}{1 \text{ mol } CaCl_2}\right] = 0.0270 \text{ mol AgCl}$

$(0.0270 \text{ mol AgCl}\left[\frac{143.32 \text{ g AgCl}}{1 \text{ mol AgCl}}\right] = 3.87 \text{ g AgCl}$

5.25 $(10.0 \text{ g } CoF_2)\left[\frac{1 \text{ mol } CoF_2}{96.93 \text{ g } CoF_2}\right]\left[\frac{1 \text{ mol } CoCl_2}{1 \text{ mol } CoF_2}\right]\left[\frac{129.83 \text{ g } CoCl_2}{1 \text{ mol } CoCl_2}\right] = 13.4 \text{ g } CoCl_2$

$(10.0 \text{ g } CoF_2)\left[\frac{1 \text{ mol } CoF_2}{96.93 \text{ g } CoF_2}\right]\left[\frac{2 \text{ mol HF}}{1 \text{ mol } CoF_2}\right]\left[\frac{20.01 \text{ g HF}}{1 \text{ mol HF}}\right] = 4.13 \text{ g HF}$

5.27 $Cl_2(g) + F_2(g) \rightarrow 2ClF(g)$

$(3.27 \text{ g } Cl_2)\left[\frac{1 \text{ mol } Cl_2}{70.90 \text{ g } Cl_2}\right]\left[\frac{1 \text{ mol } F_2}{1 \text{ mol } Cl_2}\right]\left[\frac{38.00 \text{ g } F_2}{1 \text{ mol } F_2}\right] = 1.75 \text{ g } F_2$

5.29 $Cl_2(g) + 2KBr(aq) \rightarrow 2KCl(aq) + Br_2(aq)$

$(0.289 \text{ g } Cl_2)\left[\frac{1 \text{ mol } Cl_2}{70.90 \text{ g } Cl_2}\right]\left[\frac{1 \text{ mol } Br_2}{1 \text{ mol } Cl_2}\right]\left[\frac{159.80 \text{ g } Br_2}{1 \text{ mol } Br_2}\right] = 0.651 \text{ g } Br_2$

5.31 $(1.49 \text{ g Cu})\left[\frac{1 \text{ mol Cu}}{63.55 \text{ g Cu}}\right] = 0.0234 \text{ mol Cu}$

$(0.0234 \text{ mol Cu})\left[\frac{1 \text{ mol } CuSO_4}{1 \text{ mol Cu}}\right] = 0.0234 \text{ mol } CuSO_4$

$(0.0234 \text{ mol } CuSO_4)\left[\frac{159.61 \text{ g } CuSO_4}{1 \text{ mol } CuSO_4}\right] = 3.73 \text{ g } CuSO_4$

$\frac{3.73 \text{ g } CuSO_4}{5.52 \text{ g sample}} \times 100 = 67.6 \%$

5.33 The overall equation is $TeO_2(s) + 2H^+ + 2OH^- \rightarrow H_2TeO_3(s) + H_2O(l)$

$(62.1 \text{ g } TeO_2)\left[\frac{1 \text{ mol } TeO_2}{159.60 \text{ g } TeO_2}\right]\left[\frac{1 \text{ mol } H_2TeO_3}{1 \text{ mol } TeO_2}\right]\left[\frac{177.62 \text{ g } H_2TeO_3}{1 \text{ mol } H_2TeO_3}\right]$

$= 69.1 \text{ g } H_2TeO_3$

5.35 The overall equation for the process is

$$(2)[2KClO_3 \rightarrow 2KCl + 3O_2]$$
$$(3)[CH_4 + 2O_2 \rightarrow CO_2 + 2H_2O]$$
$$3CH_4 + 4KClO_3 \rightarrow 4KCl + 3CO_2 + 6H_2O$$

$$(35.0\text{ g }CH_4)\left[\frac{1\text{ mol }CH_4}{16.05\text{ g }CH_4}\right] = 2.18\text{ mol }CH_4$$

$$(2.18\text{ mol }CH_4)\left[\frac{4\text{ mol }KClO_3}{3\text{ mol }CH_4}\right] = 2.91\text{ mol }KClO_3$$

$$(2.91\text{ mol }KClO_3)\left[\frac{122.55\text{ g }KClO_3}{1\text{ mol }KClO_3}\right] = 357\text{ g }KClO_3$$

5.37 (a) $$(2.25\text{ ton impure limestone})\left[\frac{45\text{ ton }CaCO_3}{100\text{ ton impure limestone}}\right]\left[\frac{1\text{ ton-mol }CaCO_3}{100.09\text{ ton }CaCO_3}\right] = 0.010\text{ ton-mol }CaCO_3$$

$$(0.010\text{ ton-mol }CaCO_3)\left[\frac{1\text{ ton-mol }CaO}{1\text{ ton-mol }CaCO_3}\right]\left[\frac{56.08\text{ ton }CaO}{1\text{ ton-mol }CaO}\right] = 0.56\text{ ton }CaO$$

(b) $$(0.010\text{ ton-mol }CaCO_3)\left[\frac{1\text{ ton-mol }CaO}{1\text{ ton-mol }CaCO_3}\right]\left[\frac{1\text{ ton-mol }Ca(OH)_2}{1\text{ ton-mol }CaO}\right] \times \left[\frac{74.10\text{ ton }Ca(OH)_2}{1\text{ ton-mol }Ca(OH)_2}\right] = 0.74\text{ ton }Ca(OH)_2$$

5.39 $$(0.75\text{ mol }Ca(OH)_2)\left[\frac{2\text{ mol HBr}}{1\text{ mol }Ca(OH)_2}\right]\left[\frac{1\text{ L}}{0.50\text{ mol HBr}}\right] = 3.0\text{ L}$$

5.41 $$(25.00\text{ mL})\left[\frac{1\text{ L}}{1000\text{ mL}}\right]\left[\frac{0.1037\text{ mol NaOH}}{1\text{ L}}\right]\left[\frac{1\text{ mol HCl}}{1\text{ mol NaOH}}\right] = 0.002593\text{ mol HCl}$$

$$\frac{(0.002593\text{ mol HCl})}{(23.65\text{ mL})(1\text{ L}/1000\text{ mL})} = 0.1096\text{ mol HCl/L} = 0.1096\text{ M}$$

5.43 $$(14.78\text{ mL})\left[\frac{1\text{ L}}{1000\text{ mL}}\right]\left[\frac{0.1026\text{ mol HCl}}{1\text{ L}}\right] = 0.001516\text{ mol HCl}$$

$$(0.001516\text{ mol HCl})\left[\frac{1\text{ mol }Na_2CO_3}{2\text{ mol HCl}}\right] = 0.0007580\text{ mol }Na_2CO_3$$

$$(0.0007580\text{ mol }Na_2CO_3)\left[\frac{105.99\text{ g }Na_2CO_3}{1\text{ mol }Na_2CO_3}\right] = 0.08034\text{ g }Na_2CO_3$$

$$\frac{0.08034\text{ g }Na_2CO_3}{0.1247\text{ g sample}} \times 100 = 64.43\ \%$$

5.45 $$(5.00\text{ g Na})\left[\frac{1\text{ mol Na}}{22.99\text{ g Na}}\right] = 0.217\text{ mol Na}$$

$$(7.10\ g\ Cl_2)\left[\frac{1\ mol\ Cl_2}{70.90\ g\ Cl_2}\right] = 0.100\ mol\ Cl_2$$

$$(0.217\ mol\ Na)\left[\frac{1\ mol\ Cl_2}{2\ mol\ Na}\right] = 0.109\ mol\ Cl_2\text{, more than available}$$

Therefore, Cl_2 is the limiting reactant; Na is in excess.

$$(0.100\ mol\ Cl_2)\left[\frac{2\ mol\ NaCl}{1\ mol\ Cl_2}\right] = 0.200\ mol\ NaCl$$

$$(0.200\ mol\ NaCl)\left[\frac{58.44\ g\ NaCl}{1\ mol\ NaCl}\right] = 11.7\ g\ NaCl$$

5.47 $$(45.0\ g\ PbF_2)\left[\frac{1\ mol\ PbF_2}{245.19\ g\ PbF_2}\right] = 0.184\ mol\ PbF_2$$

$$(25.0\ g\ PCl_3)\left[\frac{1\ mol\ PCl_3}{137.32\ g\ PCl_3}\right] = 0.182\ mol\ PCl_3$$

$$(0.184\ mol\ PbF_2)\left[\frac{2\ mol\ PCl_3}{3\ mol\ PbF_2}\right] = 0.123\ mol\ PCl_3$$

Therefore, PCl_3 is in excess and the PbF_2 is the limiting reactant.

$$(0.184\ mol\ PbF_2)\left[\frac{3\ mol\ PbCl_2}{3\ mol\ PbF_2}\right]\left[\frac{278.09\ g\ PbCl_2}{1\ mol\ PbCl_2}\right] = 51.2\ g\ PbCl_2$$

The mass of $PbCl_3$ unreacted is

$$(0.182\ mol\ PCl_3 - 0.123\ mol\ PCl_3)\left[\frac{137.32\ g\ PCl_3}{1\ mol\ PCl_3}\right] = 8.1\ g\ PCl_3$$

5.49 $$(8.00\ g\ HCl)\left[\frac{1\ mol\ HCl}{36.46\ g\ HCl}\right] = 0.219\ mol\ HCl$$

$$(10.00\ g\ HNF_2)\left[\frac{1\ mol\ HNF_2}{53.02\ g\ HNF_2}\right] = 0.1886\ mol\ HNF_2$$

$$(0.219\ mol\ HCl)\left[\frac{3\ mol\ HNF_2}{3\ mol\ HCl}\right] = 0.219\ mol\ HNF_2\text{, more than is available}$$

Therefore, HNF_2 is the limiting reactant and HCl is in excess.

$(0.15)(0.1886\ mol\ HNF_2) = 0.028\ mol\ HNF_2$ reacts

$0.1886\ mol - 0.028\ mol = 0.161\ mol\ HNF_2$ remains

$$(0.028\ mol\ HNF_2)\left[\frac{3\ mol\ HCl}{3\ mol\ HNF_2}\right] = 0.028\ mol\ HCl\text{ reacts}$$

$0.219\ mol - 0.028\ mol = 0.191\ mol\ HCl$ remains

$$(0.161\ mol\ HNF_2)\left[\frac{53.02\ g\ HNF_2}{1\ mol\ HNF_2}\right] = 8.54\ g\ HNF_2\text{ remains}$$

$$(0.191 \text{ mol HCl})\left[\frac{36.46 \text{ g HCl}}{1 \text{ mol HCl}}\right] = 6.96 \text{ g HCl remains}$$

$$(0.028 \text{ mol } HNF_2)\left[\frac{2 \text{ mol } ClNF_2}{3 \text{ mol } HNF_2}\right]\left[\frac{87.46 \text{ g } ClNF_2}{1 \text{ mol } ClNF_2}\right] = 1.6 \text{ g } ClNF_2$$

$$(0.028 \text{ mol } HNF_2)\left[\frac{1 \text{ mol } NH_4Cl}{3 \text{ mol } HNF_2}\right]\left[\frac{53.50 \text{ g } NH_4Cl}{1 \text{ mol } NH_4Cl}\right] = 0.50 \text{ g } NH_4Cl$$

$$(0.028 \text{ mol } HNF_2)\left[\frac{2 \text{ mol HF}}{3 \text{ mol } HNF_2}\right]\left[\frac{20.01 \text{ g HF}}{1 \text{ mol HF}}\right] = 0.37 \text{ g HF}$$

5.51 $$(38.5 \text{ g } PCl_3)\left[\frac{1 \text{ mol } PCl_3}{137.32 \text{ g } PCl_3}\right] = 0.280 \text{ mol } PCl_3$$

$$(0.280 \text{ mol } PCl_3)\left[\frac{1 \text{ mol } PCl_5}{1 \text{ mol } PCl_3}\right] = 0.280 \text{ mol } PCl_5$$

$$(0.280 \text{ mol } PCl_5)\left[\frac{208.22 \text{ g } PCl_5}{1 \text{ mol } PCl_5}\right] = 58.3 \text{ g } PCl_5$$

$$(0.85)(58.3 \text{ g } PCl_5) = 50. \text{ g } PCl_5$$

5.53 $2AgNO_3(s) \xrightarrow{\Delta} 2Ag(s) + 2NO_2(g) + O_2(g)$

$$\text{theoretical yield} = (0.575 \text{ g } AgNO_3)\left[\frac{1 \text{ mol } AgNO_3}{169.88 \text{ g } AgNO_3}\right]\left[\frac{2 \text{ mol Ag}}{2 \text{ mol } AgNO_3}\right]$$

$$\times\left[\frac{107.87 \text{ g Ag}}{1 \text{ mol Ag}}\right] = 0.365 \text{ g Ag}$$

$$\text{percent yield} = \left[\frac{0.362 \text{ g Ag}}{0.365 \text{ g Ag}}\right](100) = 99.2\ \%$$

5.55 Assume 1.00 L of solution so that the original number of moles of H_2SO_4 is 0.100 mol. If the final reaction is 100.0 % complete, then

$$(0.100 \text{ mol } H_2SO_4)\left[\frac{1 \text{ mol } H^+}{1 \text{ mol } H_2SO_4}\right] = 0.100 \text{ mol } H^+ \text{ formed}$$

$$(0.100 \text{ mol } H_2SO_4)\left[\frac{1 \text{ mol } HSO_4^-}{1 \text{ mol } H_2SO_4}\right] = 0.100 \text{ mol } HSO_4^- \text{ formed}$$

As a result of the second reaction

$$(0.100 \text{ mol } HSO_4^-)\left[\frac{1 \text{ mol } H^+}{1 \text{ mol } HSO_4^-}\right](0.100) = 0.0100 \text{ mol } H^+ \text{ formed}$$

$$(0.100 \text{ mol } HSO_4^-)\left[\frac{1 \text{ mol } SO_4^{2-}}{1 \text{ mol } HSO_4^-}\right](0.100) = 0.100 \text{ mol } SO_4^{2-} \text{ formed}$$

leaving

$0.100 \text{ mol } HSO_4^- - 0.0100 \text{ mol } HSO_4^- = 0.090 \text{ mol } HSO_4^-$

The concentrations are 0.0100 M SO_4^{2-}, 0.090 M HSO_4^-, and 0.110 M H^+.

CHAPTER 6

THE GASEOUS STATE

Solutions to Exercises

6.1 $V_2 = V_1\left[\frac{P_1}{P_2}\right] = (10.0 \text{ L})\left[\frac{145 \text{ kPa}}{75 \text{ kPa}}\right] = 19 \text{ L}$

6.2 $T_1 = 25\ °\text{C} + 273 = 298 \text{ K}$ $\quad$ $T_2 = 75\ °\text{C} + 273 = 348 \text{ K}$

$V_2 = V_1\left[\frac{T_2}{T_1}\right] = (35.2 \text{ L})\left[\frac{348 \text{ K}}{298 \text{ K}}\right] = 41.1 \text{ L}$

6.4 $T_2 = T_1\left[\frac{P_2}{P_1}\right]\left[\frac{V_2}{V_1}\right] = (298 \text{ K})\left[\frac{5.0 \text{ atm}}{15 \text{ atm}}\right]\left[\frac{75 \text{ L}}{175 \text{ L}}\right] = 43 \text{ K}$

6.5 $T_1 = 27\ °\text{C} + 273 = 300. \text{ K}$

$V_{STP} = V_1\left[\frac{P_1}{P_{STP}}\right]\left[\frac{T_{STP}}{T_1}\right] = (35.2 \text{ mL})\left[\frac{742 \text{ Torr}}{760 \text{ Torr}}\right]\left[\frac{273 \text{ K}}{300. \text{ K}}\right] = 31.3 \text{ mL}$

6.6 $V_2 = V_1\left[\frac{n_2}{n_1}\right] = (6.25 \text{ ft}^3)\left[\frac{0.116 \text{ mol}}{1.522 \text{ mol}}\right] = 0.476 \text{ ft}^3$

6.7 $(0.0036 \text{ mol})\left[\frac{22.4 \text{ L}}{1 \text{ mol}}\right] = 0.081 \text{ L} = 81 \text{ mL}$

6.9 $T = 375\ °\text{C} + 273 = 648 \text{ K}$

$P = \frac{nRT}{V} = \frac{(1.38 \text{ mol})(0.0821 \text{ L atm/K mol})(648 \text{ K})}{(5.2 \text{ L})} = 14 \text{ atm}$

6.10 $M = \frac{mRT}{PV} = \frac{(0.0218 \text{ g})(0.0821 \text{ L atm/K mol})(273 \text{ K})}{(0.0100 \text{ atm})(1.111 \text{ L})} = 44.0 \text{ g/mol}$

6.11 $T = 26.5\ °\text{C} + 273.15 = 299.7 \text{ K}$

$M = \frac{dRT}{P} = \frac{(1.979 \text{ g/L})(0.082057 \text{ L atm/K mol})(299.7 \text{ K})}{(1.013 \text{ atm})} = 48.04 \text{ g/mol}$

6.12 (a) $(27.3 \text{ Torr})\left[\frac{1 \text{ atm}}{760 \text{ Torr}}\right] = 0.0359 \text{ atm}$

$P_{He} = P_{total} - P_{H_2O} = 0.893 \text{ atm} - 0.0359 \text{ atm} = 0.857 \text{ atm}$

(b) $$X_{He} = \frac{P_{He}}{P_{total}} = \frac{0.857\ \text{atm}}{0.893\ \text{atm}} = 0.960$$

6.13 $$P = (759.3\ \text{Torr} - 22.1\ \text{Torr})\left[\frac{1\ \text{atm}}{760\ \text{Torr}}\right] = 0.9700\ \text{atm}$$

$$T = 23.8\ °\text{C} + 273.15 = 297.0\ \text{K}$$

$$V = (45.2\ \text{mL})\left[\frac{1\ \text{L}}{1000\ \text{mL}}\right] = 0.0452\ \text{L}$$

$$n = \frac{PV}{RT} = \frac{(0.9700\ \text{atm})(0.0452\ \text{L})}{(0.0821\ \text{L atm/K mol})(297.0\ \text{K})} = 0.00180\ \text{mol}$$

$$m = (0.00180\ \text{mol})\left[\frac{2.02\ \text{g}\ H_2}{1\ \text{mol}\ H_2}\right] = 0.00364\ \text{g}\ H_2$$

6.14 $$(50.\ \text{mL}\ BrF_3)\left[\frac{1\ \text{vol}\ Br_2}{2\ \text{vol}\ BrF_3}\right] = 25\ \text{mL}\ Br_2$$

$$(50.\ \text{mL}\ BrF_3)\left[\frac{3\ \text{vol}\ F_2}{2\ \text{vol}\ BrF_3}\right] = 75\ \text{mL}\ F_2$$

6.15 $$(15.0\ \text{g}\ H_2NNH_2)\left[\frac{1\ \text{mol}\ H_2NNH_2}{32.06\ \text{g}\ H_2NNH_2}\right]\left[\frac{1\ \text{mol}\ N_2}{1\ \text{mol}\ H_2NNH_2}\right] = 0.468\ \text{mol}\ N_2$$

$$V = \frac{nRT}{P} = \frac{(0.468\ \text{mol})(0.0821\ \text{L atm/K mol})(1120\ \text{K})}{(0.23\ \text{atm})} = 190\ \text{L}\ N_2$$

$$(190\ \text{L}\ N_2)\left[\frac{2\ \text{L}\ H_2O}{1\ \text{L}\ N_2}\right] = 380\ \text{L}\ H_2O$$

6.16 $$\text{time}_2 = (\text{time}_1)\frac{\sqrt{M_2}}{\sqrt{M_1}} = (23.2\ \text{s})\frac{\sqrt{30.07\ \text{g/mol}}}{\sqrt{16.04\ \text{g/mol}}} = 31.8\ \text{s}$$

6.17 $$T = 75\ °\text{C} + 273 = 348\ \text{K}$$

(a) $$P = \frac{nRT}{V} = \frac{(32.5\ \text{mol})(0.0821\ \text{L atm/K mol})(348\ \text{K})}{(1.75\ \text{L})} = 531\ \text{atm}$$

(b) $$P = \frac{nRT}{(V - nb)} - \frac{an^2}{V^2}$$

$$= \frac{(32.5\ \text{mol})(0.0821\ \text{L atm/K mol})(348\ \text{K})}{[1.75\ \text{L} - (32.5\ \text{mol})(0.05105\ \text{L/mol})]}$$

$$- \frac{(4.194\ \text{L}^2\ \text{atm/mol}^2)(32.5\ \text{atm})^2}{(1.75\ \text{L})^2}$$

$$= 10000\ \text{atm} - 1450\ \text{atm} = 9000\ \text{atm}$$

why not 10,200

Solutions to Odd-Numbered Questions and Problems

6.1 The five statements that summarize the kinetic-molecular theory for an ideal gas are

1. Molecules are relatively far apart; in comparison with their size, the spaces between them are large. This is not valid for a liquid nor for a solid, in both of which the same number of molecules occupy a much smaller volume.
2. Molecules are in constant motion, and their collisions with surfaces cause pressure. This statement is valid for liquids and solids.
3. The average speed and average kinetic energy of molecules are proportional to the temperature. This statement is valid for liquids and solids.
4. At the same temperature, the molecules in all ideal gases have the same average kinetic energy. This statement is valid for liquids and solids.
5. Collisions of molecules with each other or with the walls of a container are perfectly elastic. This statement is valid for liquids and solids.

6.3 $$\frac{E_{k,25\ °C}}{E_{k,75\ °C}} = \frac{298\ K}{348\ K} = 0.856$$

6.5 (a) The relationship between P and V is (ii) inversely proportional.
(b) The relationship between V and T is (i) directly proportional.
(c) The relationship between P and T is (i) directly proportional.
(d) The relationship between V and n is (i) directly proportional.

6.7 $$P_2 = P_1\left[\frac{V_1}{V_2}\right] = (1.00\ atm)\left[\frac{25\ L}{75\ L}\right] = 0.33\ atm$$

6.9
$$\begin{aligned} P_1V_1 &= P_2V_2 \\ (3.6\ atm)(V_1) &= (2.5\ atm)(V_1 + 4.9\ L) \\ (3.6)V_1 &= (2.5)V_1 + 12\ L \\ (1.1)V_1 &= 12\ L \\ V_1 &= 11\ L \end{aligned}$$

6.11 $$V_2 = V_1\left[\frac{T_2}{T_1}\right] = (1.000\ L)\left[\frac{274.2\ K}{273.2\ K}\right] = 1.004\ L$$

6.13 $$T_2 = T_1\left[\frac{V_2}{V_1}\right] = (298\ K)\left[\frac{126.4\ cm^3}{102.7\ cm^3}\right] = 367\ K = 94\ °C$$

6.15 $T_2 = T_1\left[\frac{P_2}{P_1}\right] = [(830 + 273)\text{K}]\left[\frac{1.0\text{ psi}}{14.0\text{ psi}}\right] = 79\text{ K}$

6.17 $T_2 = 135\ °\text{C} + 273 = 408\text{ K}$

$P_2 = P_1\left[\frac{T_2}{T_1}\right] = (1.00\text{ atm})\left[\frac{408\text{ K}}{273\text{ K}}\right] = 1.49\text{ atm}$

6.19 $V_2 = V_1\left[\frac{P_1}{P_2}\right]\left[\frac{T_2}{T_1}\right] = (16.3\text{ L})\left[\frac{0.937\text{ atm}}{5.30\text{ atm}}\right]\left[\frac{1246\text{ K}}{273\text{ K}}\right] = 13.2\text{ L}$

6.21 $T_1(°\text{C}) = (1\ \text{C}°/1.8\ \text{F}°)(85 - 32) = 29\ °\text{C}$

$T_1(\text{K}) = 29\ °\text{C} + 273 = 302\text{ K}$

$T_2(°\text{C}) = (1\ \text{C}°/1.8\ \text{F}°)(78\ °\text{F} - 32) = 26\ °\text{C}$

$T_2(\text{K}) = 26\ °\text{C} + 273 = 299\text{ K}$

$P_2 = P_1\left[\frac{V_1}{V_2}\right]\left[\frac{T_2}{T_1}\right] = (135\text{ psi})\left[\frac{375\text{ ft}^3}{37.5\text{ ft}^3}\right]\left[\frac{299\text{ K}}{302\text{ K}}\right] = 1340\text{ psi}$

6.23 $T_2 = T_1\left[\frac{P_2}{P_1}\right]\left[\frac{V_2}{V_1}\right] = (15.0\text{ K})\left[\frac{425\text{ Torr}}{763\text{ Torr}}\right]\left[\frac{83\text{ L}}{75\text{ L}}\right] = 9.2\text{ K}$

6.25 (a) $(2\text{ L H}_2)\left[\frac{1\text{ vol Cl}_2}{1\text{ vol H}_2}\right] = 2\text{ L Cl}_2$

(b) $(2\text{ L C}_2\text{H}_4)\left[\frac{1\text{ vol Cl}_2}{1\text{ vol C}_2\text{H}_4}\right] = 2\text{ L Cl}_2$

(c) $(2\text{ L CO})\left[\frac{1\text{ vol Cl}_2}{1\text{ vol CO}}\right] = 2\text{ L Cl}_2$

(d) $(2\text{ L C}_2\text{H}_2)\left[\frac{2\text{ vol Cl}_2}{1\text{ vol C}_2\text{H}_2}\right] = 4\text{ L Cl}_2$

6.27 $2H_2(g) + O_2(g) \rightarrow 2H_2O(g)$

$(250.\text{ mL H}_2)\left[\frac{1\text{ mL O}_2}{2\text{ mL H}_2}\right] = 125\text{ mL O}_2$, correct amount

$(250.\text{ mL H}_2)\left[\frac{2\text{ mL H}_2\text{O}}{2\text{ mL H}_2}\right] = 250.\text{ mL H}_2\text{O}$

6.29 $V_2 = V_1\left[\frac{n_2}{n_1}\right] = (503\text{ mL})\left[\frac{0.0256\text{ mol}}{0.0179\text{ mol}}\right] = 719\text{ mL}$

6.31 $(1.00\text{ m}^3)\left[\frac{1000\text{ L}}{1\text{ m}^3}\right]\left[\frac{1\text{ mol}}{22.4\text{ L}}\right] = 44.6\text{ mol}$

6.33 $(1.00\text{ L})\left[\frac{1\text{ mol}}{22.4\text{ L}}\right]\left[\frac{6.022 \times 10^{23}\text{ molecules}}{1\text{ mol}}\right] = 2.69 \times 10^{22}\text{ molecules}$

6.35 $\left[\frac{6.13\text{ g}}{1\text{ L}}\right]\left[\frac{22.4\text{ L}}{1\text{ mol}}\right] = 137\text{ g/mol}$

6.37 $n = \frac{PV}{RT} = \frac{(1.00\ \text{atm})(1.00\ \text{L})}{(0.0821\ \text{L atm/K mol})(298\ \text{K})} = 0.0409\ \text{mol}$

6.39 $(26.5\ \text{mL})\left[\frac{1\ \text{L}}{1000\ \text{mL}}\right]\left[\frac{1\ \text{mol}}{22.4\ \text{L}}\right] = 0.00118\ \text{mol}$

6.41 $P = \frac{nRT}{V} = \frac{(5.29\ \text{mol})(0.0821\ \text{L atm/K mol})(318\ \text{K})}{(3.45\ \text{L})} = 40.0\ \text{atm}$

6.43 $T = \frac{PV}{nR} = \frac{(45\ \text{atm})(74\ \text{L})}{(1.3\ \text{mol})(0.0821\ \text{L atm/K mol})} = 3.1 \times 10^{4}\ \text{K}$

6.45 $V = \frac{nRT}{P} = \frac{(1.00\ \text{mol})(0.0821\ \text{L atm/K mol})(198\ \text{K})}{(12.5\ \text{atm})} = 1.30\ \text{L}$

6.47 $n = \frac{PV}{RT} = \frac{(1.6 \times 10^{-9}\ \text{Torr})(1\ \text{atm/760 Torr})(1.00\ \text{L})}{(0.0821\ \text{L atm/K mol})(1475\ \text{K})} = 1.7 \times 10^{-14}\ \text{mol}$

$(1.7 \times 10^{-14}\ \text{mol})\left[\frac{6.022 \times 10^{23}\ \text{molecules}}{1\ \text{mol}}\right] = 1.0 \times 10^{10}\ \text{molecules}$

6.49 $(5.0\ \text{g})\left[\frac{1\ \text{mol}}{95\ \text{g}}\right] = 0.053\ \text{mol}$

$V = \frac{nRT}{P} = \frac{(0.053\ \text{mol})(0.0821\ \text{L atm/K mol})(298\ \text{K})}{(776\ \text{Torr})(1\ \text{atm/760 Torr})} = 1.3\ \text{L}$

6.51 $M = \frac{mRT}{PV} = \frac{(0.800\ \text{g})(0.0821\ \text{L atm/K mol})(372\ \text{K})}{(721\ \text{Torr})(1\ \text{atm/760 Torr})(103\ \text{mL})(1\ \text{L/1000 mL})} = 250.\ \text{g/mol}$

6.53 $d = \frac{PM}{RT} = \frac{(10.0\ \text{atm})(18\ \text{g/mol})}{(0.0821\ \text{L atm/K mol})(298\ \text{K})} = 7.4\ \text{g/L}$

6.55 $V = \frac{nRT}{P} = \frac{(24.0\ \text{g})(1\ \text{mol/32.00 g})(0.0821\ \text{L atm/K mol})(273\ \text{K})}{(1000.\ \text{Torr})(1\ \text{atm/760 Torr})} = 12.8\ \text{L}$

$P_{H_2} = \frac{nRT}{V} = \frac{(6.00\ \text{g})(1\ \text{mol/2.02 g})(0.0821\ \text{L atm/K mol})(273\ \text{K})}{(12.8\ \text{L})} = 5.20\ \text{atm}$

$P_{total} = (1000.\ \text{Torr})\left[\frac{1\ \text{atm}}{760\ \text{Torr}}\right] + 5.20\ \text{atm} = 6.52\ \text{atm}$

6.57 $X_{cyclopropane} = \frac{P_{cyclopropane}}{P_{total}} = \frac{150\ \text{Torr}}{150\ \text{Torr} + 550\ \text{Torr}} = 0.21$

$X_{oxygen} = \frac{P_{oxygen}}{P_{total}} = \frac{550\ \text{Torr}}{700\ \text{Torr}} = 0.79$

$\frac{n_{cyclopropane}}{n_{oxygen}} = \frac{X_{cyclopropane}}{X_{oxygen}} = \frac{0.21}{0.79} = 0.27$

6.59 P_{H_2} = 738 Torr - 23.8 Torr = 714 Torr

$$V_2 = V_1\left[\frac{P_1}{P_2}\right] = (33.3\ \text{L})\left[\frac{714\ \text{Torr}}{743\ \text{Torr}}\right] = 32.0\ \text{L}$$

6.61 $$(15.0\ \text{L}\ H_2)\left[\frac{1\ \text{volume}\ N_2}{3\ \text{volume}\ H_2}\right] = 5.00\ \text{L}\ N_2$$

6.63 $$(50.0\ \text{g}\ CaCO_3)\left[\frac{1\ \text{mol}\ CaCO_3}{100.09\ \text{g}\ CaCO_3}\right] = 0.500\ \text{mol}\ CaCO_3$$

$$(0.500\ \text{mol}\ CaCO_3)\left[\frac{1\ \text{mol}\ CO_2}{1\ \text{mol}\ CaCO_3}\right] = 0.500\ \text{mol}\ CO_2$$

$$(0.500\ \text{mol}\ CO_2)\left[\frac{22.4\ \text{L}\ CO_2}{1\ \text{mol}\ CO_2}\right] = 11.2\ \text{L}\ CO_2\ \text{at STP}$$

$$V = \frac{nRT}{P} = \frac{(0.500\ \text{mol})(0.0821\ \text{L atm/K mol})(298\ \text{K})}{(0.975\ \text{atm})} = 12.5\ \text{L}$$

6.65 $2XeF_2(s) + 2H_2O(l) \rightarrow 2Xe(g) + O_2(g) + 4HF(g)$

$$(47.2\ \text{g}\ XeF_2)\left[\frac{1\ \text{mol}\ XeF_2}{169.30\ \text{g}\ XeF_2}\right] = 0.279\ \text{mol}\ XeF_2$$

$$(0.279\ \text{mol}\ XeF_2)\left[\frac{4\ \text{mol HF}}{2\ \text{mol}\ XeF_2}\right] = 0.558\ \text{mol HF}$$

$$V = \frac{nRT}{P} = \frac{(0.558\ \text{mol HF})(0.0821\ \text{L atm/K mol})(297\ \text{K})}{(743\ \text{Torr})(1\ \text{atm}/760\ \text{Torr})} = 13.9\ \text{L HF}$$

$$(13.9\ \text{L HF})\left[\frac{1\ \text{volume}\ O_2}{4\ \text{volume HF}}\right] = 3.48\ \text{L}\ O_2$$

$$(13.9\ \text{L HF})\left[\frac{2\ \text{volume Xe}}{4\ \text{volume HF}}\right] = 6.95\ \text{L Xe}$$

6.67 $$V = \frac{(4\ \text{bottles})(250\ \text{mL/bottle})(1\ \text{L}/1000\ \text{mL})}{0.5} = 2\ \text{L}$$

$$n = \frac{PV}{RT} = \frac{(723\ \text{Torr})(1\ \text{atm}/760\ \text{Torr})(2\ \text{L})}{(0.0821\ \text{L atm/K mol})(298\ \text{K})} = 0.08\ \text{mol}\ O_2$$

$$(0.08\ \text{mol}\ O_2)\left[\frac{2\ \text{mol}\ KClO_3}{3\ \text{mol}\ O_2}\right] = 0.05\ \text{mol}\ KClO_3$$

$$(0.05\ \text{mol}\ KClO_3)\left[\frac{122.55\ \text{g}\ KClO_3}{1\ \text{mol}\ KClO_3}\right] = 6\ \text{g}\ KClO_3$$

The symbol Δ indicates that the reaction must be heated; the MnO_2 acts as a catalyst for the reaction.

6.69 $$\frac{\text{rate}_{Ne}}{\text{rate}_{Ar}} = \sqrt{\frac{\text{molar mass}_{Ar}}{\text{molar mass}_{Ne}}} = \sqrt{\frac{39.948\text{ g/mol}}{20.179\text{ g/mol}}} = 1.4070$$

6.71 $$\frac{\text{rate}_{^{235}UF_6}}{\text{rate}_{^{238}UF_6}} = \sqrt{\frac{352\text{ u}}{349\text{ u}}} = 1.004$$

6.73 $$V = \frac{nRT}{P} = \frac{(1.00\text{ mol})(0.0821\text{ L atm/K mol})(135\text{ K})}{(745\text{ Torr})(1\text{ atm/760 Torr})} = 11.3\text{ L}$$

No, the volume of the gas is less than that of an ideal gas.

6.75 (a) $$P = \frac{nRT}{V} = \frac{(3.25\text{ mol})(0.0821\text{ L atm/K mol})(388\text{ K})}{(6.25\text{ L})} = 16.6\text{ atm}$$

(b) $$P = \frac{nRT}{(V - nb)} - \frac{an^2}{V^2} = \frac{(3.25\text{ mol})(0.0821\text{ L atm/K mol})(388\text{ K})}{[6.25\text{ L} - (3.25\text{ mol})(0.1383\text{ L/mol})]} - \frac{(20.39\text{ L}^2\text{ atm/mol}^2)(3.25\text{ mol})^2}{(6.25\text{ L})^2}$$

$= 17.8\text{ atm} - 5.51 = 12.3\text{ atm}$

6.77 (a) $$(745\text{ Torr})\left[\frac{1\text{ atm}}{760\text{ Torr}}\right]\left[\frac{14.7\text{ psi}}{1\text{ atm}}\right] = 14.4\text{ psi}$$

(b) $$(745\text{ Torr})\left[\frac{1\text{ mmHg}}{1\text{ Torr}}\right] = 745\text{ mmHg}$$

(c) $$(745\text{ Torr})\left[\frac{1\text{ mmHg}}{1\text{ Torr}}\right]\left[\frac{1\text{ cm}}{10\text{ mm}}\right]\left[\frac{1\text{ in}}{2.54\text{ cm}}\right] = 29.3\text{ in Hg}$$

(d) $$(745\text{ Torr})\left[\frac{1\text{ atm}}{760\text{ Torr}}\right]\left[\frac{101{,}325\text{ Pa}}{1\text{ atm}}\right] = 99{,}300\text{ Pa}$$

(e) $$(745\text{ Torr})\left[\frac{1\text{ atm}}{760\text{ Torr}}\right] = 0.980\text{ atm}$$

(f) $$(745\text{ Torr})\left[\frac{1\text{ mmHg}}{1\text{ Torr}}\right]\left[\frac{1\text{ cm}}{10\text{ mm}}\right]\left[\frac{1\text{ in}}{2.54\text{ cm}}\right]\left[\frac{1\text{ ft}}{12\text{ in}}\right]\left[\frac{13.7\text{ ft H}_2\text{O}}{1\text{ ft Hg}}\right] = 33.5\text{ ft H}_2\text{O}$$

6.79 Assume exactly 100 g cyanogen.

	C	N
mass	46.2 g	53.8 g
molar mass	12.01 g/mol	14.01 g/mol
no. of moles	$(46.2\text{ g})\left[\frac{1\text{ mol}}{12.01\text{ g}}\right] = 3.85\text{ mol}$	$(53.8\text{ g})\left[\frac{1\text{ mol}}{14.01\text{ g}}\right] = 3.84\text{ mol}$
ratio of moles, n/n_N	$\frac{3.85}{3.84} = 1.00$	$\frac{3.84}{3.84} = 1.00$
relative ratio of atoms	1	1

Empirical formula is CN. Empirical formula molar mass is 26.0 g/mol.

$$M = \frac{mRT}{PV} = \frac{(1.00\text{ g})(0.0821\text{ L atm/K mol})(298\text{ K})}{(750\text{ Torr})(1\text{ atm/760 Torr})(0.476\text{ L})} = 52\text{ g/mol}$$

$(26)x = 52$

$x = 2$

Molecular formula is $(CN)_2$.

6.81 (a) For every 1 mL of Xe that reacts, 2 mL of F_2 would react.

(b) Dalton's law allows us to deduce partial pressure of reactants.

(c) If the pressure is increased at constant temperature, the gases in the nickel container will occupy a smaller volume.

(d) If the temperature is increased at constant pressure, the gases will occupy a larger volume.

(e) The increasing order for rates of effusion is $XeF_4 < Xe < F_2$.

(f) At higher pressures it is the volume of the molecules that determines the deviation of a real gas from ideality. Because XeF_4 is the largest molecule of the three, it should have the largest deviation from ideality.

CHAPTER 7

THERMOCHEMISTRY

Solutions to Exercises

7.1 $\Delta E = q + w$

$q = \Delta E - w = (323 \text{ kJ}) - (-111 \text{ kJ}) = 434 \text{ kJ}$

The heat flowed from the surroundings to the system.

7.2 (a) There is one mole of gaseous reactant and no gaseous product. The volume decrease signifies work done on (not by) the system.

(b) Because there is no gaseous reactant but one mole of gaseous product, work is being done by the system on the surroundings.

(c) With 2 mol of gaseous reactants and 2 mol of gaseous products, $\Delta V = 0$ and the amount of work is negligible.

(d) No gaseous reactants or products are involved and therefore, negligible work is done.

7.3 $$\text{(molar hear capacity)} = \frac{q}{(\text{no. of moles})(\Delta T)} = \frac{100.\text{ J}}{(1.00 \text{ mol})(3.98 \text{ K})}$$

$$= 25.1 \text{ J/K mol}$$

7.4 $q_{\text{lost,reaction}} = -5583.5 \text{ J}$

$q_{\text{gained,solution}} = (200.7 \text{ g})(3.97 \text{ J/K g})(6.5 \text{ K}) = 5200 \text{ J}$

$q_{\text{gained,calorimeter}} = \text{(calorimeter heat capacity)}(6.5 \text{ K})$

$q_{\text{lost,reaction}} + q_{\text{gained,solution}} + q_{\text{gained,calorimeter}} = 0$

$(-5583.5 \text{ J}) + (5200 \text{ J}) + \text{(calorimeter heat capacity)}(6.5 \text{ K}) = 0$

$$\text{calorimeter heat capacity} = \frac{(5583.5 \text{ J} - 5200 \text{ J})}{(6.5 \text{ K})} = 60 \text{ J/K}$$

7.5 $q_{\text{gained,soln}} = (200.\text{ g})(4.35 \text{ J/K g})(0.077 \text{ K}) = 67 \text{ J}$

$q_{\text{gained,calorimeter}} = (7.3 \text{ J/K})(0.077 \text{ K}) = 0.56 \text{ J}$

$q_{\text{lost,reaction}} = (0.00100 \text{ mol})q$

$q_{gained,soln} + q_{gained,calorimeter} + q_{lost,reaction} = 0$

$(67\ J) + (0.56\ J) + (0.00100\ mol)q = 0$

$$q = \frac{-67\ J - 0.56\ J}{0.00100\ mol} = -68\ kJ/mol$$

7.6 $q_{calorimeter} = (10.13\ kJ/K)(1.05\ K) = 10.6\ kJ$

$$q_{reaction} = (1.013\ g\ urea)\left[\frac{1\ mol\ urea}{60.07\ g\ urea}\right]\ \Delta E_C = (0.01686\ mol)\ \Delta E_C$$

$q_{calorimeter} + q_{reaction} = 0$

$(10.6\ kJ) + (0.01686\ mol)\Delta E_C = 0$

$$\Delta E_C = \frac{-10.6\ kJ}{0.01686\ mol} = -629\ kJ/mol$$

7.7 At 100 °C the standard state of calcium is a solid, of germanium chloride and of germanium hydride is a gas, and of octane is a liquid.

7.8 (a) $2Fe(s) + \frac{3}{2}O_2(g) \rightarrow Fe_2O_3(s)$ $\Delta H° = -824.2$ kJ

(b) $Ca(s) + C(graphite) + \frac{3}{2}O_2(g) \rightarrow CaCO_3(calcite)$ $\Delta H° = -1206.92$ kJ

7.10 $C_6H_5COOH(s) + \frac{15}{2}O_2(g) \rightarrow 7CO_2(g) + 3H_2O(l)$ $\Delta H° = -3227$ kJ

$$(1.00\ g\ C_6H_5COOH)\left[\frac{1\ mol\ C_6H_5COOH}{122.13\ g\ C_6H_5COOH}\right]\left[\frac{-3227\ kJ}{1\ mol\ C_6H_5COOH}\right] = -26.4\ kJ$$

7.11

$(4)[FeO(s) \rightarrow Fe(s) + \frac{1}{2}O_2(g)$ $\Delta H° = 272.0$ kJ]

$3Fe(s) + 2O_2(g) \rightarrow Fe_3O_4(s)$ $\Delta H° = -1118.4$ kJ

$4FeO(s) \rightarrow Fe(s) + Fe_3O_4(s)$ $\Delta H° = -30.4$ kJ

7.12 $\Delta H° = [(3\ mol)\Delta H_f°(Fe_2O_3)] - [(2\ mol)\Delta H_f°(Fe_3O_4) + (1/2\ mol)\Delta H_f°(O_2)]$

$= [(3\ mol)(-824.2\ kJ/mol)] - [(2\ mol)(-1118.4\ kJ/mol) + (1/2\ mol)(0)]$

$= -235.8$ kJ

Solutions to Odd-Numbered Questions and Problems

7.1 (a) During a process in which $q < 0$ and $w < 0$, ΔE for a system is negative.

(b) During a process in which $q = 0$ and $w > 0$, ΔE for a system is positive.

(c) During a process in which $q > 0$ and $w < 0$, ΔE for a system is positive, zero, or negative, depending on the numerical values of q and w.

7.3 $$\Delta E = q + w = (6300\ J) + (-460.\ L\ atm)\left[\frac{101.325\ J}{1\ L\ atm}\right] = -40{,}300\ J$$

7.5 $\Delta V = \frac{-w}{P_{ext}} = \frac{-(125\ J)(1\ L\ atm/101.325\ J)}{(5.2\ atm)} = -0.24\ L$

7.7 (a) For this physical change, the amount of work is negligible because the compound is never in the gaseous state.

(b) For this chemical change, there is a net decrease of 1 mol of gas and so work is done by the surroundings on the system.

(c) For this chemical change, there is a net decrease of 1 mol of gas and so work is done by the surroundings on the system.

7.9 For the formation of ozone (O_3) from oxygen gas (O_2) carried out under constant pressure conditions, heat is added to the system from the surroundings, so $q_p > 0$, and the volume of gas decreases, so $w > 0$. If the contribution of $\Delta(PV)$ is assumed to be small, $\Delta H = q_p > 0$, $w > 0$, and $\Delta E > 0$.

7.11 $\Delta T = (23\ °C) - (25\ °C) = -2\ °C = -2\ K$

number of moles $= (15.0\ g\ P)\left[\frac{1\ mol\ P}{30.97\ g\ P}\right] = 0.484\ mol\ P$

q = (number of moles)(molar heat capacity)(ΔT)

$= (0.484\ mol)(21.21\ J/K\ mol)(-2\ K) = -20\ J$

7.13 $\Delta T = 35\ °C - 25\ °C = 10.\ °C = 10.\ K$

$q_{Cl_2} = (1.00\ mol)(33.91\ J/K\ mol)(10.\ K) = 340\ J$

$q_{Na} = (1.00)(28.24)(10.) = 280\ J$

$q_{NaCl} = (1.00)(50.50)(10.) = 510\ J$

Sodium chloride requires the most heat.

7.15 molar heat capacity $= \frac{q}{(\text{number of moles})\Delta T} = \frac{413\ J}{(1.0000\ mol)(14.76\ K)}$

$= 28.0\ J/K\ mol$

sp. heat $= \frac{28.0\ J/K\ mol}{200.59\ g/mol} = 0.140\ J/K\ g$

7.17 $\Delta T = \frac{q}{(\text{number of moles})(\text{molar heat capacity})} = \frac{(-260\ J)}{(0.30\ mol)(24.435\ J/K\ mol)}$

$= -35\ K$

7.19 Al: $\left[\frac{26.98\ g}{1\ mol}\right]\left[\frac{0.900\ J}{K\ g}\right] = 24.3\ J/K\ mol$

Be: $\left[\frac{9.01\ g}{1\ mol}\right]\left[\frac{1.824\ J}{K\ g}\right] = 16.4\ J/K\ mol$

Cr: $\left[\frac{52.00\ g}{1\ mol}\right]\left[\frac{0.460\ J}{K\ g}\right] = 23.9$ J/K mol

Fe: $\left[\frac{55.85\ g}{1\ mol}\right]\left[\frac{0.452\ J}{K\ g}\right] = 25.2$ J/K mol

Pb: $\left[\frac{207.2\ g}{1\ mol}\right]\left[\frac{0.130\ J}{K\ g}\right] = 26.9$ J/K mol

Sn: $\left[\frac{118.69\ g}{1\ mol}\right]\left[\frac{0.226\ J}{K\ g}\right] = 26.8$ J/K mol

Be has a very small atomic mass and is an exception. The average value excluding Be is 25.4 J/K mol. The heat capacity relationship represents the contribution of the atoms vibrating in the crystal structure.

7.21 ΔT_{H_2O} = 19.68 °C - 16.95 °C = 2.73 °C = 2.73 K

q_{gained,H_2O} = (mass)(sp. heat)(ΔT) = (75.0 g)(4.184 J/K g)(2.73 K) = 857 J

$q_{gained,calorimeter}$ = (calorimeter heat capacity)(ΔT)
= (calorimeter heat capacity)(2.73 K)

ΔT_{Fe} = 19.68 °C - 63.14 °C = -43.46 °C = -43.46 K

$q_{lost,Fe}$ = (75.2 g)(0.450 J/K g)(-43.46 K) = -1470 J

$q_{gained,H_2O} + q_{gained,calorimeter} + q_{lost,Fe} = 0$

(857 J) + (calorimeter heat capacity)(2.73 K) + (-1470 J) = 0

calorimeter heat capacity = 225 J/K

7.23 $q_{gained,calorimeter}$ = (calorimeter heat capacity)(ΔT)
= (calorimeter heat capacity)(4.32 K)

$q_{lost,reaction}$ = (1.298 g)(-26,440 J/g) = -34,320 J

$q_{gained,calorimeter} + q_{lost,reaction} = 0$

(calorimeter heat capacity)(4.32 K) + (-34,320 J) = 0

calorimeter heat capacity = 7940 J/K

7.25 $q_{gained,calorimeter}$ = (calorimeter heat capacity)(ΔT)
= (2.74 J/K)(10.89 K) = 29.8 J

$q_{gained,soln}$ = (mass)(sp. heat)(ΔT) = (100.2 g)(4.21 J/K g)(10.89 K)
= 4590 J

$q_{gained,calorimeter} + q_{gained,soln} + q_{lost,reaction} = 0$

(29.8 J) + (4590 J) + $q_{lost,reaction}$ = 0

$q_{lost,reaction}$ = -4620 J

$$\text{no. of moles} = (0.241 \text{ g Mg})\left[\frac{1 \text{ mol Mg}}{24.31 \text{ g Mg}}\right] = 0.00991 \text{ mol Mg}$$

$$\frac{-4620 \text{ J}}{0.00991 \text{ mol}} = -466{,}000 \text{ J/mol} = -466 \text{ kJ/mol}$$

$$(1.00 \text{ mol})\left[\frac{-466 \text{ kJ}}{1 \text{ mol}}\right] = -466 \text{ kJ}$$

7.27 (a) ΔT_{H_2O} = 24.41 °C - 23.62 °C = 0.79 °C = 0.79 K

q_{gained,H_2O} = (mass)(sp. heat)(ΔT) = (100.0 g)(4.184 J/K g)(0.79 K)
= 330 J

ΔT_{metal} = 24.41 °C - 99.83 °C = -75.42 °C = -75.42 K

$q_{lost,metal}$ = (32.6 g)(sp. heat)(-75.42 K) = (-2460 g K)(sp. heat)

$q_{gained,H_2O} + q_{lost,metal} = 0$

(330 J) + (-2460 g K)(sp. heat) = 0

sp. heat = 0.13 J/K g; therefore the metal was identified as W.

(b) q_{gained,H_2O} = 330 J

$q_{gained,calorimeter}$ = (calorimeter heat capacity)(ΔT)
= (410 J/K)(0.79 K) = 320 J

$q_{lost,metal}$ = (-2460 g K)(sp. heat)

$q_{gained,H_2O} + q_{gained,calorimeter} + q_{lost,metal} = 0$

(330 J) + (320 J) + (-2460 g K)(sp. heat) = 0

(sp. heat) = 0.26 J/K g; therefore, the metal was identified as Mo.

7.29 (a) The standard state of water at -5 °C is ice (solid). That (b) at 5 °C is liquid, (c) at 93 °C is liquid, and (d) at 103 °C is steam (gas).

7.31 (a) O_2(g), (b) C(s,graphite), (c) Hg(l), (d) I_2(s).

7.33 (a) $\Delta H°$ = -(-106.12 kJ) = 106.12 kJ

(b) $\Delta H°$ = 2(-106.12 kJ) = -212.24 kJ

(c) $\Delta H°$ = -2(-106.12 kJ) = 212.24 kJ

7.35 $$(0.260 \text{ kg } C_2H_2)\left[\frac{1000 \text{ g}}{1 \text{ kg}}\right]\left[\frac{1 \text{ mol } C_2H_2}{26.04 \text{ g } C_2H_2}\right]\left[\frac{-1300. \text{ kJ}}{1 \text{ mol } C_2H_2}\right]\left[\frac{1000 \text{ J}}{1 \text{ kJ}}\right] = -1.30 \times 10^7 \text{ J}$$

7.37 (a) $$(1.00 \text{ mol } O_2)\left[\frac{285.4 \text{ kJ}}{3 \text{ mol } O_2}\right] = 95.1 \text{ kJ}$$

(b) $$(1.00 \text{ mol } O_3)\left[\frac{285.4 \text{ kJ}}{2 \text{ mol } O_3}\right] = 143 \text{ kJ}$$

(c) $(1.00\ g\ O_2)\left[\frac{1\ mol\ O_2}{32.00\ g\ O_2}\right]\left[\frac{285.4\ kJ}{3\ mol\ O_2}\right] = 2.97\ kJ$

(d) $(1.00\ g\ O_3)\left[\frac{1\ mol\ O_3}{48.00\ g\ O_3}\right]\left[\frac{285.4\ kJ}{2\ mol\ O_3}\right] = 2.97\ kJ$

7.39

$$\begin{array}{lll} 2NO_2(g) \rightarrow N_2O_4(g) & \Delta H° = & -57.20\ kJ \\ \underline{2NO(g) + O_2(g) \rightarrow 2NO_2(g)} & \underline{\Delta H° = -144.14\ kJ} \\ 2NO(g) + O_2(g) \rightarrow N_2O_4(g) & \Delta H° = & -201.34\ kJ \end{array}$$

7.41

$$\begin{array}{ll} (1/2)[2Cu(s) + Cl_2(g) \rightarrow 2CuCl(s) & \Delta H° = -274.4\ kJ] \\ \underline{(1/2)[2CuCl(s) + Cl_2(g) \rightarrow 2CuCl_2(s)} & \underline{\Delta H° = -165.8\ kJ]} \\ Cu(s) + Cl_2(g) \rightarrow CuCl_2(g) & \Delta H_f° = -220.1\ kJ \end{array}$$

7.43 (a) $\Delta H° = [(1\ mol)\Delta H_f°(PCl_5)] - [(1\ mol)\Delta H_f°(P) + (5/2\ mol)\Delta H_f°(Cl_2)]$
$= [(1\ mol)(-443.5\ kJ/mol)] - [(1\ mol)(0\ kJ/mol) + (5/2\ mol)(0)]$
$= -443.5\ kJ$

(b) $\Delta H° = [(3\ mol)\Delta H_f°(Fe_2O_3)] - [(2\ mol)\Delta H_f°(Fe_3O_4) + (1/2\ mol)\Delta H_f°(O_2)]$
$= [(3\ mol)(-824.2\ kJ/mol)] - [(2\ mol)(-1118.4\ kJ/mol) + (1\ mol)(0)]$
$= -235.8\ kJ$

(c) $\Delta H° = [(1\ mol)\Delta H_f°(K) + (1\ mol)\Delta H_f°(NaCl)]$
$- [(1\ mol)\Delta H_f°(KCl) + (1\ mol)\Delta H_f°(Na)]$
$= [(1\ mol)(0) + (1\ mol)(-411.153\ kJ/mol)]$
$- [(1\ mol)(-436.747\ kJ/mol) + (1\ mol)(0)]$
$= 25.594\ kJ$

7.45 $\Delta H° = [(1\ mol)\Delta H_f°(H_2O)] - [(1\ mol)\Delta H_f°(H^+) + (1\ mol)\Delta H_f°(OH^-)]$
$= [(1\ mol)(-285.830\ kJ/mol)] - [(1\ mol)(0) + (1\ mol)(-229.994\ kJ/mol)]$
$= -55.836\ kJ$

7.47

$$\begin{array}{ll} C(s) + O_2(g) \rightarrow CO_2(g) & \Delta H° = -394.89\ kJ \\ 2[H_2(g) + \frac{1}{2}O_2(g) \rightarrow H_2O(l) & \Delta H° = -286.10\ kJ] \\ \underline{CO_2(g) + 2H_2O(l) \rightarrow CH_4(g) + 2O_2(g)} & \underline{\Delta H° = 882.0\ kJ} \\ C(s) + 2H_2(g) \rightarrow CH_4(g) & \Delta H_f° = -85.1\ kJ \end{array}$$

7.49 (a) $Al(s) \rightarrow Al(l)$

$\Delta H°_{fus} = [(1\ mol)\Delta H_f°(Al(l))] - [(1\ mol)\Delta H_f°(Al(s))]$
$= [(1\ mol)(0)] - [(1\ mol)(-10.519\ kJ/mol)]$
$= 10.519\ kJ$

(b) $Al(l) \rightarrow Al(g)$

$$\begin{aligned}\Delta H^\circ_{vap} &= [(1\ mol)\Delta H^\circ_f(Al(g))] - [(1\ mol)\Delta H^\circ_f(Al(l))]\\ &= [(1\ mol)(310.114\ kJ/mol)] - [(1\ mol)(0)]\\ &= 310.114\ kJ\end{aligned}$$

(c) $Al(s) \rightarrow Al(g)$

$$\begin{aligned}\Delta H^\circ_{sub} &= [(1\ mol)\Delta H^\circ_f(Al(g))] - [(1\ mol)\Delta H^\circ_f(Al(s))]\\ &= [(1\ mol)(310.114\ kJ/mol)] - [(1\ mol)(-10.519\ kJ/mol)]\\ &= 320.633\ kJ\end{aligned}$$

7.51 The reaction producing liquid water will be more exothermic because additional energy is released by the condensation of the steam to liquid water.

7.53 (a) $(3\ kcal)\left[\frac{10^3\ cal}{1\ kcal}\right]\left[\frac{4.184\ J}{1\ cal}\right] = 1 \times 10^4\ J$

(b) $(2\ MJ)\left[\frac{10^6\ J}{1\ MJ}\right] = 2 \times 10^6\ J$

(c) $(14\ L\ atm)\left[\frac{101.325\ J}{1\ L\ atm}\right] = 1 \times 10^3\ J$

(d) $(7.2\ erg)\left[\frac{1\ J}{10^7\ erg}\right] = 7.2 \times 10^{-7}\ J$

(e) 15 J

(f) $(4.0\ cal)\left[\frac{4.184\ J}{1\ cal}\right] = 17\ J$

The order of increasing energy is d < e < f < c < a < b.

7.55 energy = mgd = $(1.0\ kg)(9.81\ m/s^2)(1.5\ m) = 15\ J$

$$\Delta T = \frac{q}{(mass)(sp.\ heat)} = \frac{15\ J}{(1.0\ kg)\left[\frac{1000\ g}{1\ kg}\right](4.184\ J/K\ g)} = 0.0036\ K$$

7.57 $\Delta T_{ice} = 0.0\ °C - (-5.0\ °C) = 5.0\ °C = 5.0\ K$

q_{ice} = (mass)(sp. heat)(ΔT) = (10.0 g)(2.1 J/K g)(5.0 K) = 110 J

q_{fus} = (no. of moles)(ΔH_{fus})

$$= (10.0\ g\ H_2O)\left[\frac{1\ mol\ H_2O}{18.02\ g\ H_2O}\right]\left[\frac{6009.5\ J}{1\ mol}\right] = 3330\ J$$

$\Delta T_{liquid} = 25.0\ °C - 0.0\ °C = 25.0\ °C = 25.0\ K$

Fig. 7-1

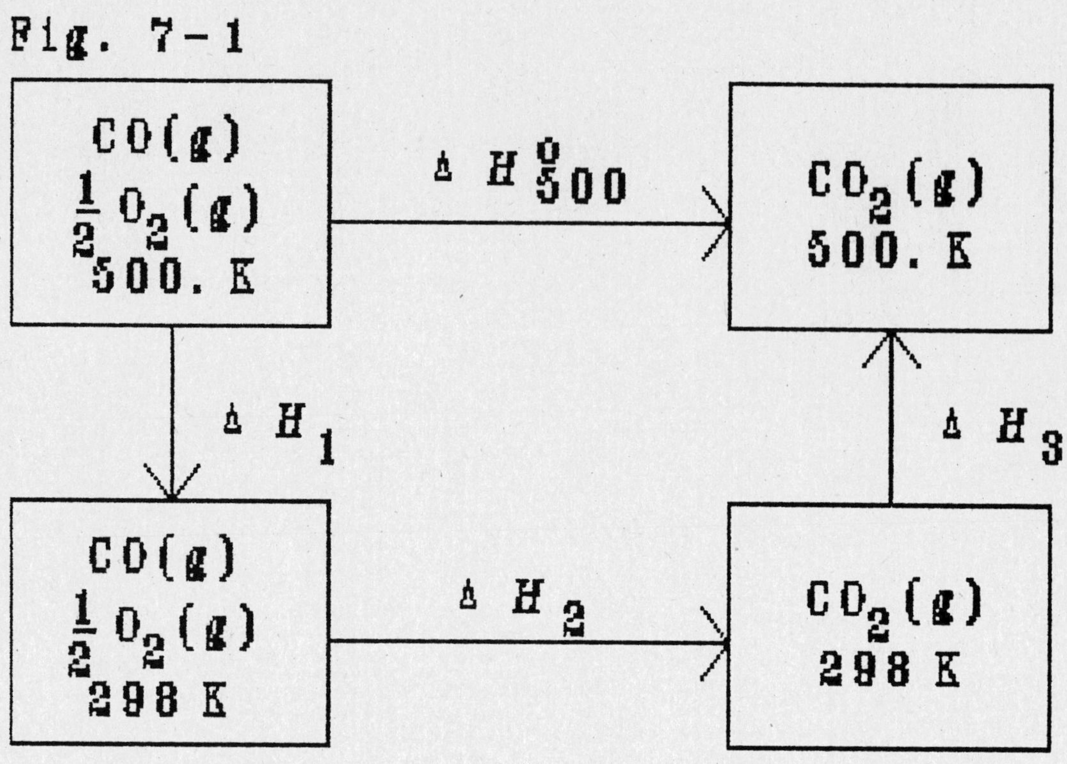

q_{liquid} = (mass)(sp. heat)(ΔT)

= (10.0 g)(4.184 J/K g)(25.0 K) = 1050 J

$\Delta H = q_{ice} + q_{fusion} + q_{liquid}$ = (110 J) + (3330 J) + (1050 J) = 4490 J

7.59 The value of ΔH°_{500} can be calculated using the series of steps shown in Fig. 7-1.

ΔT_1 = 298 K - 500. K = -202 K

q = (no. of moles)(molar heat capacity)(ΔT)

q_{CO} = (1 mol)(29.142 J/K mol)(-202 K) = -5,890 J

q_{O_2} = (1/2 mol)(29.355 J/K mol)(-202 K) = -2,960 J

$\Delta H_1 = q_{CO} + q_{O_2}$ = (-8,850 J)$\left[\frac{1 \text{ kJ}}{1000 \text{ J}}\right]$ = -8.85 kJ

ΔH_2 = -282.984 kJ

ΔT_3 = 500. K - 298 K = 202 K

q = (no. of moles)(molar heat capacity)(ΔT)

$= (1\ \text{mol})(37.11\ \text{J/K mol})(202\ \text{K})\left[\frac{1\ \text{kJ}}{1000\ \text{J}}\right] = 7.50\ \text{kJ}$

$\Delta H^{\circ}_{500} = \Delta H_1 + \Delta H_2 + \Delta H_3 = (-8.85\ \text{kJ}) + (-282.984\ \text{kJ}) + (7.50\ \text{kJ})$

$= -284.33\ \text{kJ}$

CHAPTER 8

ELECTRONIC STRUCTURE AND THE PERIODIC TABLE

Solutions to Exercises

8.1 $\lambda = \dfrac{c}{\nu} = \dfrac{(3.00 \times 10^8 \text{ m/s})(10^9 \text{ nm/1 m})}{5.2 \times 10^{14} \text{ s}^{-1}} = 580 \text{ nm}$

$\bar{\nu} = \dfrac{1}{\lambda} = \dfrac{1}{(580 \text{ nm})(10^2 \text{ cm/1 m})(1 \text{ m}/10^9 \text{ nm})} = 1.7 \times 10^4 \text{ cm}^{-1}$

8.2 $1s^2 2s^2 2p^6 3s^2 3p^6 3d^{10} 4s^2 4p^3$

8.3 (a) $[\text{Ar}]3d^{10}4s^2 4p^4$ (b) $[\text{Ar}]3d^6 4s^2$ (c) $[\text{Xe}]4f^{14}5d^{10}6s^2$

8.4 (a) $[\text{Xe}]6s^1$ (b) $[\text{Ar}]3d^1 4s^2$ (d) $[\text{Kr}]4d^{10}5s^2 5p^4$

Solutions to Odd-Numbered Questions and Problems

8.1 The two phenomena were the photoelectric effect and radiation by hot bodies.

8.3 (a) $\nu = \dfrac{c}{\lambda} = \dfrac{(2.998 \times 10^8 \text{ m/s})(10^9 \text{ nm/1 m})}{670.8 \text{ nm}} = 4.469 \times 10^{14} \text{ s}^{-1}$

(b) $\bar{\nu} = \dfrac{1}{\lambda} = \dfrac{1}{(670.8 \text{ nm})(1 \text{ m}/10^9 \text{ nm})(10^2 \text{ cm/1 m})} = 1.491 \times 10^4 \text{ cm}^{-1}$

(c) $E = h\nu = (6.626 \times 10^{-34} \text{ J s})(4.469 \times 10^{14} \text{ s}^{-1}) = 2.961 \times 10^{-19} \text{ J}$

8.5 $\nu = \dfrac{c}{\lambda} = \dfrac{(3.00 \times 10^8 \text{ m/s})(10^9 \text{ nm/1 m})}{340 \text{ nm}} = 8.8 \times 10^{14} \text{ s}^{-1}$

$E = h\nu = (6.626 \times 10^{-34} \text{ J s})(8.8 \times 10^{14} \text{ s}^{-1}) = 5.8 \times 10^{-19} \text{ J}$

$(5.8 \times 10^{-19} \text{ J/photon})(6.022 \times 10^{23} \text{ photons/mol}) = 3.5 \times 10^5 \text{ J/mol}$

8.7 $\nu = \dfrac{c}{\lambda} = \dfrac{(3.00 \times 10^8 \text{ m/s})(10^9 \text{ nm/1 m})}{550 \text{ nm}} = 5.5 \times 10^{14} \text{ s}^{-1}$

$E = h\nu = (6.626 \times 10^{-34} \text{ J s})(5.5 \times 10^{14} \text{ s}^{-1}) = 3.6 \times 10^{-19} \text{ J}$

$$\frac{10^{-17}\text{ J}}{3.6 \times 10^{-19}\text{ J/photon}} = 30\text{ photons}$$

8.9 When a low-intensity light of a frequency below the threshold energy shines on a photoelectric surface, nothing happens. If the intensity of the low-frequency light is increased by a factor of 1000, still nothing happens (except that the surface warms up faster). If a low-intensity light of a frequency above the threshold frequency is used, electrons are dislodged. If the intensity of this light is increased, the number of electrons dislodged increases.

8.11 The wave nature of particles of very small masses is important because the wavelength is in the order of magnitude of the atomic size. We neglect the wave nature of objects that have large masses because their wavelengths are too small to be observed.

8.13 $$\lambda = \frac{h}{mu} = \left[\frac{(6.626 \times 10^{-34}\text{ J s})}{(1.67 \times 10^{-27}\text{ kg})(2200\text{ m/s})}\right]\left[\frac{1\text{ kg m}^2/\text{s}^2}{1\text{ J}}\right]\left[\frac{10^9\text{ nm}}{1\text{ m}}\right] = 0.18\text{ nm}$$

8.15 The five basic parts of a spectrometer and their function are (1) a source of radiation, which excites the sample; (2) an analyzer, which divides the beam according to the property or properties being analyzed; (3) the sample holder, which contains the sample; (4) the detector, which measures the quantity or quantities being determined; and (5) a display device, which makes the results visible as a graph, chart, or photograph.

8.17 The incident radiation is absorbed by the sample to excite the sample to higher energy levels; the sample returns to the lower energy state by emitting the excess energy.

8.19 Helium was discovered by Janssen during a solar eclipse in 1868 by the detection of a new line in the solar spectrum.

8.21 $$\bar{\nu} = R\left[\frac{1}{n_1^2} - \frac{1}{n_2^2}\right] = (109{,}678\text{ cm}^{-1})\left[\frac{1}{1^2} - \frac{1}{3^2}\right] = 97{,}491.6\text{ cm}^{-1}$$

$$\lambda = \frac{1}{\bar{\nu}} = \frac{1}{(97{,}491.6\text{ cm}^{-1})(100\text{ cm/1 m})(1\text{ m}/10^9\text{ nm})} = 102.573\text{ nm}$$

This wavelength is in the ultraviolet region of the spectrum, not in the visible region.

8.23 The atomic electronic transition that emits more energy than the n = 3 to n = 1 transition is (d) the n = 4 to n = 1 transition. The amount of energy in an electronic transition is proportional to $[(1/n_1^2) - (1/n_2^2)]$. For emission, n_2 is greater than n_1. Small values of n_1 and large values of n_2 will make this value large.

8.25 $$\bar{\nu} = R\left[\frac{1}{n_1^2} - \frac{1}{n_2^2}\right] = (109{,}678\ \text{cm}^{-1})\left[\frac{1}{1^2} - \frac{1}{\infty^2}\right] = 109{,}678\ \text{cm}^{-1}$$

$$\lambda = \frac{1}{\bar{\nu}} = \frac{1}{(109{,}678\ \text{cm}^{-1})(100\ \text{cm}/1\ \text{m})(1\ \text{m}/10^9\ \text{nm})} = 91.1760\ \text{nm}$$

8.27 The three basic principles that make up the Bohr theory are (1) the electron in the hydrogen atom can move about the nucleus only in any one of several fixed circular orbits; (2) the angular momentum of the electron in a hydrogen atom is quantized as a whole-number multiple of $h/2\pi$; and (3) the electron does not radiate energy as long as it remains in one of the orbits. This theory is no longer used as the working model of the atom because it failed when applied to the spectra of atoms containing more than one electron, violated the Heisenberg uncertainty principle, and could not explain spectral changes caused by magnetic fields.

8.29 r = (0.0529 nm) n^2

r_1 = (0.0529 nm)(1^2) = 0.0529 nm $\qquad$ r_2 = (0.0529 nm)(2^2) = 0.212 nm

r_3 = (0.0529 nm)(3^2) = 0.476 nm $\qquad$ r_4 = (0.0529 nm)(4^2) = 0.846 nm

r_5 = (0.0529 nm)(5^2) = 1.32 nm

A sketch of the relative sizes of the orbits appears in Fig. 8-1. The size of the orbits increases as the square of n.

8.31 The permitted values of the principal quantum number are whole-number values 1, 2, 3, ... The range of values for the ground states of the known elements is from 1 to 7. The $n = \infty$ value corresponds to the complete removal of an electron to form an ion. A physical interpretation of n is that it defines the energy of an atomic electron and its average distance from the nucleus.

8.33 The orbital quantum number symbol is m_ℓ. Its allowed values are from $-\ell$, $-\ell+1$, ..., -1, 0, 1, 2 ..., $+\ell$. It can be interpreted as describing the orientation of orbitals in space.

Fig. 8-1

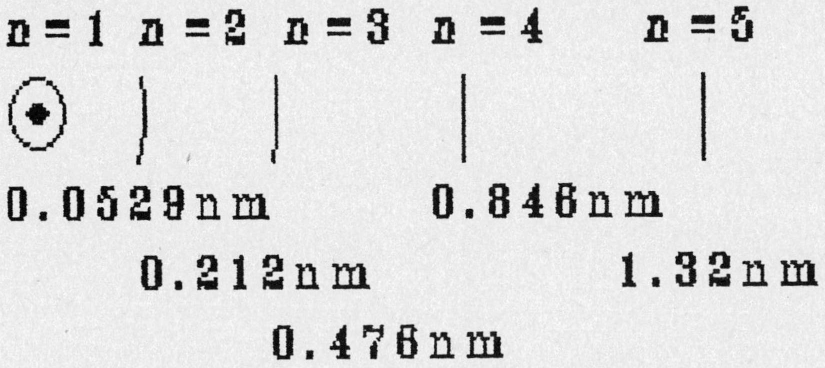

n = 1 n = 2 n = 3 n = 4 n = 5

0.0529 nm 0.848 nm

0.212 nm 1.32 nm

0.476 nm

8.35 The value of the subshell quantum number corresponding to (a) *d* is 2, (b) *f* is 3, (c) *s* is 0, (d) *p* is 1, and (e) *g* is 4.

8.37 (a) For a 4*d* orbital, m_ℓ can take the values -2, -1, 0, 1 and 2; (b) for a 1*s* orbital, m_ℓ can take the value 0; and (c) for a 3*p* orbital, m_ℓ can take the values -1, 0, and 1.

8.39 The set of quantum numbers that correctly describes an atom is (d) $n = 3$, $\ell = 1$, $m_\ell = 0$, $m_s = -\frac{1}{2}$.

8.41 The order of increasing distance of these subshells from the nucleus is 1*s*, 2*p*, 2*s*, 3*d*, 3*p*, 3*s*.

8.43 The figure shows that the most probable distance from the nucleus for the electron is 0.0529 nm, but that the electron can be found closer and farther away as well.

8.45 The Pauli exclusion principle states that no two electrons in a given atom can have the same numerical values for the four quantum numbers (n, ℓ, m_ℓ, and m_s). The electron configuration of (c) $1s^3$ violates this rule.

8.47 The Hund principle states that orbitals of equal energy are each occupied by a single electron before a second electron, which will have the opposite spin quantum number, enters any of them. The electron configuration that violates this rule is (b) $1s^2 2s^2 2p_x^2$.

8.49 (a) K: $1s^2 2s^2 2p^6 3s^2 3p^6 4s^1$; same as in Table 8.7; paramagnetic

(b) Sc: $1s^2 2s^2 2p^6 3s^2 3p^6 3d^1 4s^2$; same as in Table 8.7; paramagnetic

(c) Si: $1s^2 2s^2 2p^6 3s^2 3p^2$; same as in Table 8.7; paramagnetic

(d) F: $1s^2 2s^2 2p^5$; same as in Table 8.7; paramagnetic

(e) U: $1s^2 2s^2 2p^6 3s^2 3p^6 3d^{10} 4s^2 4p^6 4d^{10} 4f^{14} 5s^2 5p^6 5d^{10} 5f^4 6s^2 6p^6 7s^2$; it is different from that in Table 8.7 for two subshells, $5f^3 6d^1$ rather than $5f^4$; paramagnetic

(f) Ag: $1s^2 2s^2 2p^6 3s^2 3p^6 3d^{10} 4s^2 4p^6 4d^9 5s^2$; it is different from that in Table 8.7 for two subshells, $4d^{10} 5s^1$ rather than $4d^9 5s^2$; paramagnetic

(g) Mg: $1s^2 2s^2 2p^6 3s^2$; same as in Table 8.7; diamagnetic

(h) Fe: $1s^2 2s^2 2p^6 3s^2 3p^6 3d^6 4s^2$; same as in Table 8.7; paramagnetic

(i) Pr: $1s^2 2s^2 2p^6 3s^2 3p^6 3d^{10} 4s^2 4p^6 4d^{10} 4f^3 5s^2 5p^6 6s^2$; same as in Table 8.7; paramagnetic

8.51 The elements and their corresponding electron configurations are

(a) $1s^2 2s^2 2p^6 3s^2 3p^6 3d^{10} 4s^2 4p^3$ is As.

(b) $[Kr]4d^{10} 4f^{14} 5s^2 5p^6 5d^{10} 5f^{14} 6s^2 6p^6 6d^2 7s^2$ is Unq.

(c) $[Kr]4d^{10} 4f^{14} 5s^2 5p^6 5d^{10} 6s^2 6p^4$ is Po.

(d) $[Kr]4d^5 5s^2$ is Tc.

(e) $1s^2 2s^2 2p^6 3s^2 3p^6 3d^3 4s^2$ is V.

8.53 Cu, lowest energy principle: $1s^2 2s^2 2p^6 3s^2 3p^6 3d^9 4s^2$

Cu, complete 3*d*: $1s^2 2s^2 2p^6 3s^2 3p^6 3d^{10} 4s^1$

Both electron configurations show one unpaired electron, so no differences in magnetic properties are predicted.

8.55

(a)	$[Ne]3s^2 3p^6 4s^2$	no unpaired electrons
(b)	$[Xe]4f^{14} 5d^4 6s^2$	4 unpaired (5*d*) electrons
(c)	$[Ar]3d^{10} 4s^2 4p^2$	2 unpaired (4*p*) electrons
(d)	$[Xe]4f^1 5d^1 6s^2$	2 unpaired (4*f* and 5*d*) electrons
(e)	$[Kr]4d^5 5s^2$	5 unpaired (4*d*) electrons

8.57

A:	$1s^2 2s^2 2p^6 3s^2 3p^6 3d^{10} 4s^1$	B:	$[Kr]\ 4d^2 5s^2$
C:	$[Kr]\ 4d^{10} 5s^2 5p^6$	D:	$[Xe]\ 4f^{14} 5d^{10} 6s^2 6p^3$
E:	$1s^2 2s^2$		

8.59 The predicted electron configuration for element 119 is $1s^2 2s^2 2p^6 3s^2 3p^6 3d^{10} 4s^2 4p^6 4d^{10} 4f^{14} 5s^2 5p^6 5d^{10} 5f^{14} 6s^2 6p^6 6d^{10} 7s^2 7p^6 8s^1$. Element 119 would fall in the alkali metal family.

CHAPTER 9

PERIODIC PERSPECTIVE I: ATOMS AND IONS

Solutions to Exercises

9.1 (a) Cesium is in the lithium family, which is Representative Group I. Cesium has one valence electron in the n = 6 level ($6s^1$). The Lewis symbol is Cs·

(b) Silicon is in the carbon family, which is Representative Group IV. Silicon has four valence electrons: the two s and two p electrons in the n = 3 level ($3s^2 3p^2$). The Lewis symbol is ·Ṡi:

(c) Sulfur is in the oxygen family, which is Representative Group VI. Sulfur has six valence electrons: the two s and four p electrons in the n = 3 level ($3s^2 3p^4$). The Lewis symbol is :Ṩ:

9.2 (a) The element is a p-block representative element. It is a semiconducting element.

(b) The element is an f-transition element (actinide). It is a metal.

(c) The element is a d-transition element. It is a metal.

9.3 (a) The element K has a larger atomic radius than Na because size increases down a representative element family.

(b) The element Na has a larger atomic radius than Mg because size decreases across a period.

(c) The element As has a larger atomic radius than S because As has a higher occupied energy level.

9.4 Although the first ionization energy of Y in period 5 is greater than La in period 6, the order is reversed for the rest of the d-transition elements in these two periods because the intervention of the lanthanides makes the atoms virtually identical in size and the much larger nuclear charge of the period 6 elements produces larger ionization energies.

Solutions to Odd-Numbered Questions and Problems

9.1 The periodic table groups with these outer electron configurations are

(a) ns^2np^3 is the nitrogen family, Representative Group V.

(b) ns^1 is the alkali metal family, Representative Group I.

(c) $(n\text{-}2)f^{1-14}(n\text{-}1)d^{0-2}ns^2$ represents the *f*-transition elements.

9.3 The outer electronic configuration for the

(a) alkaline earth metals is ns^2.

(b) *d*-transition metals is $(n\text{-}1)d^{1-10}ns^{1,2}$.

(c) halogens is ns^2np^5.

9.5 The element with the configuration

(a) $1s^22s^22p^63s^23p^64s^2$ is calcium, an *s*-block representative element (alkaline earth element).

(b) $[Kr]4d^85s^1$ is rhodium, a *d*-transition element.

(c) $[Xe]4f^{14}5d^66s^2$ is osmium, a *d*-transition element.

(d) $[Xe]4f^{12}6s^2$ is erbium, an *f*-transition element (lanthanide).

(e) $[Kr]4d^{10}5s^25p^6$ is xenon, a noble gas element.

(f) $[Kr]4d^{10}4f^{14}5s^25p^65d^{10}6s^26p^2$ is lead, a *p*-block representative element.

9.7 (a) The alkali metal is B.

(b) The element with the outer configuration of d^8s^2 is H.

(c) The lanthanide is A.

(d) The *p*-block representative elements are C, F, G, and I.

(e) The elements with an incompletely filled *f* subshell are A and D.

(f) The halogen is I.

(g) The *s*-block representative elements are B and J.

(h) The actinide is D.

(i) The *d*-transition elements are E, H, and K.

(j) The noble gas is G.

9.9 The number of valence electrons present in atoms of these elements are

(a) Rb, 1; (b) As, 5; (c) S, 6; (d) Ne, 8; and (e) I, 7.

9.11 (a) Thallium is in Representative Group III and has three valence electrons ($6s^26p^1$). The Lewis symbol is $\cdot\mathrm{Tl}:$

(b) Arsenic is in Representative Group V and has five valence electrons ($4s^24p^3$). The Lewis symbol is $\cdot\underset{\cdot}{\ddot{\mathrm{As}}}\cdot$

(c) Magnesium is in Representative Group II and has two valence electrons ($3s^2$). The Lewis formula is Mg:

(d) Fluorine is in Representative Group VII and has seven valence electrons ($2s^22p^5$). The Lewis symbol is ·F: (with lone pairs above and below F)

(e) Oxygen is in Representative Group VI and has six valence electrons ($2s^22p^4$). The Lewis symbol is ·O: (with a pair above O and a single dot below)

9.13 The two factors that influence atomic radii are effective nuclear charge and the principal quantum number. The representative elements decrease in atomic radius across a period. In going down a family of representative elements, the atomic radii increase.

9.15 The atomic radii of Al and Ga are nearly the same because the intervention of the *d*-transition elements increases the nuclear charge of Ga enough to balance the effect of a higher occupied energy level. This similarity in atomic radii means similar chemical behavior.

9.17 Among the *p*-block representative elements, the order of increasing atomic radius is Ar < Cl < Sn < Pb because atomic radius decreases from left to right and from bottom to top in the periodic table.

(a) The atomic radius of Sn is (iii) 0.141 nm.

(b) The atomic radius of Ar is (i) 0.095 nm.

(c) The atomic radius of Cl is (ii) 0.099 nm.

(d) The atomic radius of Pb is (iv) 0.175 nm.

9.19 An anion is larger than the neutral atom from which it was formed. A cation is smaller than the neutral atom. The order of increasing radius is $Cl^+ < Cl < Cl^-$.

9.21 The loss of two electrons from the small Be atom makes Be^{2+} the smallest ion in the group. Na^+, Mg^{2+}, and F^- are isoelectronic. The ionic radius of these ions decreases as the charge becomes more positive, so Mg^{2+} is the smallest and F^- is the largest of the three. S^{2-} and Cl^- are isoelectronic, with S^{2-} being smaller than Cl^-. S^{2-} and Cl^- are larger than the Na^+, Mg^{2+}, F^- group of ions. The order of increasing ionic radius is $Be^{2+} < Mg^{2+} < Na^+ < F^- < Cl^- < S^{2-}$.

9.23 $\text{atomic radius}_{Cl} = \dfrac{0.198\text{ nm}}{2} = 0.099\text{ nm}$

$\text{atomic radius}_{P} = (0.204\text{ nm}) - (0.099\text{ nm}) = 0.105\text{ nm}$

$\text{P-F bond length} = (0.105\text{ nm}) + (0.071\text{ nm}) = 0.176\text{ nm}$

9.25 $4r = a\sqrt{2}$

$$r = \frac{a\sqrt{2}}{4} = \frac{(0.35238\text{ nm})\sqrt{2}}{4} = 0.12459\text{ nm}$$

9.27 (a) The electron configuration for Mg is [Ne]$3s^2$. It would form a +2 cation (Mg^{2+}) with an electron configuration of [Ne].

(b) The electron configuration for Al is [Ne]$3s^23p^1$. It would form a +3 cation (Al^{3+}) with an electron configuration of [Ne].

(c) The electron configuration for Sc is [Ar]$3d^14s^2$. It would form a +2 cation (Sc^{2+}) with an electron configuration of [Ar]$3d^1$. Sc will also form a +3 cation (Sc^{3+}) with an electron configuration of [Ar].

9.29 (a) The electron configuration for N is [He]$2s^22p^3$. It would form a -3 anion (N^{3-}) with an electron configuration of [Ne].

(b) The electron configuration for S is [Ne]$3s^23p^4$. It would form a -2 anion (S^{2-}) with an electron configuration of [Ar].

(c) The electron configuration for P is [Ne]$3s^23p^3$. It would form a -3 anion (P^{3-}) with an electron configuration of [Ar].

9.31 The Lewis symbol for Mg^{2+} is $[Mg]^{2+}$; for Al^{3+} is $[Al]^{3+}$; for Sc^{3+} is $[Sc]^{3+}$; for N^{3-} is $[:\ddot{\underset{..}{N}}:]^{3-}$; and for S^{2-} is $[:\ddot{\underset{..}{S}}:]^{2-}$.

9.33 The general trend of first ionization energies for atoms in the same period is an increase across the period. The general trend within a family of representative elements is a decrease down the family.

9.35 The values of the ionization energies of the metals are smaller than those of the nonmetals. These values determine the chemical behavior of metals and nonmetals in that metals form cations and nonmetals do not.

9.37 The ionization energy of nitrogen is greater than the other two and the electron affinity is less than the other two. The p^3 configuration must impart stability to the atom with respect to ionization and gain of electrons.

9.39

$$\begin{array}{lcl} F(g) \rightarrow \frac{1}{2}F_2(g) & & \Delta H^\circ_f = -78.99\text{ kJ} \\ \frac{1}{2}F_2(g) + e^- \rightarrow F^-(g) & & \Delta H^\circ_f = -255.39\text{ kJ} \\ \hline F(g) + e^- \rightarrow F^-(g) & & \Delta H^\circ_f = -334.38\text{ kJ} \end{array}$$

Fig. 9-1

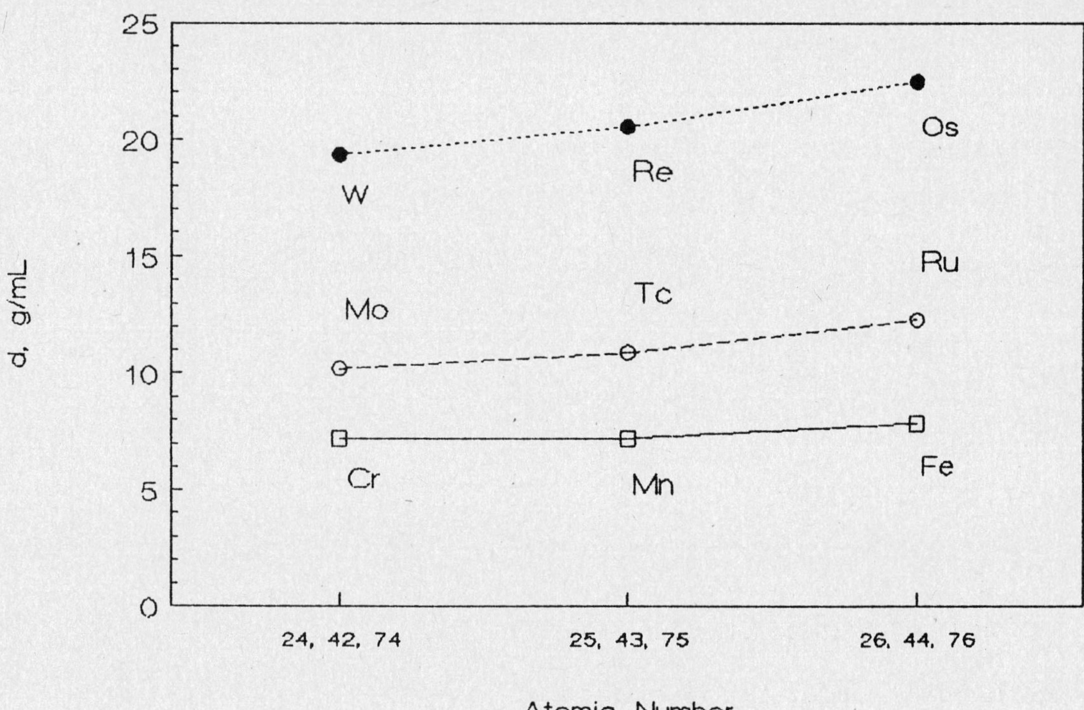

9.41 The second ionization energy is so much larger than the first because it is more difficult to remove an electron from a positively charged ion than from a neutral atom.

9.43 (a) An atom of an element with 15 electrons has an approximate atomic mass of 30 u, corresponding to the mass of 15 protons and an equal number of neutrons.

(b) The atomic number is 15 (the number of protons), which an atom of E must have in order to be a neutral atom.

(c) The total number of s electrons is 6 ($1s^2 2s^2 2p^6 3s^2 3p^3$).

(d) The element is a nonmetal (*p*-block element in the third period).

9.45

(a)		(b)		(c)	
Cr	7.20 g/mL	W	19.35 g/mL	Mo	10.2 g/mL
Mn	7.20	Re	20.53	Tc	?
Fe	7.86	Os	22.48	Ru	12.30

The data are plotted in Fig. 9-1 and the density of Tc should be about 10.9 g/mL.

CHAPTER 10

PERIODIC PERSPECTIVE II: CHEMICAL BONDS

Solutions to Exercises

10.1 (a) Ba^{2+}: [Xe] (b) Ag^{+}: [Kr]$4d^{10}$
(c) Cd^{2+}: [Kr]$4d^{10}$ (d) Pb^{2+}: [Xe]$4f^{14}5d^{10}6s^{2}$

10.2 Barium reacts with atomic chlorine to produce the ionic compound $BaCl_2$. These changes can be shown with Lewis symbols and electron configurations as

Ba: + 2 ·C̤̈l: → $[Ba]^{2+}$ 2[:C̤̈l:]$^-$
[Xe]$6s^2$ [Ne]$3s^23p^5$ [Xe] [Ar]

10.3 $\Delta H° = [(3\text{ mol})BE_{N-F} + (1\text{ mol})BE_{H-H}] - [(2\text{ mol})BE_{N-F} + (1\text{ mol})BE_{N-H} + (1\text{ mol})BE_{H-F}]$
$= [(3\text{ mol})(280\text{ kJ/mol}) + (1\text{ mol})(435\text{ kJ/mol})] - [(2\text{ mol})(280\text{ kJ/mol}) + (1\text{ mol})(389\text{ kJ/mol}) + (1\text{ mol})(569\text{ kJ/mol})]$
$= -243$ kJ

10.4 The most ionic bond is (a) Na-Cl because Na has the lowest charge-to-size ratio of the three metals. The least ionic bond is (c) Be-Cl because Be, being the smallest, has the largest charge-to-size ratio.

10.5 The order of increasing bond polarity with electronegativity differences shown is

(a) C-N < (b) S-O < (c) Si-N < (d) B-O
0.5 1.0 1.2 1.5

10.6 Metals have low electronegativities, so K has the lowest value. Sn is a *p*-block metal and will have a low value of electronegativity. The semiconducting element As will have a greater electronegativity value than the two metals, but a smaller electronegativity value than the two nonmetals N and F. Electronegativity increases from left to right within a period, and so F will be more electronegative than N. The order of increasing electronegativity of the elements, with electronegativities shown, is

$$\begin{array}{ccccccccc} K & < & Sn & < & As & < & N & < & F \\ 0.8 & & 1.8 & & 2.0 & & 3.0 & & 4.0 \end{array}$$

10.7 (a) $\overset{+1}{H}\ \overset{+5}{N}\ \overset{3(-2)}{O_3}$ (b) $\overset{+1}{H}\ \overset{+3}{N}\ \overset{2(-2)}{O_2}$ (c) $\overset{+3}{N}\ \overset{-2}{O^+}$

(d) $\overset{-3}{N^{3-}}$ (e) $\overset{0}{N_2}$ (f) $\overset{-3}{N}\ \overset{3(+1)}{H_3}$

10.8 $\overset{+2}{Ca}\ \overset{+2}{Mg}\ (\overset{2(+4)}{Si}\ \overset{6(-2)}{O_3})_2$

10.9 (a) gold(III) chloride (b) gold(I) sulfide (c) gold(III) oxide

$\overset{+3}{Au}\ \overset{3(-1)}{Cl_3}$ $\overset{2(+1)}{Au_2}\ \overset{-2}{S}$ $\overset{2(+3)}{Au_2}\ \overset{3(-2)}{O_3}$

10.10 The formulas of the following compounds named using the Stock system are
(a) arsenic(III) fluoride, AsF_3 (b) bismuth(V) oxide, Bi_2O_5
(c) cerium(IV) sulfate, $Ce(SO_4)_2$ (d) chromium(II) chloride, $CrCl_2$

Solutions to Odd-Numbered Questions and Problems

10.1 A chemical bond is a force that acts strongly enough between two atoms or groups of atoms to hold them together in a different species that has measurable properties. Atoms are held together in a chemical bond by electrostatic attraction between positive and negative charges.

10.3 An alkali metal atom can attain a noble gas electron configuration by losing the outermost electron to form a +1 ion.

10.5 The bonding in $Ba(OH)_2$ is ionic bonding between Ba^{2+} and OH^- ions and polar covalent bonding between O and H atoms.

10.7 A metallic bond is formed by atoms each giving up one or more valence electrons to become cations. All the valence electrons form an electron sea that surrounds the cations. Nonmetals do not form metallic bonds because their valence electrons are too difficult to remove.

10.9 Metals are malleable and ductile because specific bonds need not be broken as layers of cations move with respect to each other.

10.11 When an ionic bond is formed between a metal atom and a nonmetal atom, one or more electrons are transferred from the metal to the nonmetal, resulting in electrostatic attraction between the ions.

10.13 Lewis symbols and electron configurations showing the changes that occur as (a) zinc and atomic fluorine combine to form ionic zinc fluoride are

$$\begin{array}{ccccc} Zn\!: & + & 2\ \cdot\ddot{\underset{\cdot\cdot}{F}}\!: & \rightarrow & [Zn]^{2+} \quad 2\,[:\ddot{\underset{\cdot\cdot}{F}}:]^- \\ [Ar]3d^{10}4s^2 & & [He]2s^22p^5 & & [Ar]3d^{10} \quad [Ne] \end{array}$$

(b) calcium and atomic oxygen combine to form ionic calcium oxide are

$$\begin{array}{ccccc} Ca\!: & + & \cdot\ddot{\underset{\cdot}{O}}\!: & \rightarrow & [Ca]^{2+} \quad [:\ddot{\underset{\cdot\cdot}{O}}:]^{2-} \\ [Ar]4s^2 & & [He]2s^22p^4 & & [Ar] \quad [Ne] \end{array}$$

10.15 (a) The elements La and Cl_2 form the ionic compound $LaCl_3$. Likewise, the elements (b) Cu and F_2 form CuF and CuF_2 and (c) Cs and Br_2 form CsBr.

10.17 (a) Ca + N. This is simply the combination of a metal and a nonmetal. The compound would be Ca_3N_2.

(b) Al + Si. This is the combination of a metal and a semiconducting element. Silicon forms neither anions nor cations, so no ionic compound would be produced.

(c) K + Se. The representative metal K might combine with the semiconducting element Se, as Se forms a noble gas type of anion. The compound would be K_2Se.

(d) Fe + S. The transition metal Fe is known to form both Fe^{2+} and Fe^{3+} cations. Sulfur is a Group VI nonmetal and forms S^{2-} ions. Therefore, the two possible compounds would be FeS and Fe_2S_3.

10.19 Each ion is surrounded by ions of opposite charge; the geometry is determined by the charge and relative size of each type of ion.

10.21 The number of electrons shared between two atoms in (a) a single covalent bond is two, (b) a double covalent bond is four, and (c) a triple covalent bond is six.

10.23 The term used to describe a covalent bond in which both electrons come from the same atom is a coordinate covalent bond. The nitrogen atom is the donor atom and the boron atom is the acceptor atom in the example.

10.25 The species that obey the octet rule are

```
     ..   ..            ..                          
(a) :F - F:  (d) [     :O:      ]2-   (e)      :O:
     ..   ..      ..    |    ..                 ||
                 :O  -  S  -  O:           ..   ||   ..
                  ..    |    ..           :Cl - C - Cl:
                       :O:                 ..        ..
                        ..
```

10.27 Because the molecules usually remain intact upon such physical changes as melting and boiling, the properties of the substances are related to the strengths of the intermolecular forces.

10.29 In a covalently bonded substance the bond length is the distance between the nuclei of the two atoms. It is necessary to discuss an average value for the bond length because bonded atoms always vibrate back and forth within the region of bond stability. As the bonding increases from a single to a double to a triple covalent bond, the bond length decreases.

10.31 $H_2O(g) \rightarrow 2H(g) + O(g)$

$$\Delta H° = [(2\text{ mol})\Delta H_f°(H) + (1\text{ mol})\Delta H_f°(O)] - [(1\text{ mol})\Delta H_f°(H_2O)]$$

$$= [(2\text{ mol})(217.965\text{ kJ/mol}) + (1\text{ mol})(249.170\text{ kJ/mol})] - [(1\text{ mol})(-241.818\text{ kJ/mol})]$$

$$= 926.918\text{ kJ}$$

$$\text{average O-H bond energy} = \frac{926.918\text{ kJ}}{2\text{ mol}} = 463.459\text{ kJ/mol}$$

10.33 $$\text{average S-H bond energy} = \frac{381.27\text{ kJ} + 354.10\text{ kJ}}{2\text{ mol}} = 367.69\text{ kJ/mol}$$

10.35 $$\Delta H° = [(4\text{ mol})BE_{C-H} + (2\text{ mol})BE_{O=O}] - [(2\text{ mol})BE_{C=O} + (4\text{ mol})BE_{O-H}]$$

$$-802.335\text{ kJ} = [(4\text{ mol})(414\text{ kJ/mol}) + (2\text{ mol})(498\text{ kJ})] - [(2\text{ mol})BE_{C=O} + (4\text{ mol})(464\text{ kJ/mol})]$$

$$BE_{C=O} = 799\text{ kJ/mol}$$

10.37 (a) $$\Delta H° = [(2\text{ mol})BE_{C\equiv O} + (1\text{ mol})BE_{O=O}] - [(4\text{ mol})BE_{C=O}]$$

$$= [(2\text{ mol})(1075\text{ kJ/mol}) + (1\text{ mol})(498\text{ kJ/mol})] - [(4\text{ mol})(803\text{ kJ/mol})]$$

$$= -560\text{ kJ}$$

(b) $$\Delta H° = [(1\text{ mol})BE_{C\equiv O} + (1\text{ mol})BE_{Cl-Cl}] - [(1\text{ mol})BE_{C=O} + (2\text{ mol})BE_{C-Cl}]$$

$$= [(1\text{ mol})(1075\text{ kJ/mol}) + (1\text{ mol})(243\text{ kJ/mol})] - [(1\text{ mol})(728\text{ kJ/mol}) + (2\text{ mol})(326\text{ kJ/mol})]$$

$$= -62\text{ kJ}$$

(c) ΔH° = [(8 mol)BE_{C-H} + (6 mol)BE_{N-H} + (3 mol)$BE_{O=O}$]
- [(2 mol)BE_{C-H} + (2 mol)$BE_{C\equiv N}$ + (12 mol)BE_{O-H}]
= [(8 mol)(414 kJ/mol) + (6 mol)(389 kJ/mol)
+ (3 mol)(498 kJ/mol)] - [(2 mol)(414 kJ/mol)
+ (2 mol)(858 kJ/mol) + (12 mol)(464 kJ/mol)]
= -970 kJ

10.39 Yes, the value of ΔH° would be the same for both reactions because there are the same number of C-C and C-H bonds in both compounds.

10.41 When we classify the bonding in a substance as metallic, covalent, or ionic, we should keep in mind that the bonding is not fully any one of them (except for a covalent bond in a diatomic molecule of identical atoms), but lies at some intermediate point in electron arrangement and properties.

10.43 A covalent bond becomes polar when the electrons are shared unequally. The polar covalent bonds are (a) F - Xe, (d) C = O, and (e) C - O.

10.45 The cation that would be most effective in polarizing a given anion is Al^{3+} because it has the largest charge-to-radius ratio.

10.47 (a) Cu^+ $\frac{1}{0.096} = 10.$ (b) Cu^{2+} $\frac{2}{0.072} = 28$

A cation with a higher charge has a higher charge-to-radius ratio. The Cu^{2+} ion would therefore, polarize an anion more.

10.49 Electronegativity is the ability of an atom in a covalent bond to attract electrons to itself. The bonding between two atoms is nonpolar if their electronegativity difference is small or zero. Pairs of atoms with moderate differences in electronegativity form polar covalent bonds. With a large enough difference in electronegativity (≥2.0) an ionic bond can be expected to form.

10.51 The order of increasing electronegativity of the elements is

(d) Ra < (c) Bi < (b) Te < (a) I

10.53 The pair(s) of atoms whose bonding would be (a) ionic is (iii) Sr, F; (b) polar covalent are (i) Si, O and (ii) N, O; and (c) nonpolar covalent is (iv) As, As.

10.55 (a) Cu has metallic bonding. (b) O_2 has nonpolar covalent bonding.
(c) CuO has ionic bonding. (d) H_2 has nonpolar covalent bonding.

(e) H_2O has polar covalent bonding. (f) $\ddot{\underset{..}{O}} = \ddot{\underset{..}{O}}$ (g) $[Cu]^{2+}\ [:\ddot{\underset{..}{O}}:]^{2-}$

(h) H - H (i) H - $\ddot{\underset{..}{O}}$ - H

(j) H_2 and O_2 would have relatively low melting and boiling points.

(k) Cu is a good conductor of heat and electricity.

10.57 Oxidation numbers are numbers equal to the charge of ions or the charge that ions would have if the compound were ionic; no. These numbers do not represent actual charges on atoms in molecules.

10.59 The elements that have only one oxidation number other than 0 include members of the *s* block, the scandium family, zinc, cadmium, aluminum, and fluorine. The elements that have oxidation numbers equal to the group number and the group number minus 2 include the heavier metals in the boron, nitrogen, and oxygen families.

10.61 (a) $\overset{+1}{K}\ \overset{-1}{H}$ (b) $\overset{+2}{Mn}\ \overset{2(-1)}{Cl_2}$ (c) $\overset{-3}{N}\ \overset{4(+1)}{H_4^+}$ (d) $\overset{0}{P_4}$ (e) $\overset{-1}{Cl^-}$

(f) $\overset{+4}{S}\ \overset{3(-2)}{O_3^{2-}}$ (g) $\overset{2(+1)}{Na_2}\ \overset{2(-1)}{O_2}$ (h) $\overset{+3}{Mn}\ \overset{3(-1)}{F_3}$ (i) $\overset{+3}{I}\ \overset{3(-1)}{Cl_3}$

(j) $\overset{1(+1)}{H_2}\ \overset{-2}{Se}$

10.63 (a) $2\overset{-3}{N}\overset{+1}{H_3}(g) + 3\overset{+2}{Cu}\overset{-2}{O}(s) \rightarrow \overset{0}{N_2}(g) + 3\overset{0}{Cu}(s) + 3\overset{+1}{H_2}\overset{-2}{O}(g)$

N and Cu are undergoing changes in oxidation number.

(b) $\overset{+1}{H_2}\overset{+6}{S}\overset{-2}{O_4}(aq) + 2\overset{+1}{Na}\overset{-2}{O}\overset{+1}{H}(aq) \rightarrow \overset{+1}{Na_2}\overset{+6}{S}\overset{-2}{O_4}(aq) + 2\overset{+1}{H_2}\overset{-2}{O}(l)$

There are no changes in oxidation number.

10.65 The following substances are named using the Stock system:

(a) N_2O_3, nitrogen(III) oxide (b) ICl_3, iodine(III) chloride

(c) CO_2, carbon(IV) oxide (d) SO_2, sulfur(IV) oxide

(e) BF_3, boron(III) fluoride (f) N_2O_5, nitrogen(V) oxide

(g) $SiCl_4$, silicon(IV) chloride (h) CCl_4, carbon(IV) chloride

10.67 The formulas of the following compounds named using the Stock system are

(a) boron(III) nitride, BN (b) carbon(IV) selenide, CSe_2

(c) bromine(I) chloride, BrCl (d) nitrogen(III) oxide, N_2O_3

(e) oxygen(II) fluoride, OF_2 (f) sulfur(IV) fluoride, SF_4

(g) nitrogen(II) oxide, NO (h) phosphorus(V) chloride, PCl_5

CHAPTER 11

COVALENT BONDING AND PROPERTIES OF MOLECULES

Solutions to Exercises

11.2 (a) IO_3^-

The atoms are arranged with each oxygen forming a single bond to the central iodine. The total number of valence electrons is 7 + (3)(6) + 1 = 26. If 6 are used for the three covalent bonds, 20 are left for lone pairs.

```
  ..   ..   ..
[ :O - I  - O: ]-
  ..   |    ..
      :O:
       ..
```

(b) H_2O_2

The total number of valence electrons is (2)(1) + (2)(6) = 14. If 6 are used for the three single bonds, 8 are left for nonbonded electrons.

```
    ..  ..
H - O - O - H
    ..  ..
```

(c) $[B(OH)_4]^-$

The O atoms surround the central boron atom in a symmetrical arrangement. One H atom is bonded to each O atom. The total number of valence electrons is 3 + (4)(6) + (4)(1) + 1 = 32.

```
[           H           ]-
            |
           :O:
    ..      |     ..
 H - O  -   B  -  O - H
    ..      |     ..
           :O:
            |
            H           ]
```

(d) IF_5

The fluorine atoms surround the central iodine atom. The total number of valence electrons is 7 + (5)(7) = 42. If 10 are used for the five single bonds, 32 are left for nonbonded electrons.

```
         ..
        :F:
  ..     |     ..
 :F  -   I  -  F:
  ..    /..\   ..
      :F:  :F:
       ..   ..
```

(e) CS_2

Each sulfur atom is bonded to the central carbon atom. The total number of valence electrons is 4 + (2)(6) = 16. If 4 are used for the two single bonds between the carbon atom and the sulfur atoms, 12 remain for nonbonding electrons.

```
..       ..
S = C = S
..       ..
```

Under this arrangement, there is a deficiency of 4 in order for each atom to obey the octet rule. However, using two double bonds (8 electrons) between the carbon atom and the sulfur atoms leaves 8 electrons for lone pairs. With this arrangement, all atoms have an octet.

11.3 NO_2^-, nitrite ion
The two oxygen atoms are bonded to the central nitrogen atom. The total number of valence electrons is 5 + (2)(6) + 1 = 18. Resonance forms are

```
 ..   ..          ..    ..
[:O - N = O ]⁻ ↔ [ O = N - O:]
 ..        ..     ..        ..
```

11.4 The Lewis structures of (a) CO_2, (b) SF_6, (c) NH_4^+ are shown at the right. CO_2 is a linear molecule (AB_2 type). SF_6 is an octahedral molecule (type AB_6). NH_4^+ is a a tetrahedral ion (type AB_4).

```
 ..       ..
:O = C = O:

  ..    ..           [     H     ]+
 :F:   :F:           |     |     |
 ..  \ /  ..         |  H - N - H|
:F - S - F:          |     |     |
 ..  / \  ..         [     H     ]
 :F:   :F:
  ..    ..
```

11.5 The Lewis structure of $[I_3]^-$ is shown at the right. $[I_3]^-$ is a linear ion (type AB_2E_3).
The Lewis structure of ClO_2 is shown at the right. ClO_2 is a bent molecule (type AB_2E_2).
The O-Cl-O angle should be greater than 109.47° because the central Cl atom has a single unpaired electron.

```
  ..  ..  ..
[:I - I - I:]⁻
  ..  ..  ..

 ..   .    ..
:O - Cl - O:
 ..   ..   ..
```

11.6 There are four equivalent hybrid orbitals on each oxygen atom, formed by sp^3 hybridization and directed toward the four corners of a tetra-tetrahedron; two contain contain lone pairs of electrons and two contain a single electron each, which will form the σ bonds with the hydrogen atoms.

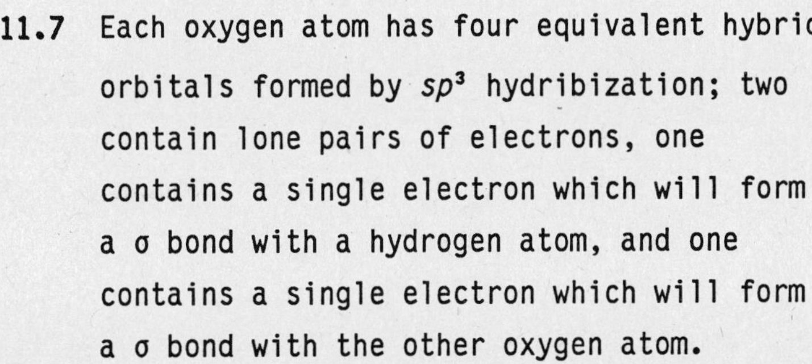

11.7 Each oxygen atom has four equivalent hybrid orbitals formed by sp^3 hydribization; two contain lone pairs of electrons, one contains a single electron which will form a σ bond with a hydrogen atom, and one contains a single electron which will form a σ bond with the other oxygen atom.

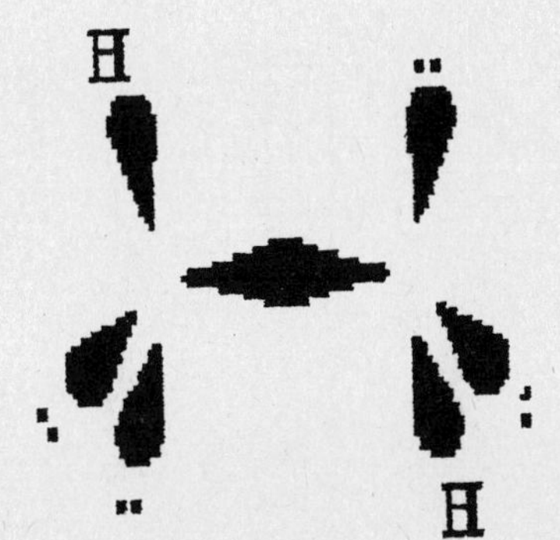

11.8 The nitrogen atom is sp^2-hybridized; the three hybridized orbitals are used for the N-H σ bond, the lone pair of electrons, and the N-O σ bond. The oxygen atom may be pictured as also sp^2 hybridized. The unhybridized $2p_z$ orbital on the N nitrogen atom is used to form the π bond by the parallel overlap of the unhybridized $2p_z$ orbital on the oxygen atom.

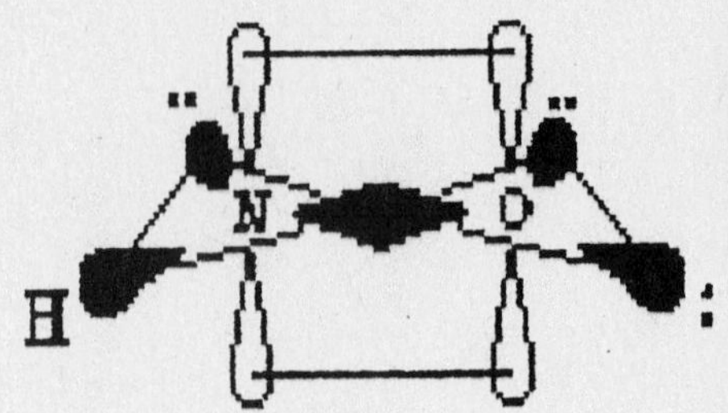

11.9 Each carbon atom is sp^2-hybridized; the three hybridized orbitals are used for the σ bonds between C and H and between C and C; the parallel unhybridized *p* orbitals form delocalized π bonds with other C atoms.

11.10 The Lewis structures of the molecules are

```
(a)    :O:             (b)     :F:            (c)  :F:
    .. .. |:.. ..          ..   |   ..             :|
  :O  -  S  -  O:        :Cl  -  B  -  Cl:           Xe:
                           ..         ..           :|
                                                   :F:
```

(a)	(b)	(c)
triangular planar (type AB_3) not polar	triangular planar (type AB_3) polar	linear (type AB_2E_3) not polar

11.11 (a) The Lewis structure of ethyl alcohol is shown at the right. The properties of ethyl alcohol will be influenced by dipole-dipole interactions, London forces, and hydrogen bonding.

```
    H   H
    |   |   ..
H - C - C - O - H
    |   |   ..
    H   H
```

(b) The Lewis structure of PH_3 is shown at the right. Because the geometry is pyramidal (type AB_3E), the properties of PH_3 will be influenced by dipole-dipole interactions and London forces.

```
    ..
H - P - H
    |
    H
```

Solutions to Odd-Numbered Questions and Problems

11.1 The Lewis structure gives information about the arrangement of the atoms in the structure and the chemical bonds that exist. It will not give quantitative information such as bond length, bond angles, bond strengths, or molecular geometry.

```
         ..                  ..    ..   ..          ..   .. ..    ..
11.3 (a) [      ..       ]-  (b) O  =  N  -  F:    (c) :F  -  Xe  -  F:
         [     :O:       ]       ..          ..         ..    / \      ..
         [  ..  |   ..   ]                                  :F:  :F:
         [ :O - Cl - O:  ]                                   ..   ..
         [  ..  |   ..   ]
         [     :O:       ]
         [      ..       ]

                                                     ..  .. ..    ..
     (d) [       ..        ]3-  (e)     :O:     (f) :F - Cl - F:
         [  ..  :Cl:  ..   ]        ..   ||   ..     ..   |    ..
         [ :Cl:  |   :Cl:  ]       :Cl - C - Cl:         :F:
         [     \ | /       ]        ..        ..          ..
         [      Cr         ]
         [     / | \       ]
         [ :Cl:  |   :Cl:  ]
         [  ..  :Cl:  ..   ]
         [       ..        ]

                                  ..     ..                          ..  ..  ..  ..
11.5 (a)      H          (b)      S  —  S     (c)    H         (d) :F - O - O - F:
         ..   |                  /..    ..\          |              ..  ..  ..  ..
     H - O - N - H            :S:        :S:    H - Si - H
         ..  ..                 |          |         |         (e) :C ≡ O:
                              :S:        :S:         H
                                 \..    ../
                                  S  —  S
                                  ..     ..

              ..         ..
     (f)     :Cl         Cl:
          ..  ..  \   /  ..  ..
         :Cl  —  Se  —  Cl:
          ..  ..  /   \  ..  ..
             :Cl         Cl:
              ..         ..

      ..         ..               ..        ..       ..
11.7 :Cl - Al - Cl:              :Cl        Cl       Cl:
      ..   |     ..               ..  \   /  ..  \  /  ..
          :Cl:                          Al         Al
           ..                     ..  /   \  ..  /  \  ..
                                 :Cl        Cl       Cl:
                                  ..        ..       ..

11.9 [        ..  ]-         [              ]-
     [ H - C - O: ]    ↔     [ H - C = O:   ]
     [     ||  .. ]          [     |    ..  ]
     [    :O:     ]          [    :O:       ]
                                   ..
```

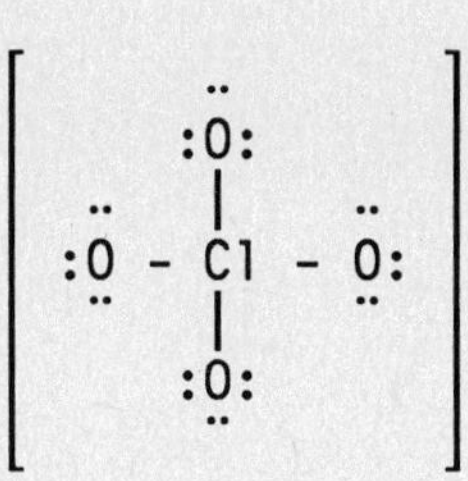

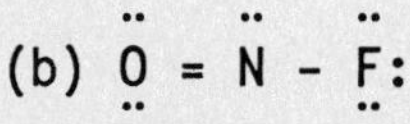

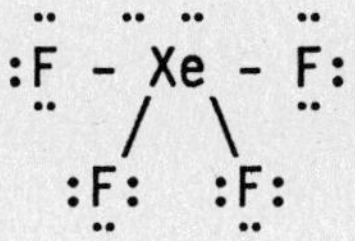

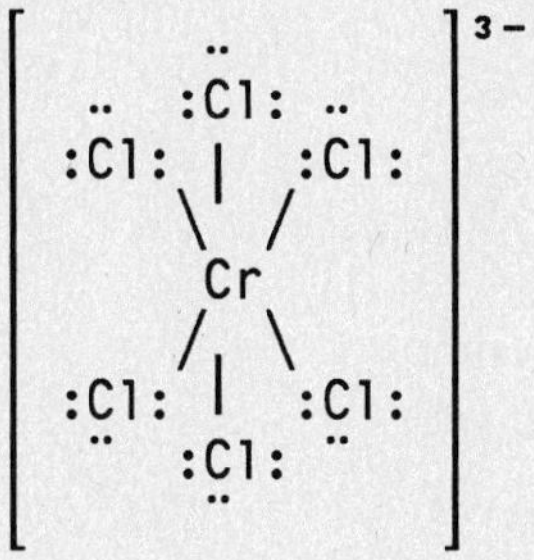

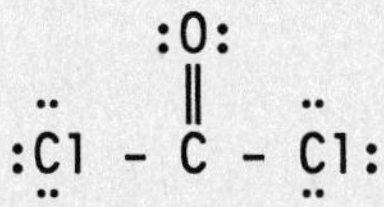

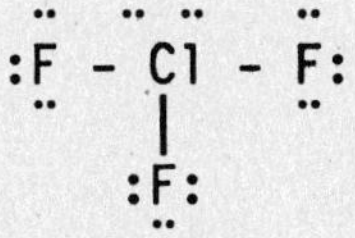

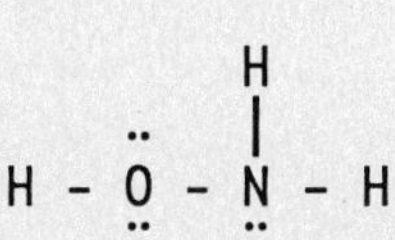

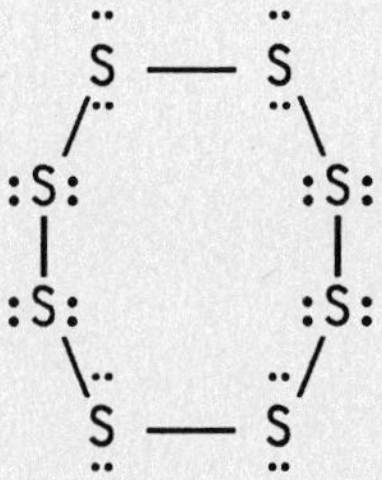

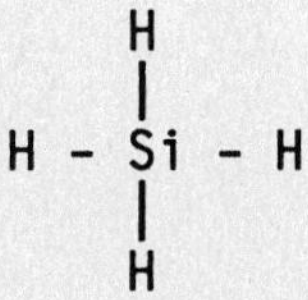

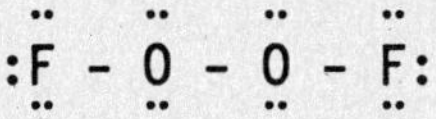

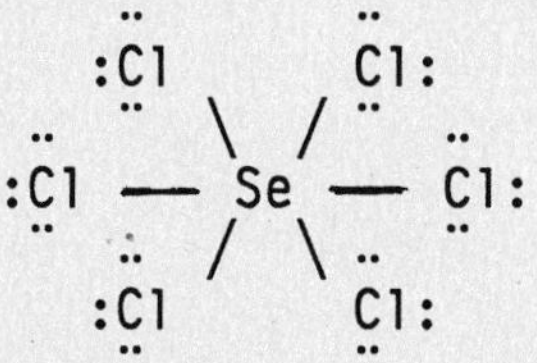

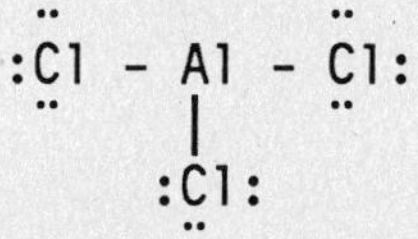

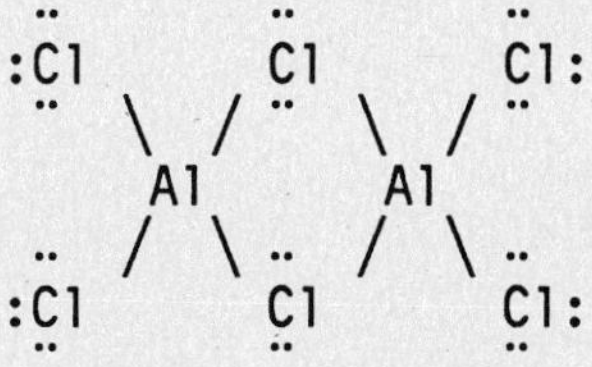

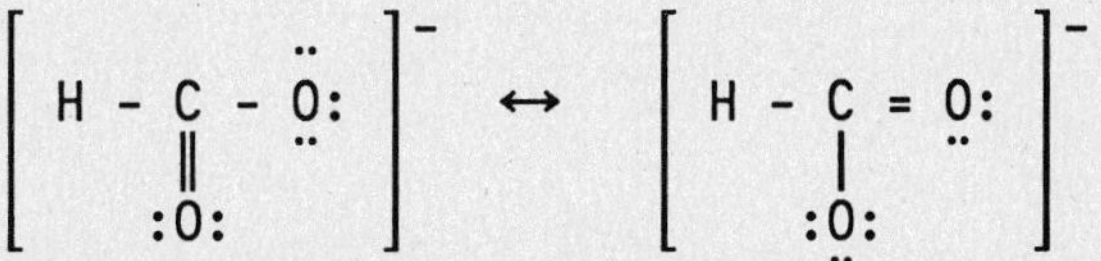

11.11 For toluene the carbon-carbon bond length in the six-member ring would be less than the carbon-carbon bond length between the CH_3 group and the carbon atom on the ring because the bond lengths in the ring are between single and double bond length.

11.13 The following molecules would have the predicted geometric structures indicated:

(a) AB, linear (b) AB_2, linear

(c) AB_3, triangular planar (d) AB_5, triangular bipyramidal

11.15 The order of decreasing strength of electron-pair repulsions involving lone pairs and bonded pairs of electrons is LP-LP > LP-BP > BP-BP.

11.17 The three different possible arrangements of the two B atoms around the central atom A for the molecule AB_2E_3 are shown in Fig. 11-1. The first sketch (linear) correctly describes the molecular geometry because it places the lone pairs of electrons as far apart as possible.

11.19 The geometries of the following species are predicted by using VSEPR theory to be

(a) $:\ddot{\underset{..}{F}} - \ddot{\underset{..}{F}}:$

linear (type A_2)

(b) $H - \ddot{\underset{..}{S}} - H$

bent (type AB_2E_2)

(c) $:\ddot{\underset{..}{F}} - \ddot{\underset{..}{O}} - \ddot{\underset{..}{F}}:$

bent (type AB_2E_2)

(d) $\left[H - \ddot{\underset{|}{\underset{}{O}}} - H \right]^+$ (with a third H bonded below O)

triangular pyramidal (type AB_3E)

(e) Xe bonded to four F atoms (above, below, left, right), each F with three lone pairs, and two lone pairs on Xe:

$:\ddot{\underset{..}{F}} - Xe - \ddot{\underset{..}{F}}:$ (with $:\ddot{F}:$ above and $:\underset{..}{F}:$ below)

square planar (type AB_4E_2)

(f) $:C \equiv O:$

linear (type AB)

(g) $\left[:\ddot{\underset{..}{I}} - \ddot{\underset{..}{I}} - \ddot{\underset{..}{I}}: \right]^-$

linear (type AB_2E_3)

11.21 The geometries of the following species are predicted by using VSEPR theory to be

(a) $:\underset{..}{O} = C = \underset{..}{O}:$

linear (type AB_2)

(b) $\left[H - Al - H \right]^-$ (with H above and H below Al)

tetrahedral (type AB_4)

Fig. 11-1

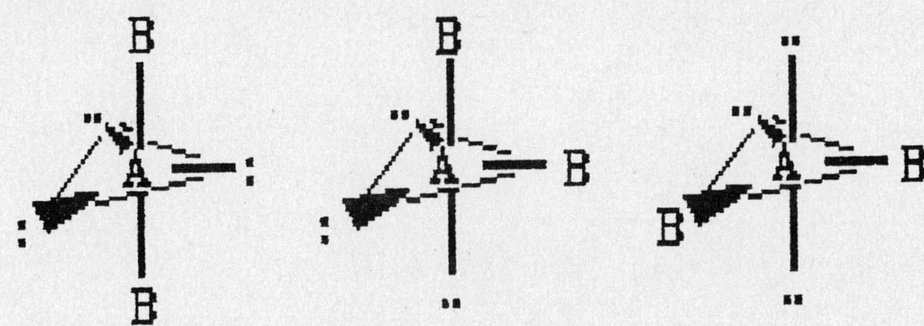


```
(c) [      H      ]+            (d) [        OH2        ]3+
    [      |      ]                 [  H2O    |   OH2   ]
    [  H - N - H  ]                 [      \     /      ]
    [      |      ]                 [        Cr         ]
    [      H      ]                 [      /     \      ]
                                    [  H2O    |   OH2   ]
    tetrahedral (type AB4)          [        OH2        ]
                                    octahedral about the central Cr (type AB6)

       ..     ..     ..                 ..          ..
(e) [ :O  -  Cl  -  O: ]-         (f) :F           F:
       ..     |      ..                 ..  \    /  ..
             :O:                     :F  -  Se  -  F:
              ..                        ..  /    \  ..
                                        :F           F:
triangular pyramidal (type AB3E)        ..           ..
                                    octahedral (type AB6)

     ..   ..   ..
(g) :O  = N  - Cl:
                ..
    bent (type AB2E)
```

11.23 The ideal angles should be 109.47°. The two lone pairs of electrons on the oxygen atom should make the N-O-H angle less than 109.47° and the lone pair on the nitrogen atom should make both the H-N-H and H-N-O angles less than 109.47°. (The observed bond angles are 102° for N-O-H, 107° for H-N-H, and 107° for H-N-O.)

11.25 See Fig. 11-2.

11.27 In terms of simple valence bond theory, the σ bond in the F_2 molecule results from the overlap of the partially filled 2*p* orbital on one F atom with the similar orbital on the other F atom.

Fig. 11-2

(a) (b) (c)

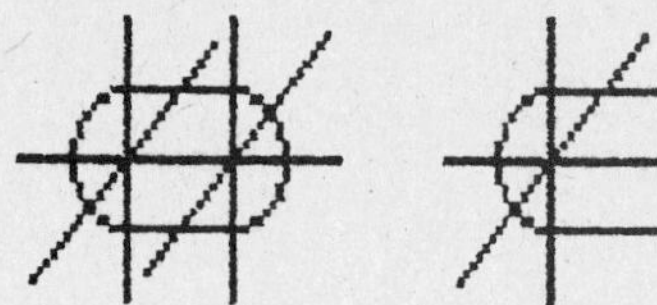

(d)

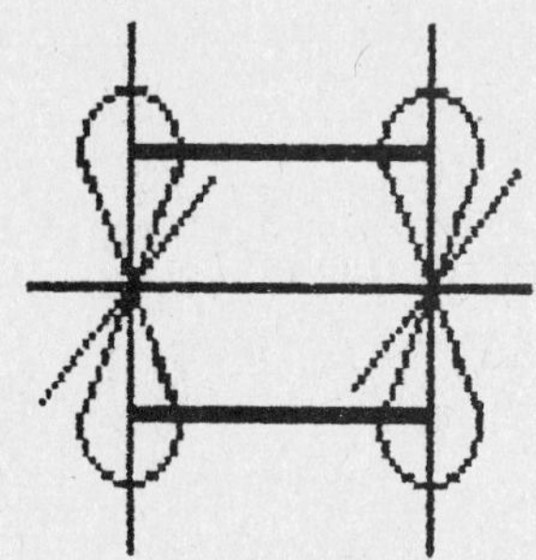

11.29 Hybridized atomic orbitals are a new set of orbitals on an atom formed by mixing of the atomic orbitals. Hybrid orbitals were introduced to explain molecular geometry in terms of atomic orbitals and valence bond theory.

11.31 The types of hybridization predicted for molecules having the following general formulas are

(a) AB_3, *sp*^{2} (b) AB_2E_2, *sp*^{3} (c) AB_3E, *sp*^{3}

(d) ABE_4, *sp*^{3}*d* (e) ABE_3, *sp*^{3}

11.33 The hybridization of the central atom in each of the following is

(a) H_2Be, *sp* *(b)* $AlCl_3$, *sp*^{2} (c) SiH_4, *sp*^{3}

(d) IO_4^-, *sp*^{3} (e) NCl_3, *sp*^{3} (f) ClO_3^-, *sp*^{3}

(g) PCl_5, *sp*^{3}*d* (h) BCl_3, *sp*^{2} (i) ClO_4^-, *sp*^{3}

11.35 The hybridization of the carbon atom in each of the following is (a) CO, *sp*; (b) CO_2, *sp*; and (c) CO_3^{2-}, *sp*^{2}. Localized π bonding is found in (a) CO and (b) CO_2. Delocalized π bonding is found in (c) CO_3^{2-}.

11.37 The Lewis structures for molecular oxygen and ozone are $:\ddot{O} = \ddot{O}:$ and $:\ddot{O} = \ddot{O} - \ddot{\underset{..}{O}}:$ ↔ $:\ddot{\underset{..}{O}} - \ddot{O} = \ddot{O}:$, respectively. The hybridization of the

Fig. 11-3

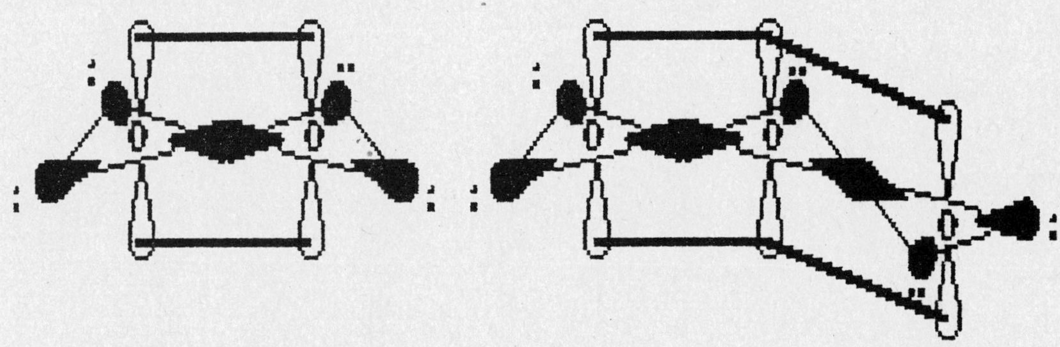

oxygen atoms in each compound is *sp*². The sketches are shown in Fig. 11-3.

11.39 Forces known as van der Waals forces are intermolecular forces. The two types of van der Waals forces considered in this chapter are dipole-dipole interactions and London forces.

11.41 London forces are caused by momentary shifts in the symmetry of the electron cloud of a molecule--the resulting slight positive charge produced at one end of one molecule induces a slight negative charge in one end of the molecule next to it, and for an instant a force of attraction exists between these molecules. The strength of London forces is influenced by the size and geometry of the molecules involved and by the ease of polarization of the electron clouds.

11.43 The Lewis structures and geometries of the molecules are

```
(a)     H
        |
    H - Si - H
        |
        H

 tetrahedral (type AB4)
 not polar
```

```
(b)  ..        ..
    :Cl - Mg - Cl:
     ..        ..

 linear (type AB2)
 not polar
```

```
      ..  ..   ..
(c) :O = N - Cl:
               ..

 nonlinear (type AB2E)
 polar
```

```
      ..    ..   ..
(d) :Cl - N - Cl:
      ..    |    ..
           :Cl:
            ..

 triangular pyramidal (type AB3E)
 polar
```

(e) $:\ddot{\underset{..}{F}} - \ddot{\underset{..}{O}} - \ddot{\underset{..}{F}}:$

bent (type AB_2E_2)
polar

11.45 The molecules which exhibit hydrogen bonding are (b) N_2H_4 and (c) CH_3CH_2OH.

11.47 Hydrogen fluoride has a lower boiling point and lower heat of vaporization than H_2O because each H_2O molecule is involved in four hydrogen bonds and each HF molecule is involved in only two.

11.49 The intermolecular forces that would be important for
(a) CO_2 are London forces.
(b) $AsCl_5$ are London forces.
(c) Cl_2CO are London forces and dipole-dipole interactions.
(d) $MgCl_2$ are London forces.
(e) SeF_4 are London forces and dipole-dipole interactions.
(f) BCl_3 are London forces.
(g) NOCl are London forces and dipole-dipole interactions.

11.51 The substance within each group that has the greatest intermolecular forces to overcome is
(a) S_8, the heaviest.
(b) SO_2, the most polar.
(c) F_2, nonspherical.
(d) *n*-octane, more nonspherical.
(e) CCl_4, the heavier.

11.53 The intermolecular forces present in liquid ammonia are dipole-dipole interactions, hydrogen bonding, and London forces; in liquid methane only London forces are present. Methane should have the lower freezing and boiling points. Ammonia would be expected to be a liquid over a larger temperature range.

11.55 The molecule expected to have the higher boiling point is the one with four atoms attached all in one continuous chain. Both molecules will have weak dipole-dipole interactions, but because of the more nearly spherical shape of the molecule with the shorter carbon chain, the London forces will be less for this molecule and hence it will have a lower boiling point.

11.57 The thermochemical equation for the S-F average bond energy in SF_4 is

$$SF_4(g) \rightarrow S(g) + 4F(g)$$

$$\begin{aligned}\Delta H^\circ &= [(1\ \text{mol})\Delta H^\circ_f(S) + (4\ \text{mol})\Delta H^\circ_f(F)] - [(1\ \text{mol})\Delta H^\circ_f(SF_4)] \\ &= [(1\ \text{mol})(278.805\ \text{kJ/mol}) + (4\ \text{mol})(78.99\ \text{kJ/mol})] \\ &\qquad - [(1\ \text{mol})(-774.9\ \text{kJ/mol})] \\ &= 1369.7\ \text{kJ}\end{aligned}$$

$$\text{average bond energy} = \frac{1369.7\ \text{kJ}}{4} = 342.4\ \text{kJ/mol}$$

The thermochemical equation for the S-F average bond energy in SF_6 is

$$SF_6(g) \rightarrow S(g) + 6F(g)$$

$$\begin{aligned}\Delta H^\circ &= [(1\ \text{mol})\Delta H^\circ_f(S) + (6\ \text{mol})\Delta H^\circ_f(F)] - [(1\ \text{mol})\Delta H^\circ_f(SF_6)] \\ &= [(1\ \text{mol})(278.805\ \text{kJ/mol}) + (6\ \text{mol})(78.99\ \text{kJ/mol})] \\ &\qquad - [(1\ \text{mol})(-1209\ \text{kJ/mol})] \\ &= 1962\ \text{kJ}\end{aligned}$$

$$\text{average bond energy} = \frac{1962\ \text{kJ}}{6} = 327\ \text{kJ/mol}$$

11.59 (a) The polar molecules would be i and iii.

(b) The structures having equal bond lengths and strengths are i, ii, and iii.

(c) The diamagnetic molecules will be i and ii.

(d) Only structure i correctly predicts all three properties.

(e) Structure ii contains a considerable amount of strain.

11.61 (a)

```
 ..   ..    ..         ..        ..                  ..        ..
:F  - Xe  - F:        :F         F:                 :F         F:
 ..   ....  ..         ..  \.. /  ..             ..  ..  \.. /  ..  ..
                             Xe                 :F  ——  Xe  ——  F:
     sp³d              ..  /.. \  ..             ..      / \        ..
                      :F         F:                  ..  /   \  ..
                       ..        ..                 :F         F:
                                                     ..        ..
                          sp³d²                        sp³d³
```

(b) The three configurations are shown in Fig. 11-4. The geometry should be linear (first sketch) because the lone pairs occupy the positions that minimize the number of lone pairs separated by 90°.

(c) The shape of XeF_4 is square planar.

(d) The important intermolecular forces in XeF_2 and XeF_4 are London forces.

Fig. 11-4

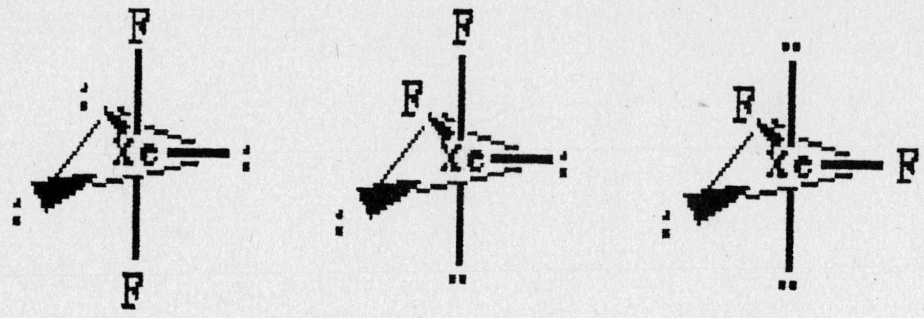

CHAPTER 12

NUCLEAR CHEMISTRY

Solutions to Exercises

12.1 calculated mass = (46 neutrons)(1.008665 u/neutron) + (35 protons)(1.007276 u/proton) + (35 electrons)(0.00054858 u/electron) = 81.67245 u

mass defect = 80.9163 u - 81.67245 u = -0.7562 u

$$(-0.7562\ \text{u})\left[\frac{1.6606 \times 10^{-27}\ \text{kg}}{1\ \text{u}}\right] = -1.256 \times 10^{-27}\ \text{kg}$$

$$E = (-1.256 \times 10^{-27}\ \text{kg})(2.9979 \times 10^{8}\ \text{m/s})^2\left[\frac{1\ \text{J}}{1\ \text{kg m}^2/\text{s}^2}\right]$$

$$= -1.129 \times 10^{-10}\ \text{J}$$

$$\left[\frac{-1.129 \times 10^{-10}\text{J}}{1\ \text{atom}}\right]\left[\frac{6.022 \times 10^{23}\ \text{atoms}}{1\ \text{mol}}\right] = -6.799 \times 10^{13}\ \text{J/mol}$$

12.2 $$k = \frac{0.693}{t_{1/2}} = \frac{0.693}{5.26\ \text{yr}} = 0.132\ \text{yr}^{-1}$$

$$\log\left[\frac{q_0}{q}\right] = \frac{kt}{2.303} = \frac{(0.132\ \text{yr}^{-1})(3.0\ \text{yr})}{2.303} = 0.17$$

$$\frac{q_0}{q} = 1.5$$

$$\frac{q}{q_0} = \frac{1}{1.5} = 0.67$$

67 % remains

12.3 $$k = \frac{0.693}{t_{1/2}} = \frac{0.693}{5730\ \text{yr}} = 1.21 \times 10^{-4}\ \text{y}^{-1}$$

$$t = \frac{(2.303)\log\left[\frac{a_0}{a}\right]}{k} = \frac{(2.303)\log\left[\frac{15.3\ \text{min}^{-1}\ (\text{g C})^{-1}}{2.93\ \text{min}^{-1}\ (\text{g C})^{-1}}\right]}{1.21 \times 10^{-4}\ \text{yr}^{-1}} = 1.37 \times 10^{4}\ \text{yr}$$

12.4 $^{10}_{5}B + ^{1}_{0}n \rightarrow ^{10}_{4}Be + ^{1}_{1}H$

The unknown particle formed is a proton.

12.5 $^{242}Cm(\alpha,n)^{245}Cf$

Solutions to Odd-Numbered Questions and Problems

12.1 The term "nucleon" is a collective term for protons and neutrons. The number of protons and the atomic number are the same. The sum of the number of protons plus the number of neutrons is the mass number.

12.3 The nuclear binding energy is the energy that would be released in the combination of nucleons to form the nucleus. This energy is calculated from the mass defect by using the mass-energy relationship.

12.5 (a) $^{14}_{7}N$

calculated mass = (7 neutrons)(1.008665 u/neutron)
+ (7 protons)(1.007276 u/proton)
+ (7 electrons)(0.00054858 u/electron)
= 14.115427 u

mass defect = 14.00307 u - 14.115427 u = -0.11236 u

$$(-0.11236\ \text{u})\left[\frac{1.6605655 \times 10^{-27}\ \text{kg}}{1\ \text{u}}\right] = -1.8658 \times 10^{-28}\ \text{kg}$$

$$E = mc^2 = (-1.8658 \times 10^{-28}\ \text{kg})(2.9979 \times 10^{8}\ \text{m/s})^2\left[\frac{1\ \text{J}}{1\ \text{kg m}^2/\text{s}^2}\right]$$
$$= -1.6769 \times 10^{-11}\ \text{J}$$

$$\text{average binding energy} = \frac{-1.6769 \times 10^{-11}\ \text{J}}{14\ \text{nucleons}} = -1.1978 \times 10^{-12}\ \text{J/nucleon}$$

(b) $^{56}_{26}Fe$

calculated mass = (30 neutrons)(1.008665 u/neutron)
+ (26 protons)(1.007276 u/proton)
+ (26 electrons)(0.00054858 u/electron)
= 56.46339 u

mass defect = 55.9349 u - 56.46339 u = -0.5285 u

$$(-0.5285\ \text{u})\left[\frac{1.6605655 \times 10^{-27}\ \text{kg}}{1\ \text{u}}\right] = -8.776 \times 10^{-28}\ \text{kg}$$

$$E = (-8.776 \times 10^{-28}\ \text{kg})(2.9979 \times 10^{8}\ \text{m/s})^2\left[\frac{1\ \text{J}}{1\ \text{kg m}^2/\text{s}^2}\right]$$

$$= -7.887 \times 10^{-11}\ \text{J}$$

$$\text{average binding energy} = \frac{-7.887 \times 10^{-11}\ \text{J}}{56\ \text{nucleons}} = -1.408 \times 10^{-12}\ \text{J/nucleon}$$

(c) $^{130}_{52}Te$

$$\begin{aligned}\text{calculated mass} &= (78\ \text{neutrons})(1.008665\ \text{u/neutron}) \\ &\quad + (52\ \text{protons})(1.007276\ \text{u/proton}) \\ &\quad + (52\ \text{electrons})(0.00054858\ \text{u/electron}) \\ &= 131.08275\ \text{u}\end{aligned}$$

$$\text{mass defect} = 129.9067\ \text{u} - 131.08275\ \text{u} = -1.1760\ \text{u}$$

$$(-1.1760\ \text{u})\left[\frac{1.6605655 \times 10^{-27}\ \text{kg}}{1\ \text{u}}\right] = -1.9528 \times 10^{-27}\ \text{kg}$$

$$E = (-1.9528 \times 10^{-27}\ \text{kg})(2.9979 \times 10^{8}\ \text{m/s})^2\left[\frac{1\ \text{J}}{1\ \text{kg m}^2/\text{s}^2}\right]$$

$$= -1.7551 \times 10^{-10}\ \text{J}$$

$$\text{average binding energy} = \frac{-1.7551 \times 10^{-10}\ \text{J}}{130\ \text{nucleons}} = -1.3501 \times 10^{-12}\ \text{J/nucleon}$$

$^{56}_{26}Fe$ has the largest binding energy per nucleon.

12.7 The very light elements have small values; the maximum stability is at mass numbers of 40 to 100; the heavier elements have slightly smaller values than the intermediate elements.

12.9 (a) The α particle will be drawn toward the negative electrode, the β particle toward the positive electrode, and the γ radiation will be unaffected by the electrical field.

(b) The α particle and β particle will be drawn toward opposite positions of the magnetic field and the γ radiation will be unaffected by the magnetic field.

(c) The piece of paper will reduce the α radiation significantly, but not the β or γ radiation; the concrete will prevent most of the α particles and β particles from passing, but not the γ radiation.

α Particles are helium-4 nuclei, β particles are electrons, and γ radiation is very high energy electromagnetic radiation.

12.11 Various isotopes of the same element undergo the same chemical reactions because they have the same number of electrons and electronic structure.

Chemical behavior is dependent upon electronic structure, not nuclear structure.

12.13 The probability of the electron to be captured is related to its location in the atom which is a function of the chemical environment.

12.15 $k = \frac{0.693}{t_{1/2}} = \frac{0.693}{29 \text{ s}} = 0.024 \text{ s}^{-1}$

$$\log\left[\frac{q_0}{q}\right] = \frac{kt}{2.303} = \frac{(0.024 \text{ s}^{-1})(5.0 \text{ s})}{2.303} = 0.052$$

$$\frac{q_0}{q} = 1.13$$

$$\frac{q}{q_0} = 0.885$$

88.5 % remains

12.17

$$k = \frac{(2.303)\log\left[\frac{q_0}{q}\right]}{t} = \frac{(2.303)\log\left[\frac{100.0}{94.5}\right]}{1.00 \text{ yr}} = 0.0566 \text{ yr}^{-1}$$

$$t_{1/2} = \frac{0.693}{k} = \frac{0.693}{0.0566 \text{ yr}^{-1}} = 12.2 \text{ yr}$$

12.19 (a) $k = \frac{0.693}{t_{1/2}} = \frac{0.693}{1.28 \times 10^{9} \text{ yr}} = 5.41 \times 10^{-10} \text{ yr}^{-1}$

(b) $$(1.00 \text{ g KCl})\left[\frac{1 \text{ mol KCl}}{74.55 \text{ g KCl}}\right]\left[\frac{6.022 \times 10^{23} \text{ KCl ion pairs}}{1 \text{ mol KCl}}\right]\left[\frac{1 \text{ K}^+ \text{ ion}}{1 \text{ KCl ion pair}}\right]$$

$$= 8.08 \times 10^{21} \text{ K}^+ \text{ ions}$$

(c) $$(8.08 \times 10^{21} \text{ K}^+ \text{ ions})\left[\frac{1.17 \ {}^{40}\text{K}^+ \text{ ions}}{100 \text{ K}^+ \text{ ions}}\right] = 9.45 \times 10^{19} \ {}^{40}\text{K}^+ \text{ ions}$$

$$\text{activity} = (5.41 \times 10^{-10} \text{ yr}^{-1})(9.45 \times 10^{19} \ {}^{40}\text{K}^+ \text{ ions})\left[\frac{1 \text{ yr}}{365 \text{ d}}\right]$$

$$\times \left[\frac{1 \text{ d}}{24 \text{ h}}\right]\left[\frac{1 \text{ h}}{3600 \text{ s}}\right] = 1620 \text{ s}^{-1}$$

12.21 $k = \frac{0.693}{t_{1/2}} = \frac{0.693}{5730 \text{ yr}} = 1.21 \times 10^{-4} \text{ yr}^{-1}$

$$t = \frac{(2.303)\log\left[\frac{q_0}{q}\right]}{k} = \frac{(2.303)\log\left[\frac{15.3 \text{ min}^{-1} (\text{g C})^{-1}}{8.3 \text{ min}^{-1} (\text{g C})^{-1}}\right]}{1.21 \times 10^{-4} \text{ yr}^{-1}} = 5.1 \times 10^{3} \text{ yr}$$

The artifact is 5100 years old.

12.23 (a)

a,cpm	10,800	9,100	7,200	6,000	3,400	3,200	2,600	1,700	280	230	160
log *a*	4.03	3.96	3.86	3.78	3.53	3.51	3.41	3.23	2.45	2.36	2.20
t,day	0	2	4	7	13	14	16	21	42	44	49

The data are plotted in the graph shown in Fig. 12-1.

(b) $k = -(2.303)(-0.038\ \text{d}^{-1}) = 0.088\ \text{d}^{-1}$

(c) $t_{1/2} = \dfrac{0.693}{k} = \dfrac{0.693}{0.088\ \text{d}^{-1}} = 7.9\ \text{d}$

12.25 Hydrogen and helium are far more abundant than other elements. Elements with even atomic numbers are more abundant than those with odd atomic numbers. Elements with lower atomic numbers are more abundant than those with higher atomic numbers.

12.27 (a) ${}^{63}_{28}\text{Ni} \rightarrow {}^{0}_{-1}e + {}^{63}_{29}\text{Cu}$ (b) $2\ {}^{2}_{1}\text{H} \rightarrow {}^{3}_{2}\text{He} + {}^{1}_{0}n$

(c) ${}^{10}_{5}\text{B} + {}^{1}_{0}n \rightarrow {}^{7}_{3}\text{Li} + {}^{4}_{2}\text{He}$ (d) ${}^{14}_{7}\text{N} + {}^{1}_{0}n \rightarrow 3\ {}^{4}_{2}\text{He} + {}^{3}_{1}\text{H}$

The unknown nuclide is ${}^{10}\text{B}$.

12.29 ${}^{235}_{92}\text{U} \rightarrow {}^{4}_{2}\text{He} + {}^{231}_{90}\text{Th}$ ${}^{231}_{90}\text{Th} \rightarrow {}^{0}_{-1}e + {}^{231}_{91}\text{Pa}$ ${}^{231}_{91}\text{Pa} \rightarrow {}^{4}_{2}\text{He} + {}^{227}_{89}\text{Ac}$

${}^{227}_{89}\text{Ac} \rightarrow {}^{0}_{-1}e + {}^{227}_{90}\text{Th}$ Radioactinium is ${}^{227}_{90}\text{Th}$.

12.31 ${}^{12}_{6}\text{C} + {}^{1}_{1}\text{H} \rightarrow {}^{13}_{7}\text{N} + \gamma$ A is ${}^{13}\text{N}$

${}^{13}_{7}\text{N} \rightarrow {}^{13}_{6}\text{C} + {}^{0}_{+1}e$ B is ${}^{13}\text{C}$

${}^{13}_{6}\text{C} + {}^{1}_{1}\text{H} \rightarrow {}^{14}_{7}\text{N} + \gamma$ C is ${}^{14}\text{N}$

${}^{14}_{7}\text{N} + {}^{1}_{1}\text{H} \rightarrow {}^{15}_{8}\text{O} + \gamma$ D is ${}^{15}\text{O}$

${}^{15}_{8}\text{O} \rightarrow {}^{15}_{7}\text{N} + {}^{0}_{+1}e$ E is ${}^{15}\text{N}$

${}^{15}_{7}\text{N} + {}^{1}_{1}\text{H} \rightarrow {}^{12}_{6}\text{C} + {}^{4}_{2}\text{He}$ F is ${}^{4}\text{He}$

12.33 (a) During isomeric transition the (i) atomic number is unchanged, (ii) mass number is unchanged, and (iii) *n*/*p* ratio is unchanged.

(b) During α decay the (i) atomic number decreases by 2, (ii) mass number decreases by 4, and (iii) *n*/*p* ratio increases.

(c) During β^- decay the (i) atomic number increases by 1, (ii) mass number is unchanged, and (iii) *n*/*p* ratio decreases.

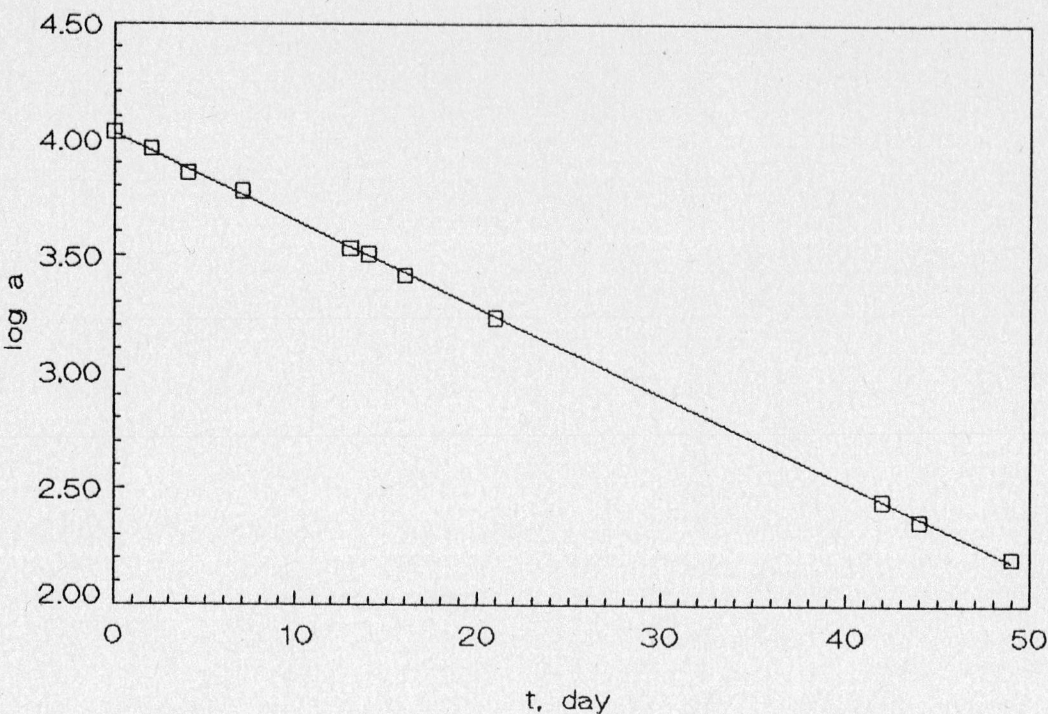

(d) $^{91m}_{39}Y \rightarrow ^{91}_{39}Y + \gamma$

(e) $^{205}_{84}Po \rightarrow ^{4}_{2}He + ^{201}_{82}Pb$

(f) $^{215}_{83}Bi \rightarrow ^{0}_{-1}e + ^{215}_{84}Po$

12.35 The mode of decay expected for a nuclide with a neutron-proton ratio larger than those of its stable isotopes is β^- decay, because by increasing the number of protons and decreasing the number of neutrons, the neutron-proton ratio is decreased.

12.37 (a) $^{13}_{5}B$ $n/p = 8/5 = 1.6$ $^{11}_{5}B$ $n/p = 6/5 = 1.2$ $^{10}_{5}B$ $n/p = 5/5 = 1.0$

Because n/p for ^{13}B is too high, it might decay by β^- emission.

(b) $^{81}_{38}Sr$ $n/p = 43/38 = 1.13$ $^{84}_{38}Sr$ $n/p = 46/38 = 1.21$

$^{88}_{38}Sr$ $n/p = 50/38 = 1.32$

Because n/p for ^{81}Sr is too low, it might decay by β^+ emission and/or by electron capture.

(c) $^{212}_{82}Pb$ $n/p = 130/82 = 1.585$ $^{204}_{82}Pb$ $n/p = 122/82 = 1.488$

$^{208}_{82}Pb$ $n/p = 126/82 = 1.537$

Because n/p for ^{212}Pb is too high, it might decay by β^- emission.

12.39 The average atomic mass of Gd is 157.25, which suggests $^{159}_{64}Gd$ has too many neutrons; therefore, it is likely a β^- emitter.

$$^{159}_{64}Gd \rightarrow {}^{0}_{-1}e + {}^{159}_{65}Tb$$

The decay scheme is shown in Fig. 12-2.

12.41 (a) $^{15}_{8}O$

calculated mass = (7 neutrons)(1.008665 u/neutron)
+ (8 protons)(1.007276 u/proton)
+ (8 electrons)(0.00054858 u/electron)
= 15.123251 u

mass defect = 15.00300 u - 15.123251 u = -0.12025 u

$$(-0.12025\ \text{u})\left[\frac{1.6606 \times 10^{-27}\ \text{kg}}{1\ \text{u}}\right] = -1.9968 \times 10^{-28}\ \text{kg}$$

$$E = mc^2 = (-1.9968 \times 10^{-28}\ \text{kg})(2.9979 \times 10^{8}\ \text{m/s})^2\left[\frac{1\ \text{J}}{1\ \text{kg m}^2/\text{s}^2}\right]$$

$$= -1.7946 \times 10^{-11}\ \text{J}$$

$$\text{average binding energy} = \frac{-1.7946 \times 10^{-11}\ \text{J}}{15\ \text{nucleons}} = -1.1964 \times 10^{-12}\ \text{J/nucleon}$$

(b) $^{16}_{8}O$

calculated mass = (8 neutrons)(1.008665 u/neutron)
+ (8 protons)(1.007276 u/proton)
+ (8 electrons)(0.00054858 u/electron)
= 16.131917 u

mass defect = 15.99491 u - 16.131917 u = -0.13701 u

$$(-0.13707\ \text{u})\left[\frac{1.6606 \times 10^{-27}\ \text{kg}}{1\ \text{u}}\right] = -2.2751 \times 10^{-28}\ \text{kg}$$

$$E = (-2.2751 \times 10^{-28}\ \text{kg})(2.9979 \times 10^{8}\ \text{m/s})^2\left[\frac{1\ \text{J}}{1\ \text{kg m}^2/\text{s}^2}\right]$$

$$= -2.0447 \times 10^{-11}\ \text{J}$$

$$\text{average binding energy} = \frac{-2.0447 \times 10^{-11}\ \text{J}}{16\ \text{nucleons}} = -1.2779 \times 10^{-12}\ \text{J/nucleon}$$

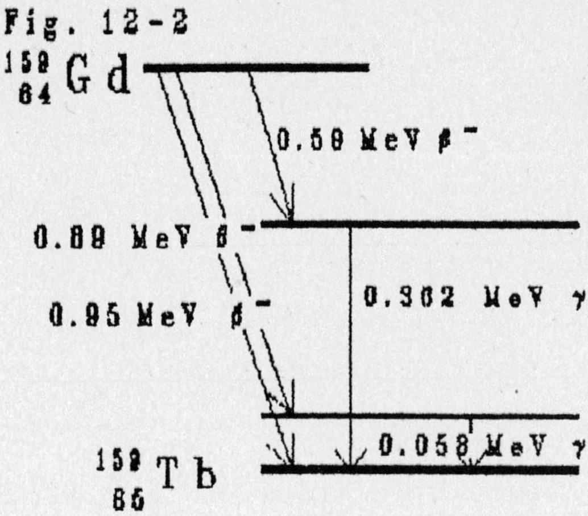

(c) $^{17}_{8}O$

calculated mass = (9 neutrons)(1.008665 u/neutron)
+ (8 protons)(1.007276 u/proton)
+ (8 electrons)(0.00054858 u/electron)
= 17.140582 u

mass defect = 16.99913 u - 17.140582 u = -0.14145 u

$$(-0.14145 \text{ u})\left[\frac{1.6606 \times 10^{-27} \text{ kg}}{1 \text{ u}}\right] = -2.3489 \times 10^{-28} \text{ kg}$$

$$E = (-2.3489 \times 10^{-28} \text{ kg})(2.9979 \times 10^{8} \text{ m/s})^2\left[\frac{1 \text{ J}}{1 \text{ kg m}^2/\text{s}^2}\right]$$

$$= -2.1111 \times 10^{-11} \text{ J}$$

$$\text{average binding energy} = \frac{-2.1111 \times 10^{-11} \text{ J}}{17 \text{ nucleons}} = -1.2418 \times 10^{-12} \text{ J/nucleon}$$

(d) $^{18}_{8}O$

calculated mass = (10 neutrons)(1.008665 u/neutron)
+ (8 protons)(1.007276 u/proton)
+ (8 electrons)(0.00054858 u/electron)
= 18.149247 u

mass defect = 17.99915 u - 18.149247 u = -0.15010 u

$$(-0.15010\ u)\left[\frac{1.6606 \times 10^{-27}\ kg}{1\ u}\right] = -2.4925 \times 10^{-28}\ kg$$

$$E = (-2.4925 \times 10^{-28}\ kg)(2.9979 \times 10^{8}\ m/s)^2\left[\frac{1\ J}{1\ kg\ m^2/s^2}\right]$$

$$= -2.2401 \times 10^{-11}\ J$$

$$\text{average binding energy} = \frac{-2.2401 \times 10^{-11}\ J}{18\ \text{nucleons}} = -1.2445 \times 10^{-12}\ J/\text{nucleon}$$

(e) $^{19}_{8}O$

calculated mass = (11 neutrons)(1.008665 u/neutron)
+ (8 protons)(1.007276 u/proton)
+ (8 electrons)(0.00054858 u/electron)
= 19.15791 u

mass defect = 19.0035 u - 19.15791 u = -0.1544 u

$$(-0.1544\ u)\left[\frac{1.6606 \times 10^{-27}\ kg}{1\ u}\right] = -2.564 \times 10^{-28}\ kg$$

$$E = (-2.564 \times 10^{-28}\ kg)(2.9979 \times 10^{8}\ m/s)^2\left[\frac{1\ J}{1\ kg\ m^2/s^2}\right]$$

$$= -2.304 \times 10^{-11}\ J$$

$$\text{average binding energy} = \frac{-2.304 \times 10^{-11}\ J}{19\ \text{nucleons}} = -1.213 \times 10^{-12}\ J/\text{nucleon}$$

The $^{16}_{8}O$ isotope would be expected to be most stable.

12.43 In bombardment reactions a "target" nucleus is struck by a moving particle that combines with the nucleus to form an unstable compound nucleus, which then decays either instantaneously or with a measurable half-life. The shorthand notation used for bombardment reactions consists of writing the symbols for the bombarding particle and the product particle between the symbols for the reactant and product nuclides.

12.45 (a) ${}^{14}_{7}N + {}^{4}_{2}He \rightarrow {}^{17}_{8}O + {}^{1}_{1}H$ (b) ${}^{106}_{46}Pd + {}^{1}_{0}n \rightarrow {}^{106}_{45}Rh + {}^{1}_{1}H$

(c) ${}^{23}_{11}Na + {}^{1}_{0}n \rightarrow {}^{0}_{-1}e + {}^{24}_{12}Mg$ The element X is ${}^{24}_{12}Mg$.

12.47 (a) ${}^{6}_{3}Li(n,\alpha){}^{3}_{1}H$ (b) ${}^{31}_{15}P(d,p){}^{32}_{15}P$ (c) ${}^{238}_{92}U(n,\beta^{-}){}^{239}_{93}Np$

12.49 (a) ${}^{14}_{7}N + {}^{4}_{2}He \rightarrow {}^{1}_{1}H + {}^{17}_{8}O$

(b) mass change = [(mass of H) + (mass of O)] − [(mass of N) + (mass of He)]

= [(1.007825 u) + (16.99913 u)] − [(14.00307 u) + (4.00260 u)]

= 0.00129 u

$$E = mc^2 = [(0.00129\ \text{u})(1.66 \times 10^{-27}\ \text{kg/u})] \times (2.9979 \times 10^{8}\ \text{m/s})^2 \left[\frac{1\ \text{J}}{1\ \text{kg m}^2/\text{s}^2}\right]$$

$$= 1.92 \times 10^{-13}\ \text{J}$$

12.51 The general penetrating abilities are $\alpha < \beta < \gamma$. α Particles that are absorbed internally by the body are particularly dangerous because they give up all their energy to the tissue in a very small distance, causing great damage.

12.53 In a nuclear fission process a heavy fissionable isotope splits into two atoms of intermediate mass and several neutrons. The two most important fissionable materials are uranium-235 and plutonium-239.

12.55 Continuous nuclear fusion processes have been observed in the sun and other stars. The main reaction taking place in these sources is the conversion of hydrogen to helium.

12.57 The five primary components of a nuclear reactor and their functions:

1. Fuel, the source of energy via fission processes
2. Moderator, to slow down neutrons to speeds at which they produce the fission reaction most efficiently
3. Control system, to allow just enough free slow neutrons to carry on the chain reaction at a safe rate
4. Cooling system, to carry away the energy of the fission reaction for transformation into electrical power and to keep the reactor from overheating
5. Shielding, to protect the walls of the reactor and the operating personnel from heat and radiation

Major ecological or environmental problems are involved in the reprocessing of spent fuel. The other components present waste storage problems when they are contaminated.

12.59 A breeder reactor produces at least as many fissionable atoms as it consumes.

$$^{238}_{92}U + ^{1}_{0}n \rightarrow ^{239}_{92}U$$

$$^{239}_{92}U \rightarrow ^{239}_{93}Np + ^{0}_{-1}e$$

$$^{239}_{93}Np \rightarrow ^{239}_{94}Pu + ^{0}_{-1}e$$

12.61 for the fission power:

$$\text{mass change} = [(\text{mass of Zn}) + (\text{mass of Ce}) + 6(\text{mass of } \beta^-) + 2(\text{mass of } n)] - [(\text{mass of U}) + (\text{mass of } n)]$$

$$\text{mass change} = [(93.9061\ u) + (139.9053\ u) + (6)(0.00055\ u) + (2)(1.00867\ u)] - [(235.0439\ u) + (1.00867\ u)]$$
$$= -0.2205\ u$$

$$E = mc^2 = (-0.2205\ u)\left[\frac{1.661 \times 10^{-27}\ kg}{1\ u}\right](2.9979 \times 10^8\ m/s)^2\left[\frac{1\ J}{1\ kg\ m^2/s^2}\right]$$
$$= -3.292 \times 10^{-11}\ J$$

$$\frac{-3.292 \times 10^{-11}\ J}{(235.0439\ u) + (1.00867\ u)} = -1.395 \times 10^{-13}\ J/u$$

for the fusion process:

$$\text{mass change} = [(\text{mass of T}) + (\text{mass of H})] - [2(\text{mass of D})]$$
$$= [(3.01605\ u) + (1.007825\ u)] - [(2)(2.0140\ u)]$$
$$= -0.0041\ u$$

$$E = (-0.0041\ u)\left[\frac{1.661 \times 10^{-27}\ kg}{1\ u}\right](2.9979 \times 10^8\ m/s)^2\left[\frac{1\ J}{1\ kg\ m^2/s^2}\right]$$
$$= -6.1 \times 10^{-13}\ J$$

$$\frac{-6.1 \times 10^{-13}\ J}{2(2.0140\ u)} = -1.5 \times 10^{-13}\ J/u$$

The fusion reaction produces the larger amount of energy per atomic mass unit of material reacting.

CHAPTER 13

LIQUID AND SOLID STATES; CHANGES OF STATE

Solutions to Exercises

13.1 Both compounds (a) and (b) are very similar in size and shape and have the same molar mass. Thus the strengths of the London forces should be similar. The intermolecular forces resulting from the dipole motion of compound (a) should be greater than that of compound (b) and so compound (a) would be predicted to have the lower vapor pressure at a given temperature. Compound (b) has the higher vapor pressure.

13.2 $P_{\text{air}} = P_1\left[\frac{V_1}{V_2}\right] = (745\ \text{Torr})\left[\frac{V_1}{V_1/5}\right] = 3730\ \text{Torr}$

$P_T = P_{\text{air}} + P_{Br_2} = 3730\ \text{Torr} + 168\ \text{Torr} = 3.90 \times 10^3\ \text{Torr}$

13.3 $\Delta H^\circ_{\text{vap}} = \dfrac{-(2.303)R}{\left[\dfrac{1}{T_2} - \dfrac{1}{T_1}\right]} \log\left[\dfrac{P_2}{P_1}\right] = \dfrac{-(2.303)(8.314\ \text{J/K mol})}{\left[\dfrac{1}{5441\ \text{K}} - \dfrac{1}{4263\ \text{K}}\right]} \log\left[\dfrac{100.0}{1.00}\right]$

$= 7.54 \times 10^5\ \text{J/mol} = 754\ \text{kJ/mol}$

13.4 Upon heating at constant pressure, dry ice originally at point *a* would become liquid. It would not become liquid from point *b* shown in Figure 13.15 of the text.

13.5 (a) $(8\ \text{corners})\left[\dfrac{(1/8)\ \text{atom}}{\text{corner}}\right] + (6\ \text{faces})\left[\dfrac{(1/2)\ \text{atom}}{\text{face}}\right] = 4\ \text{atoms}$

(b) $\dfrac{(196.97\ \text{g/mol})}{(6.022 \times 10^{23}\ \text{Au atoms/mol})} = 3.271 \times 10^{-22}\ \text{g/Au atom}$

$(4\ \text{Au atoms})(3.271 \times 10^{-22}\ \text{g/Au atom}) = 1.308 \times 10^{-21}\ \text{g}$

13.6 (a) $(0.4079\ \text{nm})\left[\dfrac{1\ \text{m}}{10^9\ \text{nm}}\right]\left[\dfrac{10^2\ \text{cm}}{1\ \text{m}}\right] = 4.079 \times 10^{-8}\ \text{cm}$

$(4.079 \times 10^{-8}\ \text{cm})^3 = 6.787 \times 10^{-23}\ \text{cm}^3$

(b) $\dfrac{1.308 \times 10^{-21}\text{ g}}{6.787 \times 10^{-23}\text{ cm}^3} = 19.27\text{ g/cm}^3$

13.7

$$\Delta H_f = \Delta H_{sub} + \Delta H_{IE} + \tfrac{1}{2}\Delta H_{BE} + \Delta H_{EA} + \Delta H_{LE}$$

$$(-562.58\text{ kJ}) = (90.00\text{ kJ}) + (424.93\text{ kJ}) + (\tfrac{1}{2})(157.99\text{ kJ}) + (-349.7\text{ kJ}) + \Delta H_{LE}$$

$$\Delta H_{LE} = -806.8\text{ kJ}$$

Solutions to Odd-Numbered Questions and Problems

13.1 Gases and liquids flow because of movement of molecules; the rigid structure of solid does not allow it to flow.

13.3 The flow of the liquid is related to viscosity. Viscosity is a function of the temperature and of the size, shape, and chemical nature of molecules. As the temperature increases, the viscosity of a liquid decreases.

13.5 The bottom of the meniscus is read for a liquid that "wets" the glass and the top of the meniscus is read for a liquid that does not "wet" the glass.

13.7 Crystalline substances may be anisotropic, are cleaved into smaller pieces with planar faces, and have sharp melting points. Amorphous substances are generally isotropic, are cleaved into smaller pieces with nonplanar faces, and soften gradually as the temperature rises.

13.9 The equilibrium that is established between two physical states of matter is an example of dynamic equilibrium because the molecules are in constant motion and the most energetic ones in the more restricted state go to the freer state, while the less energetic ones go in the opposite direction.

13.11 $BeF_2(s) \rightarrow BeF_2(g)$

$$\Delta H^\circ_{sub} = [(1\text{ mol})\Delta H^\circ_f(g)] - [(1\text{ mol})\Delta H^\circ_f(s)]$$

$$= [(1\text{ mol})(-797\text{ kJ/mol})] - [(1\text{ mol})(-1023\text{ kJ/mol})] = 226\text{ kJ}$$

The heat of sublimation is 226 kJ/mol.

13.13 ΔT = 20. °C - 35 °C = -15 °C = -15 K

$$q_{gained} = (1.00\ g)\left[\frac{1\ mol}{18.02\ g}\right]\left[\frac{44\ kJ}{1\ mol}\right]\left[\frac{1000\ J}{1\ kJ}\right] = 2400\ J$$

$$q_{lost} = -q_{gained} = -2400\ J$$

$$q_{lost} = (\text{no. of moles})(\text{molar heat capacity})\Delta T$$

(-2400 J) = (no. of moles)(75 J/K mol)(-15 K)

(no. of moles) = 2.1 mol

$$(2.1\ mol)\left[\frac{18.02\ g}{1\ mol}\right] = 38\ g$$

13.15 The substance in each pair that has the larger vapor pressure at a given temperature will be the one with the weaker intermolecular forces.

(a) The London forces in C_6Cl_6 are greater than those in C_6H_6. C_6H_6 will have the larger vapor pressure.

(b) Both molecules will have similar dipole-dipole interactions and London forces. However, CH_3OH will have hydrogen bonding that will not be present in H_2CO. H_2CO will have the larger vapor pressure.

(c) Ga and Cu will have similar London forces because they have similar masses and sizes. The metallic bonding in Cu is considerably greater than in Ga and so Ga will have the larger vapor pressure.

(d) Although the mass of He is slightly larger than that of H_2, the London forces in H_2 are stronger because of the size and shape of the molecule compared to that of the He atom. He will have the larger vapor pressure.

13.17 $P_{air} = P_{total} - P_{H_2O}$ = 627.4 Torr - 268.0 Torr = 359.4 Torr

13.19

$$\begin{array}{r} 2P_{air} + P_{SiCl_4} = 1742\ \text{Torr} \\ -[P_{air} + P_{SiCl_4} = 988\ \text{Torr}] \\ \hline P_{air} = 754\ \text{Torr} \end{array}$$

P_{SiCl_4} = 988 Torr - P_{air} = 988 Torr - 754 Torr = 234 Torr

13.21 15 °C: 26.9 mm - 22.5 mm = 4.4 mmHg = 4.4 Torr

32 °C: 47.6 mm - 1.7 mm = 45.9 mmHg = 45.9 Torr

13.23 $\frac{14.3\ Torr}{19.8\ Torr} \times 100 = 72.2\ \%$

13.25 As we can see in Figure 13.11, over large temperature ranges the vapor pressure is not a linear function of the temperature. Simply averaging two known values to predict the value at some intermediate temperature will always be in error.

13.27

$$\log\left[\frac{P_2}{P_1}\right] = \frac{-\Delta H_{vap}}{(2.303)R}\left[\frac{1}{T_2} - \frac{1}{T_1}\right]$$

$$\log\left[\frac{745\ \text{Torr}}{760\ \text{Torr}}\right] = \left[\frac{-(40{,}656\ \text{J/mol})}{(2.303)(8.314\ \text{J/K mol})}\right]\left[\frac{1}{T_2} - \frac{1}{373.15\ \text{K}}\right]$$

$$\frac{1}{T_2} = \frac{-(2.303)(8.314\ \text{J/K mol})}{(40{,}656\ \text{J/mol})}\log\left[\frac{745\ \text{Torr}}{760\ \text{Torr}}\right] + \frac{1}{373.15\ \text{K}}$$

$$= 0.002684\ \text{K}^{-1}$$

$$T_2 = 372.6\ \text{K} = 99.5\ °\text{C}$$

13.29

P, Torr	1.00	10.0	40.0	100.0	400.0	760.0
T, °C	126.2	184.0	228.8	261.7	323.0	357.0
log P	0.000	1.000	1.602	2.0000	2.6021	2.8808
T, K	399.4	457.2	502.0	534.9	596.2	630.2
$1000/T$, K^{-1}	2.504	2.187	1.992	1.869	1.677	1.586

The data are plotted in Fig. 13-1. The heat of vaporization is

$$\Delta H_{vap} = -(\text{slope})(2.303)R$$

$$= -(-3.14 \times 10^3\ \text{K})(2.303)(8.314\ \text{J/K mol})\left[\frac{1\ \text{kJ}}{1000\ \text{J}}\right]$$

$$= 60.1\ \text{kJ/mol}$$

13.31 The order of evaporation as the temperature is raised is N_2, Ar, and O_2.

13.33 The critical point is the temperature above which no amount of pressure is great enough to cause liquefaction. A substance will not always be a liquid below the critical temperature; it will be a liquid only if sufficient pressure is used.

13.35 Three phases exist at a triple point. If a small amount of heat were added under constant volume conditions to a sample of water at the triple point, some ice would melt.

13.37 When the pressure is increased under constant temperature conditions, a sample of CO_2 beginning at point *a* remains solid; a sample of H_2O beginning at point *c* melts.

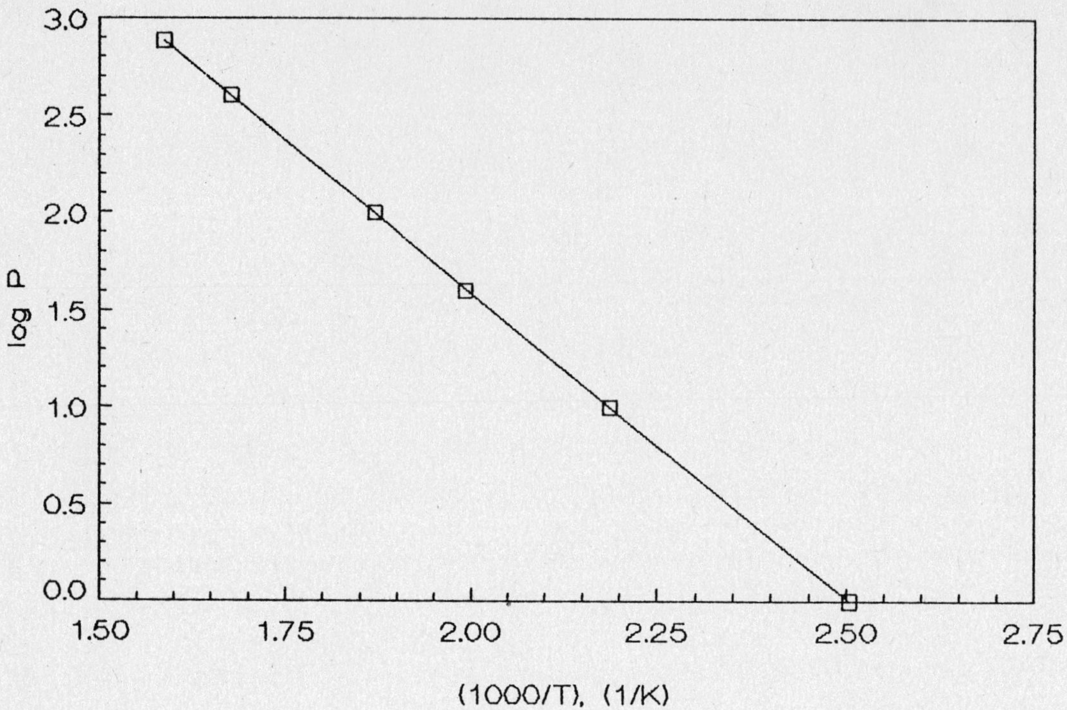

13.39 In closest packing, the atoms in a single layer are packed as closely as possible in a hexagonal arrangement. A second layer (B) is placed over the first layer (A) so that three spheres of layer B touch the same sphere of layer A. Placement of the third layer so each atom lies directly above an atom in the first layer results in hexagonal closest packing--layers of atoms arranged in an ABAB... sequence. Placement of the third layer so that its atoms are not directly above those of either layer A or layer B is cubic closest packing--layers of atoms arranged in an ABCABC... sequence.

13.41 A unit cell is the most convenient small part of a space lattice that, if repeated in three dimensions, will generate the entire lattice. See Figure 13.21 for drawings of the primitive, body-centered, and face-centered cubic unit cells.

13.43 (a) $$(1\ \text{atom})\left[\frac{209\ \text{g}}{1\ \text{mol}}\right]\left[\frac{1\ \text{mol}}{6.022 \times 10^{23}\ \text{atoms}}\right] = 3.47 \times 10^{-22}\ \text{g}$$

(b) $(0.336\ \text{nm})\left[\frac{1\ \text{m}}{10^9\ \text{nm}}\right]\left[\frac{10^2\ \text{cm}}{1\ \text{m}}\right] = 3.36 \times 10^{-8}\ \text{cm}$

$(3.36 \times 10^{-8}\ \text{cm})^3 = 3.79 \times 10^{-23}\ \text{cm}^3$

(c) $\frac{3.47 \times 10^{-22}\ \text{g}}{3.79 \times 10^{-23}\ \text{cm}^3} = 9.16\ \text{g/cm}^3$

13.45 $(4\ \text{atoms})\left[\frac{107.868\ \text{g}}{1\ \text{mol}}\right]\left[\frac{1\ \text{mol}}{6.022045 \times 10^{23}\ \text{atoms}}\right] = 7.16488 \times 10^{-22}\ \text{g}$

$\frac{7.16488 \times 10^{-22}\ \text{g}}{10.5\ \text{g/cm}^3} = 6.82 \times 10^{-23}\ \text{cm}^3$

$\sqrt[3]{68.2 \times 10^{-24}\ \text{cm}^3} = 4.09 \times 10^{-8}\ \text{cm}$

$(4.09 \times 10^{-8}\ \text{cm})\left[\frac{1\ \text{m}}{10^2\ \text{cm}}\right]\left[\frac{10^9\ \text{nm}}{1\ \text{m}}\right] = 0.409\ \text{nm}$

13.47 $(8\ \text{atoms})\left[\frac{28.0855\ \text{g}}{1\ \text{mol}}\right]\left[\frac{1\ \text{mol}}{6.022045 \times 10^{23}\ \text{atoms}}\right] = 3.73102 \times 10^{-22}\ \text{g}$

$(0.54305\ \text{nm})\left[\frac{1\ \text{m}}{10^9\ \text{nm}}\right]\left[\frac{10^2\ \text{cm}}{1\ \text{m}}\right] = 5.4305 \times 10^{-8}\ \text{cm}$

$(5.4305 \times 10^{-8}\ \text{cm})^3 = 1.6015 \times 10^{-22}\ \text{cm}^3$

$\frac{3.73102 \times 10^{-22}\ \text{g}}{1.6015 \times 10^{-22}\ \text{cm}^3} = 2.3297\ \text{g/cm}^3$

13.49 (a) See Fig. 13-2.

(b) $\frac{1}{8}(8) = 1$ cation

(c) $\frac{1}{2}(6) = 3$ anions

(d) CA_3

13.51 $(0.56402\ \text{nm})\left[\frac{1\ \text{m}}{10^9\ \text{nm}}\right]\left[\frac{10^2\ \text{cm}}{1\ \text{m}}\right] = 5.6402 \times 10^{-8}\ \text{cm}$

$(5.6402 \times 10^{-8}\ \text{cm})^3 = 1.7943 \times 10^{-22}\ \text{cm}^3$

There are 4 Na^+ and 4 Cl^- in the unit cell. The mass of the unit cell is

$$(4\ \text{Na}^+\ \text{ions})\left[\frac{22.98977\ \text{g/mol}}{6.022045 \times 10^{23}\ \text{ions/mol}}\right] + (4\ \text{Cl}^-\ \text{ions})\left[\frac{35.453\ \text{g/mol}}{6.022045 \times 10^{23}\ \text{ions/mol}}\right] = 3.8819 \times 10^{-22}\ \text{g}$$

$\frac{3.8819 \times 10^{-22}\ \text{g}}{1.7943 \times 10^{-22}\ \text{cm}^3} = 2.1635\ \text{g/cm}^3$

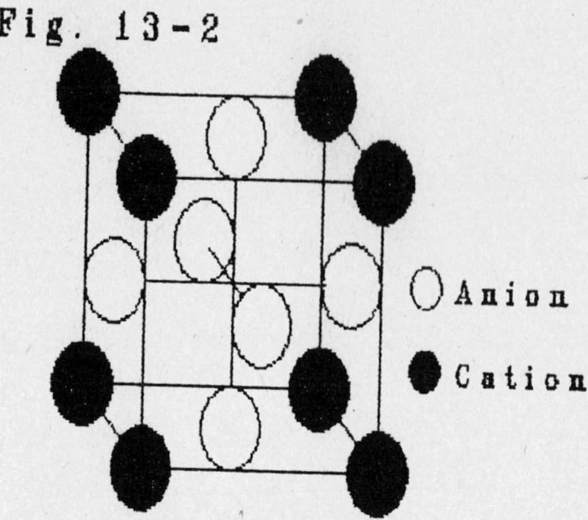

13.53 The lattice energy is the energy liberated as gaseous ions combine to give a crystalline ionic substance. The chemical equation is $M^{2+}(g) + X^{2-}(g) \rightarrow MX(s)$. The value increases with increasing ionic charge and decreases with increasing ionic radius.

13.55 $M^{+}(g) + Cl^{-}(g) \rightarrow MCl(s)$

$\Delta H° = [(1\ mol)\Delta H_f°(MCl)] - [(1\ mol)\Delta H_f°(M^{+}) + (1\ mol)\Delta H_f°(Cl^{-})]$

$\Delta H°_{NaCl} = [(1\ mol)(-411.153\ kJ/mol)] - [(1\ mol)(609.358\ kJ/mol) + (1\ mol)(-233.13\ kJ/mol)]$

$= -787.38\ kJ$

THe lattice energy is -787.38 kJ/mol for NaCl.

$\Delta H°_{KCl} = [(1\ mol)(-436.747\ kJ/mol)] - [(1\ mol)(514.26\ kJ/mol) + (1\ mol)(-233.13\ kJ/mol)]$

$= -717.88\ kJ$

The lattice energy is -717.48 kJ/mol for KCl. The value increases with decreasing interionic distance.

13.57 The theoretical density will be higher than the actual density for a pure substance because of crystal imperfections that reduce the mass and increase the volume. The theoretical density could be high or low compared to the actual density if the compound contains impurities.

13.59 Diffraction is the ability of waves to bend around corners or to spread out. In order for diffraction to occur, the wavelength of the incident radiation must be comparable to the distance between particles in a crystal. X-rays, neutrons, and electrons are suitable for diffraction studies of crystals.

CHAPTER 14

WATER AND SOLUTIONS IN WATER

Solutions to Exercises

14.4 $\dfrac{(10.\ g\ Na_2B_4O_7 \cdot 10H_2O)/(381.42\ g/mol)}{1.0\ L} = 0.026\ mol/L$

14.5 (a) partner exchange, (b) combination.

14.6 (a) $O_2(g) \xrightarrow{H_2O} O_2(aq)$ (b) $KBr(s) \xrightarrow{H_2O} K^+ + Br^-$

(c) $NaOH(s) \xrightarrow{H_2O} Na^+ + OH^-$ (d) $AgCl(s) \overset{H_2O}{\rightleftarrows} Ag^+ + Cl^-$

14.7 (a) $HCOOH(aq) + H_2O(l) \rightleftarrows HCOO^- + H_3O^+$

(b) $HBr(g) + H_2O(l) \rightarrow H_3O^+ + Br^-$

(c) $CsOH(s) \xrightarrow{H_2O} Cs^+ + OH^-$

Solutions to Odd-Numbered Questions and Problems

14.1 The Lewis structure of a water molecule is shown at the right. The bonding between the oxygen and hydrogen atoms within the molecule is polar covalent.

H - Ö - H (with two lone pairs on O)

14.3 The value of the predicted bond angle in water is 109.47°. The actual value of 104.5° is less than the predicted angle. The difference is caused by repulsions between lone pairs of electrons and between lone pairs and bonded pairs which reduce the bond angle.

14.5 Liquid water is most dense at 3.98 °C because as the temperature is increased from 0 °C, more hydrogen bonds break, which causes further collapse of the ice structure, a decrease in volume, and a corresponding increase in density. Above 3.98 °C, the normal thermal expansion caused by the increased energy of individual molecules is greater than the contraction resulting from the collapse of the ice structure.

Liquid water is more dense than ice because in ice the water molecules are built up in a big, honeycomb-like lattice containing large open spaces not found in the less organized form of liquid water.

14.7 heat released by burning 1 g of coal = $(1.00 \text{ g coal})\left[\frac{-31 \text{ kJ}}{1 \text{ g coal}}\right]\left[\frac{1000 \text{ J}}{1 \text{ kJ}}\right]$

$= -31{,}000$ J

The heat required to prepare *m* grams of steam is

$$q = m\left[\frac{4.184 \text{ J}}{\text{K g}}\right](373 \text{ K} - 298 \text{ K}) + m\left[\frac{2260 \text{ J}}{1 \text{ g}}\right] + m\left[\frac{2.1 \text{ J}}{\text{K g}}\right](398 \text{ K} - 373 \text{ K})$$

Using the law of conservation of energy gives

$$-31{,}000 \text{ J} + m\left[\frac{4.184 \text{ J}}{\text{K g}}\right](373 \text{ K} - 298 \text{ K}) + m\left[\frac{2260 \text{ J}}{1 \text{ g}}\right] + m\left[\frac{2.1 \text{ J}}{\text{K g}}\right](398 \text{ K} - 373 \text{ K}) = 0$$

$m = 12$ g

14.9 volume $= (0.4535 \text{ nm})^2(0.741 \text{ nm})(\sin 60°)\left[\frac{1 \text{ m}}{10^9 \text{ nm}}\right]^3\left[\frac{100 \text{ cm}}{1 \text{ m}}\right]^3$

$= 1.32 \times 10^{-22} \text{ cm}^3$

mass $= (4 \text{ molecules})\left[\frac{18.02 \text{ g}}{1 \text{ mol}}\right]\left[\frac{1 \text{ mol}}{6.022 \times 10^{23} \text{ molecules}}\right] = 1.197 \times 10^{-22}$ g

density $= \frac{\text{mass}}{\text{volume}} = \frac{1.197 \times 10^{-22} \text{ g}}{1.32 \times 10^{-22} \text{ cm}^3} = 0.907 \text{ g/cm}^3$

14.11 Water is a much better solvent for ionic and polar substances than for nonpolar substances because it is polar and forms favorable dipole-dipole and dipole-ion interactions. Not all ionic and highly polar substances are highly soluble in water because the forces of attraction to water are less than the interparticle forces within the pure solute.

14.13 The strength of the attraction of water molecules to ions during hydration is determined by ionic size and ionic charge. The Zn^{2+} ion will have a greater enthalpy of hydration than the Na^+ ion.

14.15 (a) For a solution having a positive heat of solution (endothermic), the combined solvent-solvent and solute-solute interactions are greater than the solvent-solute interactions.

(b) For a solution having a negative heat of solution (exothermic), the combined solvent-solvent and solute-solute interactions are not as great as the solvent-solute interactions.

14.17 $\Delta H°(\text{soln}) = [(1\text{ mol})\Delta H_f°(NH_4NO_3, 1\text{ M})] - [(1\text{ mol})\Delta H_f°(NH_4NO_3, s)]$

$= [(1\text{ mol})(-341\text{ kJ/mol})] - [(1\text{ mol})(-366\text{ kJ/mol})]$

$= 25\text{ kJ}$

The beaker would be cool to the touch.

14.19 The names of the following compounds are

(a) $Cr(CH_3COO)_3 \cdot H_2O$, chromium(III) acetate monohydrate

(b) $Cd(NO_3)_2.4H_2O$, cadmium nitrate tetrahydrate

(c) $(NH_4)_2C_2O_4 \cdot H_2O$, ammonium oxalate monohydrate

(d) $LiBr \cdot 2H_2O$, lithium bromide dihydrate

14.21 $$\text{mass \% } H_2O = \frac{\text{mass } H_2O}{\text{mass hydrate}} \times 100$$

(a) $MgSO_4 \cdot 7H_2O$

$$\text{mass \% } H_2O = \frac{(7\text{ mol } H_2O)(18.02\text{ g } H_2O/1\text{ mol } H_2O)}{246.51\text{ g}} \times 100 = 51.17\ \%$$

(b) $CuAl_6(PO_4)_4(OH)_8 \cdot 4H_2O$

$$\text{mass \% } H_2O = \frac{(4\text{ mol } H_2O)(18.02\text{ g } H_2O/1\text{ mol } H_2O)}{813.47\text{ g}} \times 100 = 8.861\ \%$$

14.23 $$(1.00\text{ lb } CaCl_2 \cdot 6H_2O)\left[\frac{454\text{ g}}{1\text{ lb}}\right]\left[\frac{110.98\text{ g } CaCl_2}{219.10\text{ g } CaCl_2 \cdot 6H_2O}\right] = 230.\text{ g } CaCl_2$$

14.25 $23.4\text{ g} - 18.5\text{ g} = 4.9\text{ g } H_2O$

$$(4.9\text{ g } H_2O)\left[\frac{1\text{ mol } H_2O}{18.02\text{ g } H_2O}\right] = 0.27\text{ mol } H_2O$$

$$(18.5\text{ g } CuCl_2)\left[\frac{1\text{ mol } CuCl_2}{134.45\text{ g } CuCl_2}\right] = 0.138\text{ mol } CuCl_2$$

$$\frac{0.27\text{ mol } H_2O}{0.138\text{ mol } CuCl_2} = 2.0$$

$CuCl_2 \cdot 2H_2O$

14.27 The chemical formula for the hydrated hydrogen ion is usually written as H_3O^+. The Lewis structure for this cation is shown at the right.

$$\left[\begin{matrix} H - \ddot{O} - H \\ | \\ H \end{matrix}\right]^+$$

14.29 There are always both H^+ and OH^- ions at equilibrium. An acidic solution simply has a greater concentration of H^+ than of OH^-.

14.31 (a) $H_2O(g) \xrightarrow{\Delta} H(g) + OH(g)$

(b) $H_2O(l) \rightleftharpoons H^+ + OH^+$

(c) $CaCl_2 \cdot H_2O(s) + 5H_2O(l) \rightarrow CaCl_2 \cdot 6H_2O(s)$

14.33 (a) iv, (b) iv, (c) i, (d) ii, (e) iii.

14.35 Water is a very weak electrolyte because relatively few ions are formed by self-ionization in pure water.

14.37 (a) K^+, NO_3^-; (b) H^+, ClO_4^-; (c) K^+, OH^-; (d) Ag^+, Cl^-; (e) HCOOH, H^+, $HCOO^-$; (f) CH_3OH.

14.39 The following chemical equations describe what happens as each of the substances is mixed with water:

(a) $H_2SO_4(l) \xrightarrow{H_2O} H^+ + HSO_4^-$ or $H_2SO_4(l) + H_2O(l) \rightarrow H_3O^+ + HSO_4^-$

and $HSO_4^- \overset{H_2O}{\rightleftharpoons} H^+ + SO_4^{2-}$ or $HSO_4^- + H_2O(l) \rightleftharpoons H_3O^+ + SO_4^{2-}$

(b) $NH_4I(s) \xrightarrow{H_2O} NH_4^+ + I^-$

(c) $Sr(OH)_2(s) \overset{H_2O}{\rightleftharpoons} Sr^{2+} + 2OH^-$

(d) $HCN(g) \overset{H_2O}{\rightleftharpoons} H^+ + CN^-$ or $HCN(g) + H_2O(l) \rightleftharpoons H_3O^+ + CN^-$

(e) $AgCl(s) \overset{H_2O}{\rightleftharpoons} Ag^+ + Cl^-$

14.41 In terms of the water-ion definition, an acid is a substance that contains hydrogen and yields hydrogen ions in aqueous solution. A base is a compound that contains hydroxide ions and that when it dissolves in water, dissociates to give hydroxide ions.

14.43 Polyprotic means more than one ionizable hydrogen atom per acid molecule. The formula of a common polyprotic acid is H_2SO_4. The chemical equations showing the stepwise ionization of this acid are

$H_2SO_4(aq) \xrightarrow{H_2O} H^+ + HSO_4^-$ or $H_2SO_4(aq) + H_2O(l) \rightarrow H_3O^+ + HSO_4^-$

$HSO_4^- \overset{H_2O}{\rightleftharpoons} H^+ + SO_4^{2-}$ or $HSO_4^- + H_2O(l) \rightleftharpoons H_3O^+ + SO_4^{2-}$

14.45 The following chemical equations show the interactions of water with each of the substances:

(a) $HOOCCOOH(s) \overset{H_2O}{\rightleftharpoons} H^+ + HOOCCOO^-$ or

$HOOCCOOH(s) + H_2O(l) \rightleftharpoons H_3O^+ + HOOCCOO^-$

$HOOCCOO^- \overset{H_2O}{\rightleftharpoons} H^+ + OOCCOO^{2-}$ or $HOOCCOO^- + H_2O(l) \rightleftharpoons H_3O^+ + OOCCOO^{2-}$

(b) $HCN(g) \overset{H_2O}{\rightleftharpoons} H^+ + CN^-$ or $HCN(g) + H_2O(l) \rightleftharpoons H_3O^+ + CN^-$

(c) $HNO_3(l) \overset{H_2O}{\rightarrow} H^+ + NO_3^-$ or $HNO_3(l) + H_2O(l) \rightarrow H_3O^+ + NO_3^-$

14.47 $HBr(aq) + NaOH(aq) \rightarrow H_2O(l) + NaBr(aq)$

$$(1.00\text{ g HBr})\left[\frac{1\text{ mol HBr}}{80.91\text{ g HBr}}\right]\left[\frac{1\text{ mol NaOH}}{1\text{ mol HBr}}\right]\left[\frac{40.00\text{ g NaOH}}{1\text{ mol NaOH}}\right] = 0.494\text{ g NaOH}$$

$HClO_4(aq) + NaOH(aq) \rightarrow H_2O(l) + NaClO_4(aq)$

$$(1.00\text{ g HClO}_4)\left[\frac{1\text{ mol HClO}_4}{100.46\text{ g HClO}_4}\right]\left[\frac{1\text{ mol NaOH}}{1\text{ mol HClO}_4}\right]\left[\frac{40.00\text{ g NaOH}}{1\text{ mol NaOH}}\right] = 0.398\text{ g NaOH}$$

The HBr requires the greater mass of NaOH.

14.49 $Ca(OH)_2(s) + 2HCl(aq) \rightarrow CaCl_2(aq) + 2H_2O(l)$

$$(16.9\text{ g Ca(OH)}_2)\left[\frac{1\text{ mol Ca(OH)}_2}{74.10\text{ g Ca(OH)}_2}\right] = 0.228\text{ mol Ca(OH)}_2$$

$$(0.228\text{ mol Ca(OH)}_2)\left[\frac{2\text{ mol HCl}}{1\text{ mol Ca(OH)}_2}\right] = 0.456\text{ mol HCl}$$

$$(0.456\text{ mol HCl})\left[\frac{1\text{ L soln}}{0.1123\text{ mol HCl}}\right] = 4.06\text{ L}$$

14.51 A complex ion consists of a central metal atom or cation to which are bonded one or more molecules or anions. The molecules or ions that bond to the central metal atom or cation are called ligands.

14.53 The charge on the following complex ions is

(a)	(b)	(c)
+3 4(-1)	+3 4(0) 2(-1)	+3 4(-1)
$[AuCl_4]^-$	$[Cr(H_2O)_4Cl_2]^+$	$[Al(OH)_4]^-$
(+3) + (-4) = -1	(+3) + (-2) = +1	(+3) + (-4) = -1

14.55 (a) Dissolved gases are flushed out by air. (b) Solids settle in pools and lakes and are filtered as water seeps through soil. (c) Dissolved solids can be precipitated. (d) Bacteria and other microorganisms decompose animal and plant byproducts.

14.57 $$\left[\frac{19{,}000\text{ mg Cl}^-}{1.00\text{ L}}\right]\left[\frac{1\text{ g}}{1000\text{ mg}}\right]\left[\frac{1\text{ mol Cl}^-}{35.45\text{ g Cl}^-}\right] = 0.54\text{ mol/L}$$

14.59 Hard water contains metal ions (principally Ca^{2+}, Mg^{2+}, and Fe^{2+}) that form precipitates with soap or upon boiling. Hard water that also contains HCO_3^- anions displays temporary hardness; such water can be softened by boiling, which drives off carbon dioxide, causing precipitation of metal carbonates. Hard water that contains no HCO_3^- anion displays permanent hardness, which cannot be softened by boiling.

14.61 For the removal of Mg^{2+}:

$$(1.00\ \text{L})\left[\frac{7.6\ \text{mg Mg}^{2+}}{1\ \text{L}}\right]\left[\frac{1\ \text{g}}{1000\ \text{mg}}\right]\left[\frac{1\ \text{mol Mg}^{2+}}{24.31\ \text{g Mg}^{2+}}\right] = 3.1 \times 10^{-4}\ \text{mol Mg}^{2+}$$

$$(3.1 \times 10^{-4}\ \text{mol Mg}^{2+})\left[\frac{1\ \text{mol Ca(OH)}_2}{1\ \text{mol Mg}^{2+}}\right] = 3.1 \times 10^{-4}\ \text{mol Ca(OH)}_2$$

$$(3.1 \times 10^{-4}\ \text{mol Ca(OH)}_2)\left[\frac{74.10\ \text{g Ca(OH)}_2}{1\ \text{mol Ca(OH)}_2}\right] = 0.023\ \text{g Ca(OH)}_2$$

$$(3.1 \times 10^{-4}\ \text{mol Mg}^{2+})\left[\frac{1\ \text{mol Na}_2\text{CO}_3}{1\ \text{mol Mg}^{2+}}\right] = 3.1 \times 10^{-4}\ \text{mol Na}_2\text{CO}_3$$

$$(3.1 \times 10^{-4}\ \text{mol Na}_2\text{CO}_3)\left[\frac{105.99\ \text{g Na}_2\text{CO}_3}{1\ \text{mol Na}_2\text{CO}_3}\right] = 0.033\ \text{g Na}_2\text{CO}_3$$

For the removal of Ca^{2+}:

$$(1.00\ \text{L})\left[\frac{34\ \text{mg Ca}^{2+}}{1\ \text{L}}\right]\left[\frac{1\ \text{g}}{1000\ \text{mg}}\right]\left[\frac{1\ \text{mol Ca}^{2+}}{40.08\ \text{g Ca}^{2+}}\right] = 8.5 \times 10^{-4}\ \text{mol Ca}^{2+}$$

$$(8.5 \times 10^{-4}\ \text{mol Ca}^{2+})\left[\frac{1\ \text{mol Ca(OH)}_2}{1\ \text{mol Ca}^{2+}}\right] = 8.5 \times 10^{-4}\ \text{mol Ca(OH)}_2$$

$$(8.5 \times 10^{-4}\ \text{mol Ca(OH)}_2)\left[\frac{74.10\ \text{g Ca(OH)}_2}{1\ \text{mol Ca(OH)}_2}\right] = 0.063\ \text{g Ca(OH)}_2$$

The total $Ca(OH)_2$ required is 0.023 g + 0.063 g = 0.086 g.

14.63 The common classes are pathogens such as viruses and bacteria; oxygen-demanding organic waste from dead or live animals or plants; and specific chemicals such as metals, detergents, pesticides, fertilizers, and radioactive wastes.

CHAPTER 15

SOLUTIONS AND COLLOIDS

Solutions to Exercises

15.1 (a) Potassium ferricyanide is an ionic compound containing K^+ cations and the $[Fe(CN)_6]^{3-}$ anion. It is more likely to be soluble in water than in *n*-hexane.

(b) Carbon tetrachloride is nonpolar because of its tetrahedral geometry. Thus it is more likely to be soluble in *n*-hexane than in water.

(c) *n*-Decane is a nonpolar hydrocarbon, so it is more likely to be soluble in *n*-hexane than in water.

(d) Calcium nitrate is an ionic compound, so it is more likely to be soluble in water than in *n*-hexane.

15.2
$$C = \frac{P_{Ar}}{k} = \frac{2.16 \text{ atm}}{396 \text{ L atm/mol}} = 5.45 \times 10^{-3} \text{ mol/L}$$

15.3
$$\text{mass percent} = \frac{\text{mass of solute}}{\text{mass of solution}} \times 100 = \frac{36.2 \text{ g}}{136.2 \text{ g}} \times 100 = 26.6 \text{ mass \%}$$

15.4
$$n = (30.0 \text{ g sugar})\left[\frac{1 \text{ mol sugar}}{342.3 \text{ g sugar}}\right] = 0.0876 \text{ mol sugar}$$

$$\text{molality} = \frac{\text{no. of moles solute}}{\text{solvent mass in kg}}$$

$$= \left[\frac{0.0876 \text{ mol sugar}}{70.0 \text{ g water}}\right]\left[\frac{1000 \text{ g}}{1 \text{ kg}}\right] = 1.25 \text{ mol sugar/kg} = 1.25 \text{ m}$$

15.5 (a) 0.500 g + 99.500 g = 100.000 g soln

$$(100.000 \text{ g soln})\left[\frac{1 \text{ mL soln}}{0.9993 \text{ g soln}}\right] = 100.1 \text{ mL soln}$$

$$(0.500 \text{ g acetone})\left[\frac{1 \text{ mol acetone}}{58.09 \text{ g acetone}}\right] = 0.00861 \text{ mol acetone}$$

$$\text{molarity} = \frac{\text{no. of moles acetone}}{\text{soln vol. in L}} = \left[\frac{0.00861\text{ mol acetone}}{100.1\text{ mL soln}}\right]\left[\frac{1000\text{ mL}}{1\text{ L}}\right]$$

$$= 0.0860\text{ mol acetone/L} = 0.0860\text{ M acetone}$$

(b) $$\text{molality} = \frac{\text{no. of moles acetone}}{\text{solvent mass in kg}} = \left[\frac{0.00861\text{ mol acetone}}{99.500\text{ g water}}\right]\left[\frac{1000\text{ g}}{1\text{ kg}}\right]$$

$$= 0.0865\text{ mol acetone/kg} = 0.0865\text{ m}$$

(c) $$(99.500\text{ g water})\left[\frac{1\text{ mol water}}{18.015\text{ g water}}\right] = 5.5232\text{ mol water}$$

$$X_{\text{acetone}} = \frac{n_{\text{acetone}}}{n_{\text{acetone}} + n_{\text{water}}} = \frac{0.00861\text{ mol}}{0.00861\text{ mol} + 5.5232\text{ mol}} = 1.56 \times 10^{-3}$$

15.6 $$(1.75\text{ L})\left[\frac{6.0\text{ mol }HNO_3}{1\text{ L}}\right] = 11\text{ mol }HNO_3$$

$$(11\text{ mol }HNO_3)\left[\frac{1\text{ L}}{15.6\text{ mol }HNO_3}\right] = 0.71\text{ L}$$

15.7 $$(14.6\text{ g NaCl})\left[\frac{1\text{ mol NaCl}}{58.44\text{ g NaCl}}\right] = 0.250\text{ mol NaCl}$$

$$\text{molality} = \left[\frac{0.250\text{ mol NaCl}}{996\text{ g }H_2O + 225\text{ g }H_2O}\right]\left[\frac{1000\text{ g}}{1\text{ kg}}\right] = 0.205\text{ m}$$

15.8 $P_{\text{benz}} = X_{\text{benz}}\, P^{\circ}_{\text{benz}} = (0.70)(73\text{ Torr}) = 51\text{ Torr}$

$P_{\text{tol}} = X_{\text{tol}}\, P^{\circ}_{\text{tol}} = (0.30)(27\text{ Torr}) = 8.1\text{ Torr}$

$P_{\text{soln}} = P_{\text{benz}} + P_{\text{tol}} = 51\text{ Torr} + 8.1\text{ Torr} = 59\text{ Torr}$

15.9 $$(1000.0\text{ g }H_2O)\left[\frac{1\text{ mol }H_2O}{18.015\text{ g }H_2O}\right] = 55.509\text{ mol }H_2O$$

$$X_{\text{urea}} = \frac{n_{\text{urea}}}{n_{\text{urea}} + n_{H_2O}} = \frac{0.083\text{ mol}}{0.083\text{ mol} + 55.509\text{ mol}} = 0.0015$$

$\Delta P = X_{\text{urea}}\, P^{\circ}_{H_2O} = (0.0015)(71.88\text{ Torr}) = 0.11\text{ Torr}$

$P_{\text{soln}} = P^{\circ}_{H_2O} - \Delta P = 71.88\text{ Torr} - 0.11\text{ Torr} = 71.77\text{ Torr}$

15.10 $\Delta T_b = K_b\, m = (0.512\text{ °C kg/mol})(0.125\text{ mol/kg}) = 0.0640\text{ °C}$

$T_{b,\text{soln}} = T_{b,\text{solvent}} + \Delta T_b = 100.0000\text{ °C} + 0.0640\text{ °C} = 100.0640\text{ °C}$

$\Delta T_f = K_f\, m = (1.86\text{ °C kg/mol})(0.125\text{ mol/kg}) = 0.233\text{ °C}$

$T_{f,\text{soln}} = T_{f,\text{solvent}} - \Delta T_f = 0.000\text{ °C} - 0.233\text{ °C} = -0.233\text{ °C}$

15.11 $\Delta T_f = T_{f,solvent} - T_{f,soln} = 0.00\ °C - (-5.02\ °C) = 5.02\ °C$

$$m = \frac{\Delta T_f}{K_f} = \frac{5.02\ °C}{1.86\ °C\ kg/mol} = 2.70\ mol/kg$$

$$(2.70\ mol/kg)(85.00\ g)\left[\frac{1\ kg}{1000\ g}\right] = 0.230\ mol$$

$$\frac{15.00\ g}{0.230\ mol} = 65.2\ g/mol$$

(The actual molar mass of urea is 60.06 g/mol.)

15.12 $\Pi = MRT = (0.25\ mol/L)(0.0821\ L\ atm/K\ mol)(298\ K) = 6.1\ atm$

15.13 $$\left[\frac{0.100\ mol\ C_6H_5COOH}{1\ kg}\right]\left[\frac{2.6\ mol\ H^+}{100\ mol\ C_6H_5COOH}\right] = 0.0026\ mol\ H^+/kg$$

$$\left[\frac{0.100\ mol\ C_6H_5COOH}{1\ kg}\right]\left[\frac{2.6\ mol\ C_6H_5COO^-}{100\ mol\ C_6H_5COOH}\right] = 0.0026\ mol\ C_6H_5COO^-/kg$$

$$\left[\frac{0.100\ mol\ C_6H_5COOH}{1\ kg}\right]\left[\frac{97.4\ mol\ unionized\ C_6H_5COOH}{100\ mol\ C_6H_5COOH}\right]$$
$$= 0.0974\ mol\ unionized\ C_6H_5COOH/kg$$

total no. of particles = 0.0026 mol/kg + 0.0026 mol/kg + 0.0974 mol/kg
= 0.1026 mol/kg

$\Delta T_f = K_f\ m = (1.86\ °C\ kg/mol)(0.1026\ mol/kg) = 0.191\ °C$

Solutions to Odd-Numbered Questions and Problems

15.1 (a) salt water, NaCl (solute) in H_2O (solvent); (b) air, O_2 (solute) in N_2 (solvent); (c) hydrochloric acid, HCl (solute) in H_2O (solvent); (d) vinegar, CH_3COOH (solute) in H_2O (solvent); and (e) brass, Cu (solute) in Zn (solvent).

15.3 The equilibrium between a saturated solution and undissolved solute is a dynamic one--equal amounts of solute are dissolving and recrystallizing continually. Such a description might be proved by putting "tagged" solute (containing different isotopes of one of the elements) into the saturated solution and observing it in the solution after a period of time.

15.5 The forces of attraction between the oil and water molecules are not great enough to overcome the intermolecular forces in the oil and the very strong hydrogen bonding in the water.

15.7 The solvent-solvent, solute-solute, and solvent-solute intermolecular forces should be as identical as possible. Solution (c) best represents a solution with equal forces.

15.9 $X = \frac{P}{k} = \frac{0.15\ \text{atm}}{3.02 \times 10^{4}\ \text{atm}} = 5.0 \times 10^{-6}$

15.11

$$C_{N_2} = \frac{P_{N_2}}{k} = \frac{608\ \text{Torr}}{4.34 \times 10^{5}\ \text{Torr/(g N}_2\text{/100 g H}_2\text{O)}} = 1.40 \times 10^{-3}\ \text{g N}_2\text{/100 g H}_2\text{O}$$

$$C_{O_2} = \frac{152\ \text{Torr}}{1.93 \times 10^{5}\ \text{Torr/(g O}_2\text{/100 g H}_2\text{O)}} = 7.88 \times 10^{-4}\ \text{g O}_2\text{/100 g H}_2\text{O}$$

$$\frac{m_{O_2}}{m_{N_2}} = \frac{7.88 \times 10^{-4}\ \text{g O}_2\text{/100 g H}_2\text{O}}{1.40 \times 10^{-3}\ \text{g N}_2\text{/100 g H}_2\text{O}} = 0.563 \text{ in aqueous solution}$$

For masses in air, assume 1 L:

$$n_{N_2} = \frac{PV}{RT} = \frac{(608\ \text{Torr})(1\ \text{atm/760 Torr})(1.00\ \text{L})}{(0.0821\ \text{L atm/K mol})(298\ \text{K})} = 0.0327\ \text{mol}$$

$$m_{N_2} = (0.0327\ \text{mol})\left[\frac{28.01\ \text{g N}_2}{1\ \text{mol N}_2}\right] = 0.916\ \text{g N}_2$$

$$n_{O_2} = \frac{(152\ \text{Torr})(1\ \text{atm/760 Torr})(1.00\ \text{L})}{(0.0821\ \text{L atm/K mol})(298\ \text{K})} = 0.0817\ \text{mol}$$

$$m_{O_2} = (0.0817\ \text{mol O}_2)\left[\frac{32.00\ \text{g O}_2}{1\ \text{mol O}_2}\right] = 0.261\ \text{g O}_2$$

$$\frac{m_{O_2}}{m_{N_2}} = \frac{0.261\ \text{g}}{0.916\ \text{g}} = 0.285 \text{ in air}$$

The ratio of the mass of O_2 to N_2 is greater in water than in air.

15.13 The solubility generally increases with increasing temperature. Many salts containing the SO_4^{2-}, SeO_4^{2-}, SO_3^{2-}, AsO_4^{3-}, and PO_4^{3-} anions do not follow the general trend.

15.15 The concentration unit dependent on temperature is molarity.

15.17 Let s be the solubility in units of (g solute/100 g H_2O). Then

$$\text{mass percent} = \frac{\text{mass of solute}}{\text{mass of solution}} \times 100 = \frac{s}{s + 100} \times 100$$

15.19

$$(40.0\ \text{mL ether})\left[\frac{0.714\ \text{g ether}}{1\ \text{mL ether}}\right] = 28.6\ \text{g ether}$$

$$(28.6\ \text{g ether})\left[\frac{1\ \text{mol ether}}{74.1\ \text{g ether}}\right] = 0.386\ \text{mol ether}$$

$$\text{molarity} = \frac{\text{no. of moles ether}}{\text{soln vol. in L}}$$

$$= \left[\frac{0.386 \text{ mol ether}}{250. \text{ mL soln}}\right]\left[\frac{1000 \text{ mL}}{1 \text{ L}}\right] = 1.54 \text{ mol/L} = 1.54 \text{ M}$$

15.21 The number of moles of each component is

$$(30.0 \text{ g sucrose})\left[\frac{1 \text{ mol sucrose}}{342.3 \text{ g}}\right] = 0.0876 \text{ mol sucrose}$$

$$(70.0 \text{ g water})\left[\frac{1 \text{ mol water}}{18.02 \text{ g water}}\right] = 3.88 \text{ mol water}$$

The mole fraction of the sugar is

$$X_{\text{sucrose}} = \frac{n_{\text{sucrose}}}{n_{\text{sucrose}} + n_{\text{water}}} = \frac{0.0876 \text{ mol}}{0.0876 \text{ mol} + 3.88 \text{ mol}} = 0.0221$$

15.23 $$(160. \text{ g soln})\left[\frac{12 \text{ g NaCl}}{100 \text{ g soln}}\right] = 19 \text{ g NaCl}$$

$160. \text{ g } H_2O - 19 \text{ g NaCl} = 141 \text{ g } H_2O$

15.25 $$(25 \text{ g } K_2ZrF_6)\left[\frac{1 \text{ mol } K_2ZrF_6}{283.42 \text{ g } K_2ZrF_6}\right] = 0.088 \text{ mol } K_2ZrF_6$$

$$\text{molality} = \frac{\text{no. of moles } K_2ZrF_6}{\text{solv mass in kg}} = \left[\frac{0.088 \text{ mol } K_2ZrF_6}{100.0 \text{ g } H_2O}\right]\left[\frac{1000 \text{ g}}{1 \text{ kg}}\right]$$

$$= 0.88 \text{ mol } K_2ZrF_6/\text{kg} = 0.88 \text{ m}$$

15.27 $$(1.000 \text{ L soln})\left[\frac{1.011 \text{ g soln}}{1 \text{ mL soln}}\right]\left[\frac{1000 \text{ mL}}{1 \text{ L}}\right] = 1011 \text{ g soln}$$

$$(1011 \text{ g soln})\left[\frac{1 \text{ kg}}{1000 \text{ g}}\right]\left[\frac{0.250 \text{ mol NaCl}}{1 \text{ kg soln}}\right] = 0.253 \text{ mol NaCl}$$

$$(0.253 \text{ mol NaCl})\left[\frac{58.44 \text{ g NaCl}}{1 \text{ mol NaCl}}\right] = 14.8 \text{ g NaCl}$$

$1011 \text{ g soln} - 14.8 \text{ g NaCl} = 996 \text{ g } H_2O$

Dissolve 14.8 g NaCl in 996 g H_2O.

15.29 $$(250. \text{ mL soln})\left[\frac{1 \text{ L}}{1000 \text{ mL}}\right]\left[\frac{0.876 \text{ mol NaOH}}{1 \text{ L soln}}\right] = 0.219 \text{ mol NaOH}$$

$$(0.219 \text{ mol NaOH})\left[\frac{1 \text{ L soln}}{4.38 \text{ mol NaOH}}\right]\left[\frac{1000 \text{ mL}}{1 \text{ L}}\right] = 50.0 \text{ mL soln}$$

Measure out 50.0 mL of 4.38 M NaOH and dilute to exactly 250 mL with H_2O.

15.31 $\Delta H°(\text{dilution}) = [(1\ \text{mol})\Delta H_f°(H_2SO_4, 1\ M)] - [(1\ \text{mol})\Delta H_f°(H_2SO_4, 16\ M)]$

$= [(1\ \text{mol})(-887\ \text{kJ/mol})] - [(1\ \text{mol})(-866\ \text{kJ/mol})] = -21\ \text{kJ}$

The beaker would be warm to the touch.

15.33 $$(100.00\ \text{g soln})\left[\frac{1\ \text{mL soln}}{1.0201\ \text{g soln}}\right] = 98.030\ \text{mL soln}$$

$$(6.50\ \text{g}\ NH_4Cl)\left[\frac{1\ \text{mol}\ NH_4Cl}{53.50\ \text{g}\ NH_4Cl}\right] = 0.121\ \text{mol}\ NH_4Cl$$

$$\text{molarity} = \frac{\text{no. of moles}\ NH_4Cl}{\text{soln vol. in L}} = \left[\frac{0.121\ \text{mol}\ NH_4Cl}{98.030\ \text{mL soln}}\right]\left[\frac{1000\ \text{mL}}{1\ \text{L}}\right]$$

$$= 1.23\ \text{mol}\ NH_4Cl/\text{L} = 1.23\ \text{M}$$

$$100.00\ \text{g soln} - 6.50\ \text{g}\ NH_4Cl = 93.50\ \text{g}\ H_2O$$

$$\text{molality} = \frac{\text{no. of moles}\ NH_4Cl}{\text{solv mass in kg}} = \left[\frac{0.121\ \text{mol}\ NH_4Cl}{93.50\ \text{g}\ H_2O}\right]\left[\frac{1000\ \text{g}}{1\ \text{kg}}\right]$$

$$= 1.29\ \text{mol}\ NH_4Cl/\text{kg} = 1.29\ m\ NH_4Cl$$

$$(93.50\ \text{g}\ H_2O)\left[\frac{1\ \text{mol}\ H_2O}{18.02\ \text{g}\ H_2O}\right] = 5.189\ \text{mol}\ H_2O$$

$$X_{NH_4Cl} = \frac{n_{NH_4Cl}}{n_{NH_4Cl} + n_{H_2O}} = \frac{0.121\ \text{mol}}{0.121\ \text{mol} + 5.189\ \text{mol}} = 0.0228$$

15.35 Assume 1.000 L of solution.

$$(1.000\ \text{L soln})\left[\frac{3.75\ \text{mol}\ H_2SO_4}{1\ \text{L soln}}\right] = 3.75\ \text{mol}\ H_2SO_4$$

$$(3.75\ \text{mol}\ H_2SO_4)\left[\frac{98.09\ \text{g}\ H_2SO_4}{1\ \text{mol}\ H_2SO_4}\right] = 368\ \text{g}\ H_2SO_4$$

$$(1.000\ \text{L soln})\left[\frac{1.225\ \text{g soln}}{1\ \text{mL soln}}\right]\left[\frac{1000\ \text{mL}}{1\ \text{L}}\right] = 1225\ \text{g soln}$$

$$1225\ \text{g soln} - 368\ \text{g}\ H_2SO_4 = 857\ \text{g}\ H_2O$$

$$\text{molality} = \frac{\text{no. of moles}\ H_2SO_4}{\text{solv mass in kg}} = \left[\frac{3.75\ \text{mol}\ H_2SO_4}{857\ \text{g}\ H_2O}\right]\left[\frac{1000\ \text{g}}{1\ \text{kg}}\right]$$

$$= 4.38\ \text{mol}\ H_2SO_4/\text{kg} = 4.38\ m\ H_2SO_4$$

$$(857\ \text{g}\ H_2O)\left[\frac{1\ \text{mol}\ H_2O}{18.02\ \text{g}\ H_2O}\right] = 47.6\ \text{mol}\ H_2O$$

$$X_{H_2SO_4} = \frac{n_{H_2SO_4}}{n_{H_2SO_4} + n_{H_2O}} = \frac{3.75\ \text{mol}}{3.75\ \text{mol} + 47.6\ \text{mol}} = 0.0730$$

$$\text{mass \%} = \frac{\text{mass } H_2O}{\text{mass soln}} \times 100 = \frac{857 \text{ g}}{1225 \text{ g}} \times 100 = 70.0 \text{ mass \% } H_2O$$

15.37 (a) The composition of the liquid and vapor phases will be the same unless azeotrope is formed.

(b) The composition of the vapor phase will be richer in A than the liquid phase.

(c) The composition of the vapor phase will be richer in B than the liquid phase.

15.39 A flask containing the immiscible liquids toluene and water will boil at a temperature below the boiling point of either because the system boils when the total vapor pressure equals the atmospheric pressure. The two liquids exert vapor pressures independently; when the total vapor pressure equals atmospheric pressure, boiling occurs.

15.41

$$X_{\text{acetone}} = \frac{n_{\text{acetone}}}{n_{\text{acetone}} + n_{\text{chloroform}}} = \frac{0.300 \text{ mol}}{0.300 \text{ mol} + 0.200 \text{ mol}} = 0.600$$

$$P_{\text{acetone}} = X_{\text{acetone}} P^{\circ}_{\text{acetone}} = (0.600)(345 \text{ Torr}) = 207 \text{ Torr}$$

$$X_{\text{chloroform}} = 1 - X_{\text{acetone}} = 1 - 0.600 = 0.400$$

$$P_{\text{chloroform}} = X_{\text{chloroform}} P^{\circ}_{\text{chloroform}} = (0.400)(295 \text{ Torr}) = 118 \text{ Torr}$$

$$P_{\text{soln}} = P_{\text{acetone}} + P_{\text{chloroform}} = 207 \text{ Torr} + 118 \text{ Torr} = 325 \text{ Torr}$$

15.43

$$(10.0 \text{ g urea})\left[\frac{1 \text{ mol urea}}{60.07 \text{ g urea}}\right] = 0.166 \text{ mol urea}$$

$$(100.00 \text{ g } H_2O)\left[\frac{1 \text{ mol } H_2O}{18.02 \text{ g } H_2O}\right] = 5.549 \text{ mol } H_2O$$

$$X_{\text{urea}} = \frac{n_{\text{urea}}}{n_{\text{urea}} + n_{H_2O}} = \frac{0.166 \text{ mol}}{0.166 \text{ mol} + 5.549 \text{ mol}} = 0.0290$$

$$\Delta P = P^{\circ}_{H_2O} X_{\text{urea}} = (23.76 \text{ Torr})(0.0290) = 0.689 \text{ Torr}$$

$$P = P^{\circ} - \Delta P = 23.76 \text{ Torr} - 0.689 \text{ Torr} = 23.07 \text{ Torr}$$

15.45 The changes in the boiling point and freezing point are

$$\Delta T_b = K_b\, m = (0.512 \text{ °C kg/mol})(0.125 \text{ mol/kg}) = 0.0640 \text{ °C}$$

$$\Delta T_f = K_f\, m = (1.86 \text{ °C kg/mol})(0.125 \text{ mol/kg}) = 0.233 \text{ °C}$$

The boiling point of the solution is

$$T_{b,solution} = T_{b,solvent} + \Delta T_b = 100.0000\ °C + 0.0640\ °C = 100.0640\ °C$$

and the freezing point of the solution is

$$T_{f,solution} = T_{f,solvent} - \Delta T_f = 0.000\ °C - 0.233\ °C = -0.233\ °C$$

15.47 (a) $$(25\ g\ CH_3OH)\left[\frac{1\ mol\ CH_3OH}{32.05\ g\ CH_3OH}\right] = 0.78\ mol\ CH_3OH$$

$$(25\ g\ HOH_2CCH_2OH)\left[\frac{1\ mol\ HOH_2CCH_2OH}{62.08\ g\ HOH_2CCH_2OH}\right] = 0.40\ mol\ HOH_2CCH_2OH$$

For a given radiator the mass of water for both solutions will be the same, so the molality of the methyl alcohol will be greater. Because $\Delta T_f = K_f\ m$, methyl alcohol will be more effective.

(b) The effectiveness of both solutions will be the same because the freezing point depression is directly proportional to molality.

15.49 $$\Delta T_b = T_{b,soln} - T_{b,solvent} = 56.58\ °C - 55.95\ °C = 0.63\ °C$$

$$m = \frac{\Delta T_b}{K_b} - \frac{0.63\ °C}{1.71\ °C\ kg/mol} = 0.37\ m$$

$$(95.0\ g\ acetone)\left[\frac{1\ kg}{1000\ g}\right]\left[\frac{0.37\ mol\ solute}{1\ kg\ acetone}\right] = 0.035\ mol\ solute$$

$$\frac{3.75\ g\ solute}{0.035\ mol\ solute} = 110\ g/mol$$

15.51 $\Pi = MRT = (0.001\ mol/L)(0.0821\ L\ atm/K\ mol)(348\ K) = 0.03\ atm$

15.53 $$T = \frac{\Pi}{MR} = \frac{1.00\ atm}{(1.00\ mol/L)(0.0821\ L\ atm/K\ mol)} = 12.2\ K$$

No, this is not a reasonable answer because the solution would be frozen.

15.55 $$M = \frac{\Pi}{RT} = \frac{(539\ Torr)(1\ atm/760\ Torr)}{(0.0821\ L\ atm/K\ mol)(298\ K)} = 0.0290\ mol/L$$

$$0.0290\ mol/L = \frac{\left[[(1.00 - x)g\ drug]\left[\frac{1\ mol\ drug}{369\ g\ drug}\right] + (x\ g\ lactose)\left[\frac{1\ mol\ lactose}{342\ g\ lactose}\right]\right]}{0.10000\ L}$$

$$X = 0.89\ g\ lactose$$

$$mass\ \%\ lactose = \frac{0.89\ g}{1.00\ g} \times 100 = 89\ mass\ \%$$

15.57 In solutions of equal concentrations, the number of solute particles present is

strong electrolytes >> weak electrolytes > nonelectrolytes

15.59 $$m = \frac{\text{no. of mol of particles}}{\text{solv. mass in kg}} = \frac{(8.5\text{ g NaCl})\left[\frac{1\text{ mol NaCl}}{58.44\text{ g NaCl}}\right]\left[\frac{2\text{ mol particles}}{1\text{ mol NaCl}}\right]}{1.00\text{ kg}}$$

$$= 0.29\text{ mol/kg} \approx 0.29\text{ mol/L}$$

$$\Pi = MRT = (0.29\text{ mol/L})(0.0821\text{ L atm/K mol})(310.\text{ K}) = 7.4\text{ atm}$$

15.61 $$\Delta T_f = T_{f,\text{solvent}} - T_{f,\text{soln}} = 0.0000\ °\text{C} - (-0.2800\ °\text{C}) = 0.2800\ °\text{C}$$

$$m = \frac{\Delta T_f}{K_f} = \frac{0.2800\ °\text{C}}{1.86\ °\text{C kg/mol}} = 0.151\text{ mol/kg}$$

The concentration increase is slightly over threefold, which indicates that $K_3Fe(CN)_6$ forms four ions (three K^+ ions and one $Fe(CN)_6^{3-}$ ion) in solution. The concentration increase would be near tenfold if the substance formed three K^+, one Fe^{3+}, and six CN^- ions in solution.

15.63 $$(1.00\text{ g NaCl})\left[\frac{1\text{ mol NaCl}}{58.44\text{ g NaCl}}\right]\left[\frac{2\text{ mol ions}}{1\text{ mol NaCl}}\right] = 0.0342\text{ mol ions}$$

$$(1.00\text{ g NaBr})\left[\frac{1\text{ mol NaBr}}{102.89\text{ g NaBr}}\right]\left[\frac{2\text{ mol ions}}{1\text{ mol NaBr}}\right] = 0.0194\text{ mol ions}$$

$$(1.00\text{ g NaI})\left[\frac{1\text{ mol NaI}}{149.89\text{ g NaI}}\right]\left[\frac{2\text{ mol ions}}{1\text{ mol NaI}}\right] = \underline{0.0133\text{ mol ions}}$$

$$0.0669\text{ mol ions}$$

$$(1.00\text{ g H}_2\text{O})\left[\frac{1\text{ mol H}_2\text{O}}{18.02\text{ g H}_2\text{O}}\right] = 5.549\text{ mol ions}$$

$$X_{H_2O} = \frac{n_{H_2O}}{n_{\text{ions}} + n_{H_2O}} = \frac{5.549\text{ mol}}{0.0669 + 5.5549\text{ mol}} = 0.9881$$

$$P_{H_2O} = X_{H_2O}\,P^\circ_{H_2O} = (0.9881)(760.0\text{ Torr}) = 751.0\text{ Torr}$$

15.65 $$\Delta T_f = T_{f,\text{solvent}} - T_{f,\text{soln}} = 0.00\ °\text{C} - (-0.31\ °\text{C}) = 0.31\ °\text{C}$$

$$m = \frac{\Delta T_f}{K_f} = \frac{0.31\ °\text{C}}{1.86\ °\text{C kg/mol}} = 0.17\text{ mol/kg}$$

$$(99.00\text{ g H}_2\text{O})\left[\frac{1\text{ kg}}{1000\text{ g}}\right]\left[\frac{0.17\text{ mol acetic acid}}{1\text{ kg H}_2\text{O}}\right] = 0.017\text{ mol acetic acid}$$

The approximate molar mass in H_2O is

$$\frac{1.00\ g}{0.017\ mol} = 59\ g/mol$$

$$m = \frac{0.441\ °C}{4.90\ °C\ kg/mol} = 0.0900\ mol/kg$$

$$(99.00\ g\ benzene)\left[\frac{1\ kg}{1000\ g}\right]\left[\frac{0.0900\ mol\ acetic\ acid}{1\ kg\ benzene}\right] = 0.00891\ mol\ acetic\ acid$$

The approximate molar mass in benzene is

$$\frac{1.00\ g}{0.00891\ mol} = 112\ g/mol$$

Acetic acid is a weak acid in aqueous solution, so only a small amount ionizes and it behaves as a nonelectrolyte. In benzene, the acetic acid dimerizes because of hydrogen bonding and so it behaves as if it had a molar mass of two times the value in water.

```
       O•••H-O-C-CH₃
       ‖       ‖
CH₃-C-O-H•••O
```

15.67 Assume exactly 10^6 g soln.

1,000,000. g soln - 18,980. g Cl^- - 10,561 g Na^+ - 2652 g SO_4^{2-}
- 1272 g Mg^{2+} - 400. g Ca^{2+} - 380. g K^+ - 142 g HCO_3^- - 65 g Br^-
= 965,548 g H_2O = 965.548 kg H_2O

$$\frac{(18{,}980.\ g\ Cl^-)\left[\frac{1\ mol\ Cl^-}{35.453\ g\ Cl^-}\right]}{965.548\ kg} = 0.55446\ mol\ Cl^-/kg$$

$$\frac{(10{,}561\ g\ Na^+)\left[\frac{1\ mol\ Na^+}{22.990\ g\ Na^+}\right]}{965.548\ kg} = 0.47576\ mol\ Na^+/kg$$

$$\frac{(2652\ g\ SO_4^{2-})\left[\frac{1\ mol\ SO_4^{2-}}{96.06\ g\ SO_4^{2-}}\right]}{965.548\ kg} = 0.02859\ mol\ SO_4^{2-}/kg$$

$$\frac{(1272\ g\ Mg^{2+})\left[\frac{1\ mol\ Mg^{2+}}{24.31\ g\ Mg^{2+}}\right]}{965.548\ kg} = 0.05419\ mol\ Mg^{2+}/kg$$

$$\frac{(400.\ g\ Ca^{2+})\left[\frac{1\ mol\ Ca^{2+}}{40.08\ g\ Ca^{2+}}\right]}{965.548\ kg} = 0.0103\ mol\ Ca^{2+}/kg$$

$$\frac{(380.\ \text{g } K^+)\left[\frac{1\ \text{mol } K^+}{39.10\ \text{g } K^+}\right]}{965.548\ \text{kg}} = 0.0101\ \text{mol } K^+/\text{kg}$$

$$\frac{(142\ \text{g } HCO_3^-)\left[\frac{1\ \text{mol } HCO_3^-}{61.02\ \text{g } HCO_3^-}\right]}{965.548\ \text{kg}} = 0.00241\ \text{mol } HCO_3^-/\text{kg}$$

$$\frac{(65\ \text{g } Br^-)\left[\frac{1\ \text{mol } Br^-}{79.90\ \text{g } Br^-}\right]}{965.548\ \text{kg}} = 0.00084\ \text{mol } Br^-/\text{kg}$$

total molality = 1.1367 mol/kg

$\Delta T_f = K_f\, m = (1.86\ °\text{C kg/mol})(1.1367\ \text{mol/kg}) = 2.11\ °\text{C}$

$T_{f,soln} = T_{f,solvent} - \Delta T_f = 0.00\ °\text{C} - 2.11\ °\text{C} = -2.11\ °\text{C}$

$\Delta T_b = K_b\, m = (0.512\ °\text{C kg/mol})(1.1367\ \text{mol/kg}) = 0.582\ °\text{C}$

$T_{b,soln} = T_{b,solvent} + \Delta T_b = 100.000\ °\text{C} + 0.582\ °\text{C} = 100.582\ °\text{C}$

$\Pi = MRT = (1.1367\ \text{mol/L})(0.0821\ \text{L atm/K mol})(298\ \text{K}) = 27.8\ \text{atm}$

15.69 A colloidal dispersion differs from a true solution in having suspended particles larger than the solute molecules or ions.

15.71 Gases exist as molecules and form only true solutions upon mixing. The colloid can be precipitated by adding a colloid of opposite charge or by using a Cottrell precipitator.

15.73 (a) Colloids in which the dispersing phase is a gas are called aerosols.
(b) Colloids in which the dispersed phase is a gas are called foams.

15.75

$$k_{CO} = \frac{P_{CO}}{C_{CO}} = \frac{1.000\ \text{atm}}{0.002603\ \text{g CO}/100\ \text{g } H_2O} = 384.2\ \text{atm/(g CO/100 g } H_2O)$$

$$k_{SO_2} = \frac{P_{SO_2}}{C_{SO_2}} = \frac{1.000\ \text{atm}}{9.41\ \text{g } SO_2/100\ \text{g } H_2O} = 0.106\ \text{atm/(g } SO_2/100\ \text{g } H_2O)$$

$$(0.002603\ \text{g CO})\left[\frac{1\ \text{mol CO}}{28.01\ \text{g CO}}\right] = 9.293 \times 10^{-5}\ \text{mol CO}$$

$$\text{molality} = \frac{\text{no. of moles CO}}{\text{solv mass in kg}} = \left[\frac{9.293 \times 10^{-5}\ \text{mol CO}}{100.0\ \text{g } H_2O}\right]\left[\frac{1000\ \text{g}}{1\ \text{kg}}\right] = 9.293 \times 10^{-4}\ \text{mol/kg}$$

$$(9.41\ \text{g } SO_2)\left[\frac{1\ \text{mol } SO_2}{64.06\ \text{g } SO_2}\right] = 0.147\ \text{mol } SO_2$$

$$\text{molality} = \frac{\text{no. of moles } SO_2}{\text{solv mass in kg}} = \left[\frac{0.147 \text{ mol } SO_2}{100.0 \text{ g } H_2O}\right]\left[\frac{1000 \text{ g}}{1 \text{ kg}}\right] = 1.47 \text{ mol/kg}$$

$$\Delta T_f = K_f\, m = (1.86\ °C\ \text{kg/mol})(9.293 \times 10^{-4}\ \text{mol/kg}) = 0.00173\ °C$$

$$T_{f,soln} = T_{f,solvent} - \Delta T_f = 0.00000\ °C - 0.00173\ °C = -0.00173\ °C$$

$$\Delta T_{f,SO_2} = (1.86\ °C\ \text{kg/mol})(1.47\ \text{mol/kg}) = 2.73\ °C$$

$$T_{f,SO_2} = 0.00\ °C - 2.73\ °C = -2.73\ °C \quad \text{assuming no ionization}$$

There is such a large difference in the solubilities because SO_2 reacts with water to form H_2SO_3.

CHAPTER 16

HYDROGEN AND OXYGEN; OXIDATION-REDUCTION REACTIONS

Solutions to Exercises

16.1 (a)

$$\overset{-1}{I^-} + \overset{+1\ -2}{ClO^-} \rightarrow \overset{+1\ -2}{IO^-} + \overset{-1}{Cl^-}$$

This is a redox reaction--iodine is being oxidized (-1 → +1) and chlorine is being reduced (+1 → -1).

(b)

$$\overset{+1\ -2}{H_2O(l)} \rightleftharpoons \overset{+1}{H^+} + \overset{-2\ +1}{OH^-}$$

This is not a redox reaction--no changes in oxidation number have occurred.

(c)

$$\overset{+1\ +5\ -2}{4HNO_3(l)} \rightarrow \overset{+4\ -2}{2N_2O_4(g)} + \overset{0}{O_2(g)} + \overset{+1\ -2}{2H_2O(l)}$$

This is a redox reaction--oxygen is being oxidized (-2 → 0) and nitrogen is being reduced (+5 → +4).

(d)

$$\overset{-1\ +1}{C_6H_6(l)} \rightarrow \overset{-1\ +1}{C_6H_6(s)}$$

This is not a redox reaction--no changes in oxidation number have occurred.

16.2 Step A. Unbalanced equation showing changes in oxidation numbers.

$$\overset{+4\ -2}{MnO_2(s)} + \overset{+4\ -2}{PbO_2(s)} \xrightarrow{H^+} \overset{+7\ -2}{MnO_4^-} + \overset{+2}{Pb^{2+}}$$

$$\overset{+4}{MnO_2} \rightarrow \overset{+7}{MnO_4^-} \quad (1)(+3) = +3 \qquad \overset{+4}{PbO_2} \rightarrow \overset{+2}{Pb^{2+}} \quad (1)(-2) = -2$$

Step B. Coefficients needed to balance oxidation number change.

(2)(+3)

$$2MnO_2 + 3PbO_2 \rightarrow 2MnO_4^- + 3Pb^{2+}$$

(3)(-2)

Step C. Balance charges by adding H^+.

$$2MnO_2 + 3PbO_2 + 4H^+ \rightarrow 2MnO_4^- + 3Pb^{2+}$$

Step D. Balance O by adding H_2O.

$$2MnO_2(s) + 3PbO_2(s) + 4H^+ \rightarrow 2MnO_4^- + 3Pb^{2+} + 2H_2O(l)$$

Step E. Check. On each side of the equation there are 2 Mn atoms, 3 Pb atoms, 4 H atoms, 10 O atoms, and +4 charge.

16.3 Step A. Unbalanced equation showing changes in oxidation numbers.

$$\overset{+4\ -2}{SO_3^{2-}} + \overset{+6\ -2}{CrO_4^{2-}} \rightarrow \overset{+3\ -2\ +1}{[Cr(OH)_4]^-} + \overset{+6\ -2}{SO_4^{2-}}$$

$$\overset{+4}{SO_3^{2-}} \rightarrow \overset{+6}{SO_4^{2-}} \quad (1)(+2) = +2 \qquad \overset{+6}{CrO_4^{2-}} \rightarrow \overset{+3}{[Cr(OH)_4]^-} \quad (1)(-3) = -3$$

Step B. Coefficients needed to balance oxidation number change.

(3)(+2)

$$3SO_3^{2-} + 2CrO_4^{2-} \rightarrow 2[Cr(OH)_4]^- + 3SO_4^{2-}$$

(2)(-3)

Step C. Balance charges by adding OH^-.

$$3SO_3^{2-} + 2CrO_4^{2-} \rightarrow 2[Cr(OH)_4]^- + 3SO_4^{2-} + 2OH^-$$

Step D. Balance O by adding H_2O.

$$3SO_3^{2-} + 2CrO_4^{2-} + 5H_2O(l) \rightarrow 2[Cr(OH)_4]^- + 3SO_4^{2-} + 2OH^-$$

Step E. Check. On each side of the equation there are 3 S atoms, 2 Cr atoms, 10 H atoms, 22 O atoms, and -10 charge.

16.4 The nonmetal oxide, P_4O_{10}, is the acidic oxide. Because Cs is a Representative Group I metal, Cs_2O is a basic oxide. This leaves ZnO as the amphoteric oxide.

Solutions to Odd-Numbered Questions and Problems

16.1 The changes in oxidation number that represent oxidation are (a) 0 → +2, (b) -3 → -2, and (f) -2 → +1.

16.3 (a)

$$\overset{+1\ -1}{2Na_2O_2(s)} + \overset{+1\ -2}{2H_2O(l)} \rightarrow \overset{0}{O_2(g)} + \overset{+1\ -2\ +1}{4NaOH(aq)}$$

This is a redox reaction--oxygen in O_2^{2-} is both the oxidizing agent and the reducing agent.

(b)

$$\overset{0}{Sn(s)} + \overset{+1\ -1}{2HCl(aq)} \rightarrow \overset{+2\ -1}{SnCl_2(aq)} + \overset{0}{H_2(g)}$$

This is a redox reaction--tin is the reducing agent and hydrogen ion is the oxidizing agent.

(c) +2 −2   +1 −2   +2 −2 +1

$CaO(s) + H_2O(l) \rightarrow Ca(OH)_2(aq)$

This is not a redox reaction--no changes in oxidation number have occurred.

(d) +5 −2   +1 −2   +1 +5 −2

$N_2O_5(s) + H_2O(l) \rightarrow 2HNO_3(aq)$

This is not a redox reaction--no changes in oxidation number have occurred.

16.5 (a) +2   +3   +4   +2

$Sn^{2+} + 2Fe^{3+} \rightarrow Sn^{4+} + 2Fe^{2+}$

Tin(II) ion is the reducing agent; iron(III) ion is the oxidizing agent.

(b) +4 −2   +1 −1   +2 −1   0   +1 −2

$MnO_2(s) + 4HCl(aq) \rightarrow MnCl_2(aq) + Cl_2(g) + 2H_2O(l)$

Manganese(IV) oxide is the oxidizing agent; hydrochloric acid is the reducing agent.

(c) +2 −1   +1 −2   0   0   +1 −1

$2XeF_2(s) + 2H_2O(l) \rightarrow 2Xe(g) + O_2(g) + 4HF(g)$

Xenon(II) fluoride is the oxidizing agent; water is the reducing agent.

(d) −1   +1 −2   +1 −2   −1

$I^- + ClO^- \rightarrow IO^- + Cl^-$

I^- is the reducing agent; ClO^- is the oxidizing agent.

(e) +1 +5 −2   +4 −2   0   +1 −2

$4HNO_3(l) \rightarrow 2N_2O_4(g) + O_2(g) + 2H_2O(l)$

HNO_3 is both the oxidizing agent and the reducing agent.

16.7 (a) Step A. +4 −2   +1 −2   0   +1 −2

$SO_2(g) + H_2S(g) \rightarrow S_8(s) + H_2O(g)$

+4   0   −2   0

$8SO_2 \rightarrow S_8$ $(8)(-4) = -32$ $8H_2S \rightarrow S_8$ $(8)(+2) = +16$

Step B.

(1)(−32)

$8SO_2 + 16H_2S \rightarrow 3S_8$

(2)(+16)

Step D. $8SO_2(g) + 16H_2S(g) \rightarrow 3S_8(s) + 16H_2O(l)$

Step E. On each side of the equation there are 24 S atoms, 16 O atoms, and 32 H atoms. There is no charge on either side.

(b) Step A.

$$\overset{+2\ +5\ -2}{Ca_3(PO_4)_2(s)} + \overset{0}{C(s)} + \overset{+4\ -2}{SiO_2(s)} \rightarrow \overset{+2\ +4\ -2}{CaSiO_3(l)} + \overset{0}{P_4(g)} + \overset{+2\ -2}{CO(g)}$$

$$\overset{0}{C} \rightarrow \overset{+2}{CO} \quad (1)(+2) = +2 \qquad 2Ca_3(\overset{+5}{P}O_4)_2 \rightarrow \overset{0}{P_4} \quad (4)(-5) = -20$$

Step B.

$$2Ca_3(PO_4)_2 + 10C \rightarrow P_4 + 10CO$$

(1)(-20) [from $2Ca_3(PO_4)_2$ to P_4]; (10)(+2) [from 10C to 10CO]

Step D. $2Ca_3(PO_4)_2(s) + 10C(s) + 6SiO_2(s) \rightarrow$
$6CaSiO_3(l) + P_4(g) + 10CO(g)$

Step E. On each side of the equation there are 6 Ca atoms, 4 P atoms, 28 O atoms, 10 C atoms, and 6 Si atoms. There is no charge on either side of the equation.

(c) Step A.

$$\overset{+4\ -2}{SO_2(g)} + \overset{+6\ -2}{Cr_2O_7^{2-}} + \overset{+1}{H^+} \rightarrow \overset{+3}{Cr^{3+}} + \overset{+1\ +6\ -2}{H_2SO_4(aq)} + \overset{+1\ -2}{H_2O(l)}$$

$$\overset{+4}{SO_2} \rightarrow \overset{+6}{H_2SO_4} \quad (1)(+2) = +2 \qquad \overset{+6}{Cr_2O_7^{2-}} \rightarrow \overset{+3}{2Cr^{3+}} \quad (2)(-3) = -6$$

Step B.

$$3SO_2 + Cr_2O_7^{2-} \rightarrow 3H_2SO_4 + 2Cr^{3+}$$

(3)(+2) [from $3SO_2$ to $3H_2SO_4$]; (1)(-6) [from $Cr_2O_7^{2-}$ to $2Cr^{3+}$]

Step C. $3SO_2 + Cr_2O_7^{2-} + 8H^+ \rightarrow 3H_2SO_4 + 2Cr^{3+}$

Step D. $3SO_2(g) + Cr_2O_7^{2-} + 8H^+ \rightarrow 2Cr^{3+} + 3H_2SO_4(aq) + H_2O(l)$

Step E. On each side of the equation there are 3 S atoms, 13 O atoms, 2 Cr atoms, 8 H atoms, and +6 charge.

(d) Step A.

$$\overset{0}{Cl_2(g)} + \overset{-2\ +1}{OH^-} \rightarrow \overset{+5\ -2}{ClO_3^-} + \overset{-1}{Cl^-} + \overset{+1\ -2}{H_2O(l)}$$

$$\overset{0}{Cl_2} \rightarrow \overset{+5}{2ClO_3^-} \quad (2)(+5) = +10 \qquad \overset{0}{Cl_2} \rightarrow \overset{-1}{2Cl^-} \quad (2)(-1) = -2$$

Step B.

$$6Cl_2 \rightarrow 2ClO_3^- + 10Cl^-$$

(1)(+10) [from $6Cl_2$ to $2ClO_3^-$]; (5)(-2) [from $6Cl_2$ to $10Cl^-$]

Step C. $6Cl_2 + 12OH^- \rightarrow 2ClO_3^- + 10Cl^-$

Step D. $6Cl_2(g) + 12OH^- \rightarrow 2ClO_3^- + 10Cl^- + 6H_2O(l)$

$3Cl_2(g) + 6OH^- \rightarrow ClO_3^- + 5Cl^- + 3H_2O(l)$

Step E. On each side of the equation there are 6 Cl atoms, 6 O atoms, 6 H atoms, and -6 charge.

(e) Step A.
0 -2 +1 -2 +4 -2 +1 -2
$S(s) + OH^- \rightarrow S^{2-} + SO_3^{2-} + H_2O(l)$

0 -2 0 +4
$S \rightarrow S^{2-}$ (1)(-2) = -2 $S \rightarrow SO_3^{2-}$ (1)(+4) = +4

Step B.
(2)(-2)
$3S \rightarrow 2S^{2-} + SO_3^{2-}$
(1)(+4)

Step C. $3S + 6OH^- \rightarrow 2S^{2-} + SO_3^{2-}$

Step D. $3S(s) + 6OH^- \rightarrow 2S^{2-} + SO_3^{2-} + 3H_2O(l)$

Step E. On each side of the equation there are 3 S atoms, 6 O atoms, 6 H atoms, and -6 charge.

(f) Step A.
+7 -2 +5 -2 +1 -2 +4 -2 +7 -2 -2 +1
$MnO_4^- + IO_3^- + H_2O(l) \rightarrow MnO_2(s) + IO_4^- + OH^-$

+7 +4 +5 +7
$MnO_4^- \rightarrow MnO_2$ (1)(-3) = -3 $IO_3^- \rightarrow IO_4^-$ (1)(+2) = +2

Step B.
(2)(-3)
$2MnO_4^- + 3IO_3^- \rightarrow 2MnO_2 + 3IO_4^-$
(3)(+2)

Step C. $2MnO_4^- + 3IO_3^- \rightarrow 2MnO_2 + 3IO_4^- + 2OH^-$

Step D. $2MnO_4^- + 3IO_3^- + H_2O(l) \rightarrow 2MnO_2(s) + 3IO_4^- + 2OH^-$

Step E. On each side of the equation there are 2 Mn atoms, 3 I atoms, 18 O atoms, 2 H atoms, and -5 charge.

(g) Step A.
0 -2 +1 +1 -2 +2 -2 +1 0
$Sn(s) + OH^- + H_2O(l) \rightarrow [Sn(OH)_4]^{2-} + H_2(g)$

0 +2 +1 0
$Sn \rightarrow [Sn(OH)_4]^-$ (1)(+2) = +2 $H_2O \rightarrow H_2$ (2)(-1) = -2

Step B.
(1)(+2)
$Sn(s) + H_2O \rightarrow [Sn(OH)_4]^- + H_2$
(1)(-2)

Step C. $Sn + H_2O + 2OH^- \rightarrow [Sn(OH)_4]^{2-} + H_2$

Step D. $Sn(s) + 2H_2O(l) + 2OH^- \rightarrow [Sn(OH)_4]^{2-} + H_2(g)$

Step E. On each side of the equation there are 1 Sn atom, 6 H atoms, 4 O atoms, and -2 charge.

16.9 The electron configuration for atomic hydrogen is $1s^1$. The predicted oxidation states of hydrogen are 0 and ±1.

16.11 $1.00797\ u = (1 - x)(1.007825\ u) + x(2.0140\ u)$

$0.00015 = (1.0062)x$

$x = 0.00015$

The percentage of natural abundance of ^2_1H is 0.015 %.

16.13 $H_2(g) + 2Li(s) \rightarrow 2LiH(s)$, $H_2(g) + CuO(s) \xrightarrow{\Delta} Cu(s) + H_2O(g)$

16.15 The electronegativity of hydrogen is considerably larger than that of the active metals, and so the bonding is mainly ionic with the hydrogen becoming the anion. The electronegativity of hydrogen is slightly less than that of the nonmetals, and so the bonding is mainly polar covalent with hydrogen having a +1 oxidation state.

16.17 (a) $N_2(g) + 3H_2(g) \rightarrow 2NH_3(g)$ Hydrogen acts as a reducing agent.

(b) $I_2(g) + H_2(g) \rightarrow 2HI(g)$ Hydrogen acts as a reducing agent.

(c) $Ca(s) + H_2(g) \rightarrow CaH_2(s)$ Hydrogen acts as an oxidizing agent.

(d) $SnO(s) + H_2(g) \rightarrow Sn(s) + H_2O(g)$ Hydrogen acts as a reducing agent.

16.19 steam reforming of hydrocarbons: $CH_4(g) + H_2O(g) \xrightarrow{\Delta} CO(g) + 3H_2(g)$

water gas reaction: $C(s) + H_2O(g) \xrightarrow{\Delta} H_2(g) + CO(g)$

electrolysis of water: $2H_2O(l) \rightarrow 2H_2(g) + O_2(g)$

16.21
$$\begin{aligned}\Delta H^\circ &= [(1\ mol)\Delta H^\circ_f(H_2O)] - [(2\ mol)\Delta H^\circ_f(H) + (\tfrac{1}{2}\ mol)\Delta H^\circ_f(O_2)] \\ &= [(1\ mol)(-242\ kJ/mol)] - [(2\ mol)(218\ kJ/mol) + (1\ mol)(0)] \\ &= -678\ kJ\end{aligned}$$

$$\frac{-678\ kJ}{-242\ kJ} = 2.80 \text{ times as great}$$

16.23 The electron configuration for atomic oxygen is $1s^2 2s^2 2p^4$. The predicted oxidation states for oxygen are 0, -2, +4, and +6.

16.25 Fluorine is the element. Oxygen is less electronegative than fluorine.

16.27 Oxides will form with the element in a lower oxidation state. The regular oxides with the element in higher oxidation state will form.

16.29 (a) $P_4(s) + 5O_2(g) \rightarrow P_4O_{10}(s)$

(b) $C(s) + O_2(g) \rightarrow CO_2(g)$

(c) $2SO_2(g) + O_2(g) \rightarrow 2SO_3(g)$

(d) $2SnO(s) + O_2(g) \rightarrow 2SnO_2(s)$

16.31 (a) $MgO(s) + H_2O(l) \rightarrow Mg(OH)_2(aq)$

(b) $P_4O_6(s) + 6H_2O(l) \rightarrow 4H_3PO_3(aq)$

(c) $CO_2(g) + H_2O(l) \rightarrow H_2CO_3(aq)$

(d) $SO_2(g) + H_2O(l) \rightarrow H_2SO_3(aq)$

(e) $Na_2O(s) + H_2O(l) \rightarrow 2NaOH(aq)$

16.33 The oxide that is different is MO. It is the most alkaline in nature because the element M has the lowest oxidation state.

16.35 $$(25\text{ g }KClO_3)\left[\frac{1\text{ mol }KClO_3}{122.55\text{ g }KClO_3}\right] = 0.20\text{ mol }KClO_3$$

$$(0.20\text{ mol }KClO_3)\left[\frac{3\text{ mol }O_2}{2\text{ mol }KClO_3}\right] = 0.30\text{ mol }O_2$$

$$V = \frac{nRT}{P} = \frac{(0.30\text{ mol})(0.0821\text{ L atm/K mol})(310.\text{ K})}{(1.007\text{ atm})} = 7.6\text{ L}$$

16.37 The Lewis structures of ozone are shown at the right. The geometry of the molecule is bent (or nonlinear).

$:\ddot{O} = \ddot{O} - \ddot{\underset{..}{O}}: \leftrightarrow :\ddot{\underset{..}{O}} - \ddot{O} = \ddot{O}:$

16.39 The equations for the formation are

$O_2(g) \xrightarrow{h\nu} 2O(g)$

$O(g) + O_2(g) + M(g) \rightarrow O_3(g) + M(g)$

and for the decomposition is

$O_3(g) \xrightarrow{h\nu} O_2(g) + O(g)$

The amount of ultraviolet radiation absorbed would change significantly.

16.41 $2KI(aq) + O_3(g) + H_2O(l) \rightarrow 2KOH(aq) + I_2(aq) + O_2(g)$

$$(101.6\text{ mg }I_2)\left[\frac{1\text{ g}}{1000\text{ mg}}\right]\left[\frac{1\text{ mol }I_2}{253.80\text{ g }I_2}\right] = 4.003 \times 10^{-4}\text{ mol }I_2$$

$$(4.003 \times 10^{-4}\text{ mol }I_2)\left[\frac{1\text{ mol }O_3}{1\text{ mol }I_2}\right] = 4.003 \times 10^{-4}\text{ mol }O_3$$

$$(4.003 \times 10^{-4}\text{ mol }O_3)\left[\frac{48.00\text{ g }O_3}{1\text{ mol }O_3}\right]\left[\frac{1000\text{ mg}}{1\text{ g}}\right] = 19.21\text{ mg }O_3$$

16.43 The Lewis structure for the hydrogen peroxide molecule is shown at the right. The geometry of the molecule can be described as two O-O-H planes roughly perpendicular to each other (see Figure 16.2 of the text). The intermolecular forces present are London forces, dipole-dipole interactions, and hydrogen bonding.

$H - \ddot{\underset{..}{O}} - \ddot{\underset{..}{O}} - H$

16.45 The decomposition of H_2O_2 is an example of an oxidation-reduction reaction. Oxygen is being oxidized from -1 to 0 and reduced from -1 to -2.

$$\overset{+1\ -2}{2H_2O_2(l)} \rightarrow \overset{+1\ -2}{2H_2O(l)} + \overset{0}{O_2(g)}$$

16.47 $X(l) \rightarrow X(g)$

$$\Delta H^\circ_{vap} = [(1\ mol)\Delta H^\circ_f(g)] - [(1\ mol)\Delta H^\circ_f(l)]$$

$$\Delta H^\circ_{vap}(H_2O) = [(1\ mol)(-241.818\ kJ/mol)] - [(1\ mol)(-285.83\ kJ/mol)] = 44.01\ kJ$$

$$\Delta H^\circ_{vap}(H_2O_2) = [(1\ mol)(-136.31\ kJ/mol)] - [(1\ mol)(-187.78\ kJ/mol)] = 51.47\ kJ$$

The strength of the intermolecular forces of H_2O_2 are stronger.

16.49 The student decided oxygen was in the first tube (because O_2 is needed for combustion of the wood) and hydrogen was in the second tube (because H_2 combines explosively with O_2 in the air at the mouth of the test tube).

16.51 The partial pressures are

$$P_{H_2} = \frac{n_{H_2}RT}{V} = \frac{(9.90\ mol)(0.0821\ L\ atm/K\ mol)(298\ K)}{(20.0\ L)} = 12.1\ atm$$

$$P_{O_2} = \frac{(2.00)(0.0821)(298)}{20.0} = 2.45\ atm$$

The total pressure before ignition was

$$P_t = P_{H_2} + P_{O_2} = 12.1\ atm + 2.45\ atm = 14.6\ atm$$

Upon ignition and cooling,

$$2H_2(g) + O_2(g) \rightarrow 2H_2O(l)$$

all of the O_2 will be consumed (the limiting reactant) and the amount of H_2 remaining will be

$$(9.90\ mol\ H_2) - (2.00\ mol\ O_2)\left[\frac{2\ mol\ H_2}{1\ mol\ O_2}\right] = 5.90\ mol\ H_2$$

The pressure exerted by the H_2 is

$$P = \frac{(5.90\ mol)(0.0821\ L\ atm/K\ mol)(298\ K)}{(20.0\ L)} = 7.22\ atm$$

In addition to the pressure of the H_2, the liquid water exerts a pressure of 23.756 Torr, giving a total pressure of

$$P_t = 7.22\ atm + (23.756\ Torr)\left[\frac{1\ atm}{760\ Torr}\right] = 7.25\ atm$$

The pressure after the ignition was 7.25 atm. [In taking 7.25 atm as the final pressure, two assumptions are made--the solubility of H_2 in H_2O (0.0002 g H_2/100 g H_2O) is negligible and the volume of the container occupied by the liquid water (~ 72 mL) is negligible.]

CHAPTER 17

CHEMICAL REACTIONS IN PERSPECTIVE

Solutions to Exercises

17.1 (a) This is a partner-exchange reaction that is likely to go to completion because HCN is a gaseous product.

(b) This is a partner-exchange reaction that is likely to go to completion because Ag_2CrO_4 is only slightly soluble.

(c) This reaction will not occur because all reactants and products are strong electrolytes.

17.2 (a) The partner-exchange reaction will occur because of the formation of a gas.

(b) The partner-exchange reaction will occur because of the formation of a precipitate.

(c) The partner-exchange reaction will not occur because the products are strong electrolytes (a soluble salt and a strong acid) and the reactants are a weak electrolyte (a weak acid) and a very slightly soluble salt.

17.3 (a) Partner-exchange reaction driven by the formation of a gaseous product.

$$H_2SO_4(conc) + NaCl(s) \rightarrow NaHSO_4(s) + HCl(g)$$

(b) No reaction occurs (the reverse reaction is favorable because $PbSO_4$ is slightly soluble).

(c) Partner-exchange reaction driven by the formation of a more stable complex ion.

$$[Zn(NH_3)_4]^{2+} + 4CN^- \rightarrow [Zn(CN)_4]^{2-} + 4NH_3(aq)$$

17.4 (a) and (b) The metals are the reducing agents (undergoing oxidation) and the nonmetals are the oxidizing agents (undergoing reduction).

$$3Mg(s) + N_2(g) \xrightarrow{\Delta} Mg_3N_2(s)$$

$$2K(s) + Br_2(l) \rightarrow 2KBr(s)$$

(c) The more electronegative of the two nonmetals, Cl_2, is the oxidizing agent; P is the reducing agent and would be oxidized to either the +3 or +5 oxidation state.

$$2P(s) + 3Cl_2(g) \rightarrow 2PCl_3(l) \quad \text{or} \quad 2P(s) + 5Cl_2(g) \rightarrow 2PCl_5(s)$$

17.5 (a) This reaction will not occur because Cl_2 is a stronger oxidizing agent than I_2.

(b) This reaction will not occur because Zn is a stronger reducing agent than Pb.

(c) This reaction will occur because Cu is a stronger reducing agent than Ag.

17.6 From reaction (a) Cu is a stronger reducing agent than Pd. Reaction (b) shows that Pd is stronger than Pt. Reaction (c) shows that Rh is stronger than Pd. Reaction (d) shows that Cu is stronger than Rh. Thus the order of decreasing strength as reducing agents is Cu > Rh > Pd > Pt.

17.7 (a) PCl_5. Phosphorus can have either a +3 or a +5 oxidation state. Chlorine is a strong enough oxidizing agent to react with PCl_3 to form PCl_5.

(b) $FeBr_2$ or $FeBr_3$. Iron has two common oxidation states.

(c) No reaction will occur. Bromine is not as strong an oxidizing agent as chlorine.

17.8 (a) This is a redox reaction, further classified as displacement of one element from a compound by another element.

(b) This is a redox reaction, further classified as combination of elements.

(c) This is a nonredox reaction, further classified as displacement of a more volatile oxide by a less volatile oxide.

(d) This is a redox reaction (more complex type) involving an oxygen-containing anion.

17.9 (a) Nitrous acid is an oxidizing acid that usually forms NO or NO_2. Iodide is a reducing agent that forms I_2.

$$2HNO_2(aq) + 2HI(aq) \rightarrow 2NO(g) + I_2(aq) + 2H_2O(l)$$

(b) Dichromate ion is a strong oxidizing agent that forms Cr^{3+}. Iron(II) ion is a reducing agent that forms Fe^{3+}.

$$Cr_2O_7^{2-} + 14H^+ + 6Fe^{2+} \rightarrow 2Cr^{3+} + 6Fe^{3+} + 7H_2O(l)$$

(c) Lead dioxide is an oxidizing agent that forms Pb^{2+}. Sulfite ion is a reducing agent that forms SO_4^{2-}. The $PbSO_4$ that forms is quite insoluble.

$$PbO_2(s) + SO_3^{2-} + 2H^+ \rightarrow PbSO_4(s) + H_2O(l)$$

17.10 (a) The first "yes" answer to the questions in Table 17.8 is for question 3. The answer to question 4 is "no", but the reaction can be a redox reaction of ions in aqueous solution (more complex type).

$$BrO_3^- + 5Br^- + 6H^+ \rightarrow 3Br_2(aq) + 3H_2O(l)$$

(b) The first "yes" answer is for question 2. It is a redox reaction of elements (combination). Iodine is not a strong enough oxidizing agent to oxidize P to the +5 oxidation state.

$$P_4(s) + 6I_2(g) \xrightarrow{\Delta} 4PI_3(s)$$

(c) The first "yes" answer is for question 3. The next "yes" answer is for question 5, in which a redox reaction of ions in aqueous solution will occur. The permanganate ion is a good oxidizing agent and H_2S is a reducing agent.

$$2KMnO_4(aq) + 3H_2S(g) \rightarrow 3S(s) + 2MnO_2(s) + 2H_2O(l) + 2OH^- + 2K^+$$

17.11 (a) Question 2 is answered "yes." However, even though Cu is a better reducing agent than Ag, AgCl is highly insoluble and no reaction will occur.

(b) Question 3 is answered "yes," as is question 5. Electron transfer could possibly occur to give $Fe^{2+} + I_2$.

(c) Question 6 is answered "yes." A nonredox partner-exchange reaction could occur in which HCl and $Si(OH)_4$ or SiO_2 would be produced.

Solutions to Odd-Numbered Questions and Problems

17.1 When a chemical reaction reaches equilibrium, the amounts of reactants and products may be equal but usually are not--the equilibrium "lies" to the reactants side (greater amounts of reactants) or to the products side (greater amounts of products). Reactions do not stop at equilibrium because an equilibrium system is a dynamic system (forward and reverse reactions continuing at equal rates).

17.3 The reactions likely to go to completion include (a) because of the formation of a gas, (c) because of the formation of a precipitate that removes ions from solution, and (d) because H_2O_2 is a strong oxidizing agent and the water that is formed is not a reducing agent. Reaction (b) cannot occur because there are two reductions and no oxidation.

17.5 Because AgCl reacts with NH_3 to form $[Ag(NH_3)_2]^+$, $[Ag(NH_3)_2]^+$ is more stable than AgCl in the presence of excess ammonia.

17.7 (a) $2NH_3(g) + H_2SO_4(aq) \rightarrow (NH_4)_2SO_4(aq)$
(b) $CaO(s) + SiO_2(l) \rightarrow CaSiO_3(l)$

17.9 (a) $Na_2CO_3(l) + SiO_2(s) \rightarrow Na_2SiO_3(l) + CO_2(g)$
(b) $[Cu(H_2O)_4]^{2+} + 4NH_3(aq) \rightarrow [Cu(NH_3)_4]^{2+} + 4H_2O(l)$
(c) $[AgCl_2]^- + 2NH_3(aq) \rightarrow [Ag(NH_3)_2]^+ + 2Cl^-$

17.11 (a) $Na_2S(aq) + 2HCl(aq) \rightarrow 2NaCl(aq) + H_2S(g)$
(b) $CdSO_4(aq) + Na_2S(aq) \rightarrow Na_2SO_4(aq) + CdS(s)$
(c) $(NH_4)_2CrO_4(aq) + Pb(CH_3COO)_2(aq) \rightarrow 2NH_4CH_3COO(aq) + PbCrO_4(s)$
(d) $2AsCl_3(aq) + 3H_2S(aq) \rightarrow As_2S_3(s) + 6HCl(aq)$

17.13 (a) nonredox decomposition
$Ca(OH)_2(s) \xrightarrow{\Delta} CaO(s) + H_2O(g)$
(b) nonredox displacement
$[Cu(H_2O)_4]Cl_2(aq) + 4NH_3(aq) \rightarrow [Cu(NH_3)_4]Cl_2(aq) + 4H_2O(l)$
(c) nonredox partner exchange
$CuCl_2(aq) + 2NaOH(aq) \rightarrow Cu(OH)_2(s) + 2NaCl(aq)$
(d) nonredox partner exchange
$PCl_3(l) + 3H_2O(l) \rightarrow 3HCl(aq) + H_3PO_3(aq)$
(e) nonredox combination
$SO_2(g) + CaO(s) \rightarrow CaSO_3(s)$
(f) nonredox decomposition
$Be(OH)_2(s) \xrightarrow{\Delta} BeO(s) + H_2O(g)$
(g) nonredox combination
$CO_2(g) + LiOH(aq) \rightarrow LiHCO_3(aq)$

17.15

	Oxidizing agent	Reducing agent
(a) Nonredox reaction		
(b) Nonredox reaction		
(c) Redox reaction	$TiCl_4$	Mg
(d) Nonredox reaction		

17.17 (a) $C(s) + O_2(g) \rightarrow CO_2(g)$

(b) $P_4(s) + 10Cl_2(g) \rightarrow 4PCl_5(s)$

(c) $Ti(s) + 2Cl_2(g) \rightarrow TiCl_4(l)$

(d) $3Mg(s) + N_2(g) \rightarrow Mg_3N_2(s)$

(e) $4FeO(s) + O_2(g) \rightarrow 2Fe_2O_3(s)$

(f) $2NO(g) + O_2(g) \rightarrow 2NO_2(g)$

(g) $P_4O_6(s) + 2O_2(g) \rightarrow P_4O_{10}(s)$

17.19 (a) The reaction will not occur because there cannot be two oxidations with no reduction.

(b) The reaction will not occur because there cannot be two reductions with no oxidation.

(c) The reaction will not occur because Zn is a stronger reducing agent than H_2.

(d) The reaction will occur because Br_2 is a stronger oxidizing agent than I_2.

17.21 (a) $SO_3^{2-} + Br_2(aq) + H_2O(l) \rightarrow SO_4^{2-} + 2Br^- + 2H^+$

(b) $2SO_3^{2-} + O_2(g) \rightarrow 2SO_4^{2-}$

(c) $SO_3^{2-} + H_2O_2(aq) \rightarrow SO_4^{2-} + H_2O(l)$

(d) $5SO_3^{2-} + 2MnO_4^- + 6H^+ \rightarrow 5SO_4^{2-} + 3H_2O(l) + 2Mn^{2+}$

17.23 (a) $6Cl_2(g) + P_4(s) \rightarrow 4PCl_3(l)$

(b) $Cl_2(g) + PCl_3(l) \rightarrow PCl_5(s)$ or $10Cl_2(g) + P_4(s) \rightarrow 4PCl_5(s)$

(c) $Cl_2(g) + 2CuCl(s) \rightarrow 2CuCl_2(s)$

(d) $Cl_2(g) + 2I^- \rightarrow 2Cl^- + I_2(aq)$

17.25 (a) more complex

$MnO_2(s) + 2Cl^- + 4H^+ \rightarrow Mn^{2+} + Cl_2(g) + 2H_2O(l)$

(b) electron transfer between monatomic ions

$Sn^{2+} + 2Fe^{3+} \rightarrow Sn^{4+} + 2Fe^{2+}$

(c) combination of elements

$16K(s) + S_8(s) \rightarrow 8K_2S(s)$

(d) more complex (combustion)

$2N_2H_2(l) + O_2(g) \rightarrow 2N_2(g) + 2H_2O(l)$

(e) decomposition (disproportionation)

$Cu_2SO_4(s) \xrightarrow{H_2O} Cu(s) + CuSO_4(aq)$

(f) decomposition (disproportionation)

$Hg_2Cl_2(s) \rightarrow Hg(l) + HgCl_2(s)$

17.27 $3HCHO(aq) + Cr_2O_7^{2-} + 8H^+ \rightarrow 3HCOOH(aq) + 2Cr^{3+} + 4H_2O(l)$

$$(50.0\text{ g HCHO})\left[\frac{1\text{ mol HCHO}}{30.03\text{ g HCHO}}\right]\left[\frac{1\text{ mol }Cr_2O_7^{2-}}{3\text{ mol HCHO}}\right]\left[\frac{1\text{ mol }K_2Cr_2O_7}{1\text{ mol }Cr_2O_7^{2-}}\right]$$

$$\times\left[\frac{294.20\text{ g }K_2Cr_2O_7}{1\text{ mol }K_2Cr_2O_7}\right] = 163\text{ g }K_2Cr_2O_7$$

17.29 (a) This is a nonredox reaction, further classified as a partner-exchange reaction between ions in aqueous solution (formation of water).

(b) This is a redox reaction, further classified as displacement of one element from a compound by another element.

(c) This is a nonredox reaction, further classified as decomposition to give compounds.

(d) This is a nonredox reaction, further classified as combination of compounds.

(e) This is a redox reaction, further classified as a more complex type.

(f) This is a nonredox reaction, further classified as displacement.

(g) This is a redox reaction that does not fit into any of the categories listed.

(h) This is a redox reaction, further classified as combination of an element with a compound to give another compound.

(i) This is a redox reaction, further classified as decomposition to give compounds and elements.

(j) This is a redox reaction, further classified as electron transfer between a metal and an ion in solution.

17.31 (a) redox displacement: $2Li(s) + H_2O(l) \rightarrow 2LiOH(aq) + H_2(g)$

(b) redox decomposition: $2Ag_2O(s) \xrightarrow{\Delta} 4Ag(s) + O_2(g)$

(c) nonredox combination: $Li_2O(s) + H_2O(l) \rightarrow 2LiOH(aq)$

(d) redox decomposition: $2H_2O(l) \xrightarrow{\text{electrolysis}} 2H_2(g) + O_2(g)$

(e) $I_2(s) + Cl^-(aq) \rightarrow$ no reaction

(f) $Cu(s) + HCl(aq) \rightarrow$ no reaction

(g) redox decomposition: $2NaNO_3(s) \xrightarrow{\Delta} 2NaNO_2(s) + O_2(g)$

17.33 (a) $NH_4NO_3(s) \xrightarrow{\Delta} N_2O(g) + 2H_2O(g)$

(b) $H_2S(g) + Pb(CH_3COO)_2(aq) \rightarrow PbS(s) + 2CH_3COOH(aq)$

(c) $Ba(OH)_2(aq) + H_2SO_4(aq) \rightarrow BaSO_4(s) + 2H_2O(l)$

(d) $2Fe^{3+} + Sn^{2+} \rightarrow 2Fe^{2+} + Sn^{4+}$

(e) $2FeCl_2(aq) + Cl_2(g) \rightarrow 2FeCl_3(aq)$

(f) $2Hg(l) + O_2(g) \rightarrow 2HgO(s)$

(g) $CaCl_2(aq) + H_2SO_4(l) \rightarrow CaSO_4(s) + 2HCl(aq)$

(h) $SO_2(g) + 2NaOH(aq) \rightarrow Na_2SO_3(aq) + H_2O(l)$

17.35 (a) No reaction will occur. The nonredox partner-exchange reaction would form two strong electrolytes.

(b) No reaction will occur. The redox displacement reaction is not spontaneous because copper metal is below Fe in the activity series.

(c) No reaction will occur. The nonredox partner-exchange reaction would form two strong electrolytes.

(d) The nonredox partner-exchange reaction will occur because water is formed: $Ba(OH)_2(aq) + 2HCl(aq) \rightarrow BaCl_2(aq) + H_2O(l)$

(e) No reaction will occur. The redox displacement reaction is not spontaneous because Zn is below Al in the activity series.

17.37 $2Mg(s) + O_2(g) \xrightarrow{\Delta} 2MgO(s)$

redox, combination of two elements to give a compound

$3Mg(s) + N_2(g) \xrightarrow{\Delta} Mg_3N_2(s)$

redox, combination of two elements to give a compound

$Mg_3N_2(s) + 6H_2O(l) \rightarrow 3Mg(OH)_2(aq) + 2NH_3(g)$

nonredox, partner-exchange reaction

$Mg(OH)_2(s) \xrightarrow{\Delta} MgO(s) + H_2O(g)$

nonredox, decomposition to give compounds

CHAPTER 18

CHEMICAL KINETICS

Solutions to Exercises

18.1 The "rate of reaction" could be defined as

$$\text{Rate} = -\Delta[O_3]/\Delta t \quad \text{or} \quad \Delta[O_2]/\Delta t$$

18.2 The rate equation for the decomposition of hydrogen peroxide is

$$\text{Rate} = k[H_2O_2][I^-]$$

The reaction is second order overall (1 + 1 = 2).

18.3 The concentration of N_2O_5 unreacted at time t is 5 % (= 100 % - 95 %) of the initial concentration, or

$$[N_2O_5] = (0.05)(0.40 \text{ mol/L}) = 0.02 \text{ mol/L}$$

The time is

$$t = \frac{(2.303)(\log[N_2O_5]_0 - \log[N_2O_5])}{k}$$

$$= \frac{(2.303)[\log(0.40) - \log(0.02)]}{6.32 \times 10^{-4} \text{ s}^{-1}}$$

$$= 4.74 \times 10^3 \text{ s} = 79 \text{ min}$$

18.4
$$k = \frac{(2.303)(\log P_{PH_3,0} - \log P_{PH_3})}{t} = \frac{(2.303)[\log(262) - \log(16)]}{120.\text{ s}}$$

$$= 0.0233 \text{ s}^{-1}$$

$$t_{1/2} = \frac{0.693}{k} = \frac{0.693}{0.0233 \text{ s}^{-1}} = 29.7 \text{ s}$$

18.5 (a) Rate = $k[Fe^{3+}][HO_2^-]$ (b) Rate = $k[Fe^{3+}][HO_2]$

(c) Rate = $k[Fe^{3+}][O_2^-]$ (d) Rate = $k[Fe^{2+}][H_2O_2]$

(e) Rate = $k[Fe^{2+}][OH]$

18.6 Rate = $k[A]^m[B]^n$

Experiments (a) and (b) show that the reaction rate essentially doubles as the concentration of A doubles. Thus the reaction is first order with respect to [A] and $m = 1$. Likewise we can see from the data for experiments (b) and (c) that for constant [A], the rate is independent of [B]. Thus the reaction is zero order with respect to [B] and $n = 0$.

Rate = k[A]

$$k = \frac{\text{Rate}}{[\text{A}]} \qquad k_a = \frac{0.030 \text{ mol/L s}}{0.10 \text{ mol/L}} = 0.30 \text{ s}^{-1}$$

$$k_b = \frac{0.059 \text{ mol/L s}}{0.20 \text{ mol/L}} = 0.30 \text{ s}^{-1} \qquad k_c = \frac{0.060 \text{ mol/L s}}{0.20 \text{ mol/L}} = 0.30 \text{ s}^{-1}$$

$$\text{average } k = \frac{0.30 \text{ s}^{-1} + 0.30 \text{ s}^{-1} + 0.30 \text{ s}^{-1}}{3} = 0.30 \text{ s}^{-1}$$

18.7 T_1 = 462 °C = 735 K $\qquad$ T_2 = 487 °C = 760. K

$$\log\left[\frac{k_2}{k_1}\right] = \frac{E_a}{(2.303)R}\left[\frac{T_2 - T_1}{T_2T_1}\right]$$

$$\log\left[\frac{k_2}{5.8 \times 10^{-6} \text{ s}^{-1}}\right] = \frac{(260 \text{ kJ/mol})(1000 \text{ J/1 kJ})}{(2.303)(8.314 \text{ J/K mol})}\left[\frac{760. \text{ K} - 735 \text{ K}}{(760. \text{ K})(735 \text{ K})}\right] = 0.61$$

$$\frac{k_2}{5.8 \times 10^{-6} \text{ s}^{-1}} = 4.1$$

$$k_2 = 2.4 \times 10^{-5} \text{ s}^{-1}$$

18.8 T_1 = 429 °C = 702 K $\qquad$ T_2 = 522 °C = 795 K

$$E_a = (2.303)R\left[\frac{T_1T_2}{T_2 - T_1}\right]\log\left[\frac{k_2}{k_1}\right]$$

$$= (2.303)(8.314 \text{ J/K mol})\left[\frac{(702 \text{ K})(795 \text{ K})}{795 \text{ K} - 702 \text{ K}}\right]\left[\frac{1 \text{ kJ}}{1000 \text{ J}}\right]\log\left[\frac{1.7 \times 10^{-4} \text{ s}^{-1}}{7.9 \times 10^{-7} \text{ s}^{-1}}\right]$$

$$= 270 \text{ kJ/mol}$$

Answers to Odd-Numbered Questions and Problems

18.1 An elementary reaction is a reaction that occurs in a single step exactly as written. A reaction mechanism consists of all of the elementary steps in a single reaction pathway. These steps show all of the changes that take place at the molecular level.

18.3 The three conditions for effective collisions between reacting species are proper orientation, sufficient penetration of outer electron shells, and a resulting structure with stable bonds. The condition most influenced by (a) temperature is sufficient penetration of outer electron shells; (b) molecular geometry is proper orientation; and (c) relative bond energies is a resulting structure with stable bonds.

18.5 From the diagram the
(a) reactant(s) are (i)
(b) product(s) are (v)
(c) energy change of reaction is (iv)
(d) activation energy for the forward reaction is (ii)
(e) activation energy for the reverse reaction is (iii)

18.7 See Fig. 18-1.

18.9 Both elementary reactions are bimolecular reactions.

18.11 The rate of reaction might be defined for the following reactions as

(a) Rate = $\Delta[O_2]/\Delta t$
(b) Rate = $-\Delta[HCl]/\Delta t$ or Rate = $\Delta[CaCl_2]/\Delta t = \Delta[CO_2]/\Delta t$
(c) Rate = $-\Delta[S^{2-}]/\Delta t = \Delta[HS^-]/\Delta t = \Delta[OH^-]/\Delta t$
(d) Rate = $\Delta[O_2]/\Delta t$
(e) Rate = $-\Delta[H_2O]/\Delta t = \Delta[H_2]/\Delta t$ or Rate = $\Delta[O_2]/\Delta t$

18.13 The rate equation for the overall first-order reaction of A is

$$\text{Rate} = k[A]$$

The units of k are (time)$^{-1}$. The integrated rate equation is

$$\log [A] = -(k/2.303)t + \log [A]_0$$

A plot of log [A] versus t would result in a linear plot of concentration-time data.

18.15 The rate reaction will take the general form

$$\text{Rate} = k[NO]^m[H_2]^n$$

when m is the order of reaction with respect to NO and n is the order of reaction with respect to H_2. We have been told that $m = 2$ and $n = 1$, so the rate equation is

$$\text{Rate} = k[NO]^2[H_2]$$

The reaction is an overall third-order reaction ($m + n = 2 + 1 = 3$).

Fig. 18-1

Energy

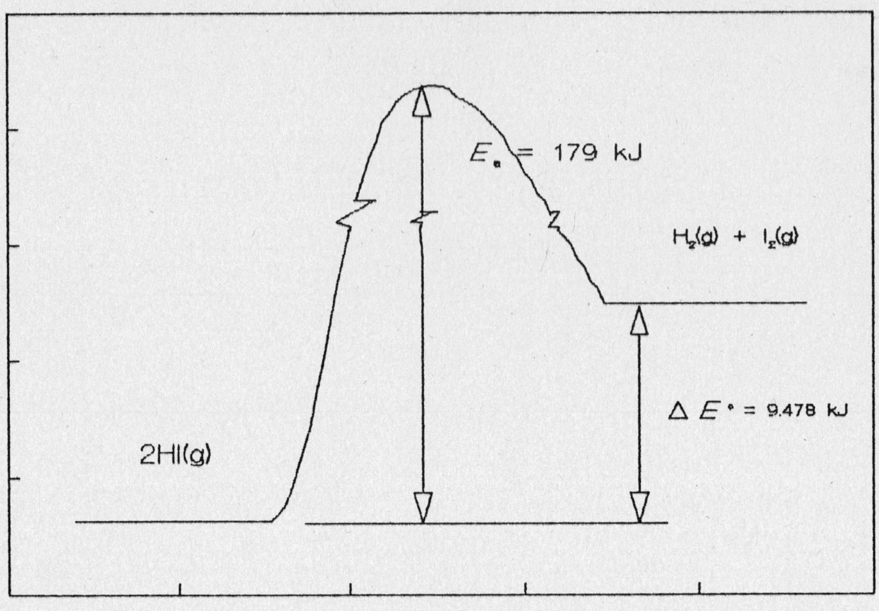

Progress of Reaction

18.17 The rate equation for the hydrogenation of ethene is

$$\text{Rate} = k[H_2][C_2H_4]^{-1}$$

18.19 Rate = $k[N_2O_5]$

(a) Rate = $(6.32 \times 10^{-4}\ \text{min}^{-1})(0.10\ \text{mol/L}) = 6.3 \times 10^{-5}\ \text{mol/L min}$

(b) Rate = $(6.32 \times 10^{-4}\ \text{min}^{-1})(0.010\ \text{mol/L}) = 6.3 \times 10^{-6}\ \text{mol/L min}$

18.21
$$\log\,[CS_2] = \log\,[CS_2]_0 - \frac{kt}{2.303}$$

$$= \log(1.30 \times 10^{-2}) - \frac{(2.8 \times 10^{-7}\ \text{s}^{-1})(2.0\ \text{day})}{2.303}\left[\frac{24\ \text{h}}{1\ \text{day}}\right]\left[\frac{3600\ \text{s}}{1\ \text{h}}\right]$$

$$= -1.907$$

$$[CS_2] = 1.24 \times 10^{-2}\ \text{mol/L}$$

$$t_{1/2} = \frac{0.693}{k} = \left[\frac{0.693}{2.8 \times 10^{-7}\ \text{s}^{-1}}\right]\left[\frac{1\ \text{h}}{3600\ \text{s}}\right]\left[\frac{1\ \text{day}}{24\ \text{h}}\right] = 29\ \text{day}$$

18.23
$$t_{1/2} = \frac{1}{k[A]} = \frac{1}{(6.49\ \text{L/mol min})(0.0100\ \text{mol/L})} = 15.4\ \text{min}$$

$[A] = (0.05)(0.0100\ \text{mol/L}) = 0.0005\ \text{mol/L}$

$$\frac{1}{[A]} = \frac{1}{[A]_0} + kt$$

$$t = \frac{\frac{1}{[A]} - \frac{1}{[A]_0}}{k} = \frac{\frac{1}{0.0005\ \text{mol/L}} - \frac{1}{0.0100\ \text{mol/L}}}{6.49\ \text{L/mol min}} = 290\ \text{min}$$

18.25 Comparison of experiments (a) and (b) shows that doubling P_{NO} essentially quadruples the rate, so the reaction is second order with respect to NO. Comparison of experiments (c) and (d) shows that doubling P_{H_2} essentially doubles the rate, so the reaction is first order with respect to H_2. The rate equation is

$$\text{Rate} = k\, P^2_{NO}\, P_{H_2}$$

18.27 Rate = $k[A]^m[B]^n$

Experiments (a) and (b) show that the rate quadruples for [A] doubling and [B] constant. Thus the reaction is second order with respect to A and m = 2. Experiments (a) and (c) show that the rate doubles for [B] doubling and [A] constant. Thus the reaction is first order with respect to B and n = 1. The reaction is third order overall.

18.29

t, min	0	5	10	15	20	25
$[CH_2N_2]$, mol/L	0.100	0.076	0.058	0.044	0.033	0.025
log $[CH_2N_2]$	-1.000	-1.12	-1.24	-1.36	-1.48	-1.60

See Fig. 18-2 The rate constant is

$k = -(2.303)(\text{slope}) = -(2.303)(-0.0241\ \text{min}^{-1}) = 0.0555\ \text{min}^{-1}$

18.31 (a) See Fig. 18-3.

(b) From Fig. 18-3, $[N_2O_5]_{600}$ = 0.978 mol/L, $[N_2O_5]_{1200}$ = 0.685 mol/L, and $[N_2O_5]_{1800}$ = 0.480 mol/L.

(c) The rate of reaction is given by Rate = $-\ \Delta[N_2O_5]/\Delta t$ = -(slope). The slope at t = 600 s is -5.80×10^{-4} mol/L s giving Rate = 5.80×10^{-4} mol/L s.

(d) Likewise at t = 1200 s, the slope is -4.06×10^{-4} mol/L s giving Rate = 4.06×10^{-4} mol/L s and at 1800 s, Rate = 2.85×10^{-4} mol/L s. The rate does seem to depend on concentration.

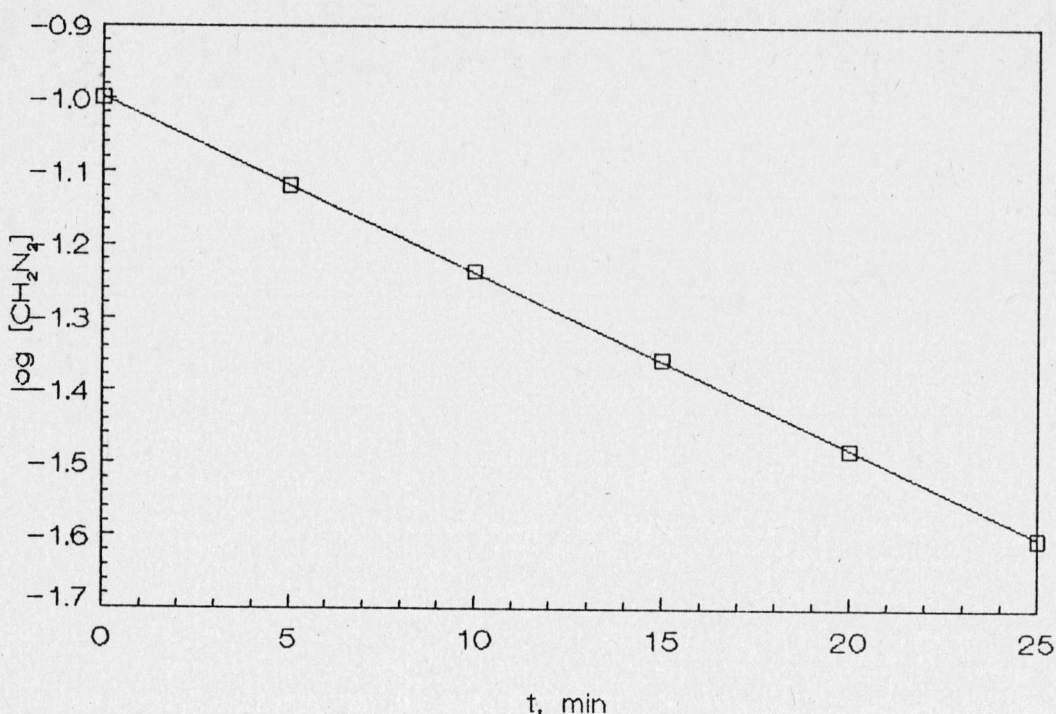

(e) Rate = $k_1[N_2O_5]$ or Rate = $k_2[N_2O_5]^2$

rate	$[N_2O_5]$	k_1 = rate/$[N_2O_5]$	k_2 = rate/$[N_2O_5]^2$
mol/L s	mol/L	s^{-1}	L/mol s
5.80×10^{-4}	0.978	5.93×10^{-4}	6.06×10^{-4}
4.06×10^{-4}	0.685	5.93×10^{-4}	8.65×10^{-4}
2.85×10^{-4}	0.480	5.94×10^{-4}	1.24×10^{-3}

The reaction is first order with respect to N_2O_5 because k_1 is constant.

18.33 If the rate determining step is the first step, then the rate law would be

$$\text{Rate} = k_1[NO_2Cl]$$

which agrees with the experimental rate equation.

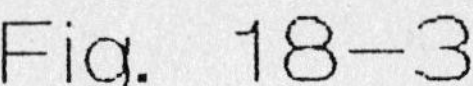

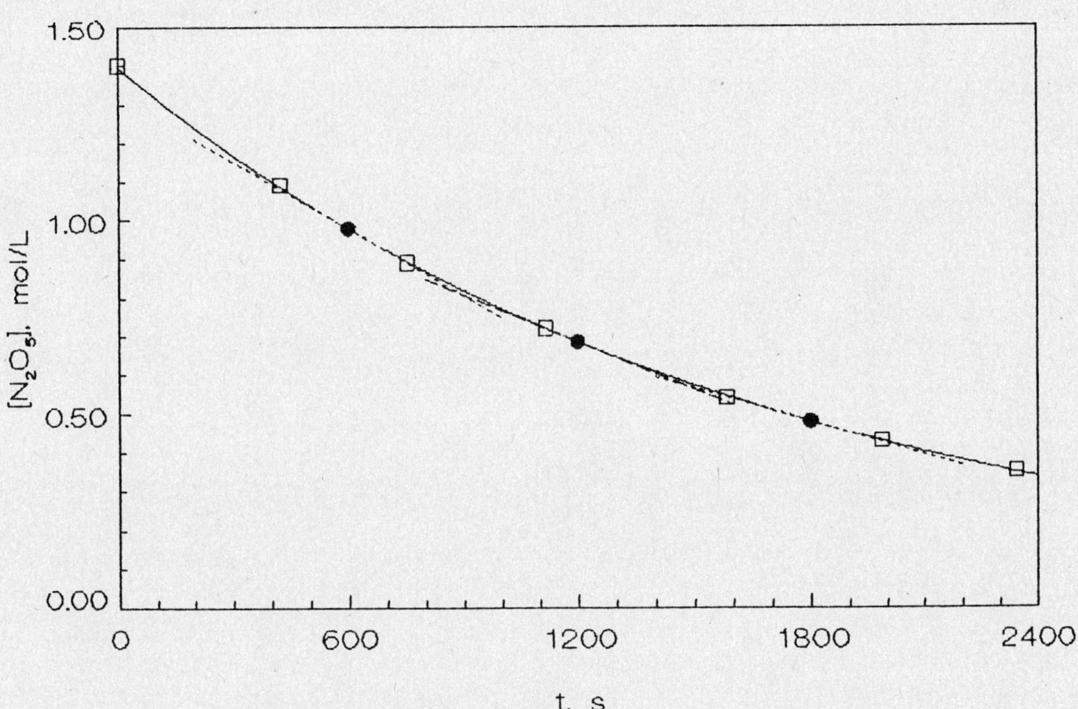

18.35 The overall stoichiometric equation is

$$\begin{array}{lll} A + B & \rightarrow & C \qquad \text{(fast)} \\ B + C & \rightarrow & D + E \qquad \text{(slow)} \\ D + F & \rightarrow & A + E \qquad \text{(fast)} \\ \hline 2B + F & \rightarrow & 2E \end{array}$$

(a) Species A is a catalyst.

(b) Species C and D are intermediates.

(c) The rate equation for the rate-determining step is

$$\text{Rate} = k[B][C]$$

(d) The rate equation including concentration terms only of reactants and products:

$$[C] = k'[A]$$

$$\text{Rate} = k[B](k'[A])$$

$$= k''[B][A]$$

(e) The reaction is second order overall, 1 + 1 = 2.

Fig. 18–4

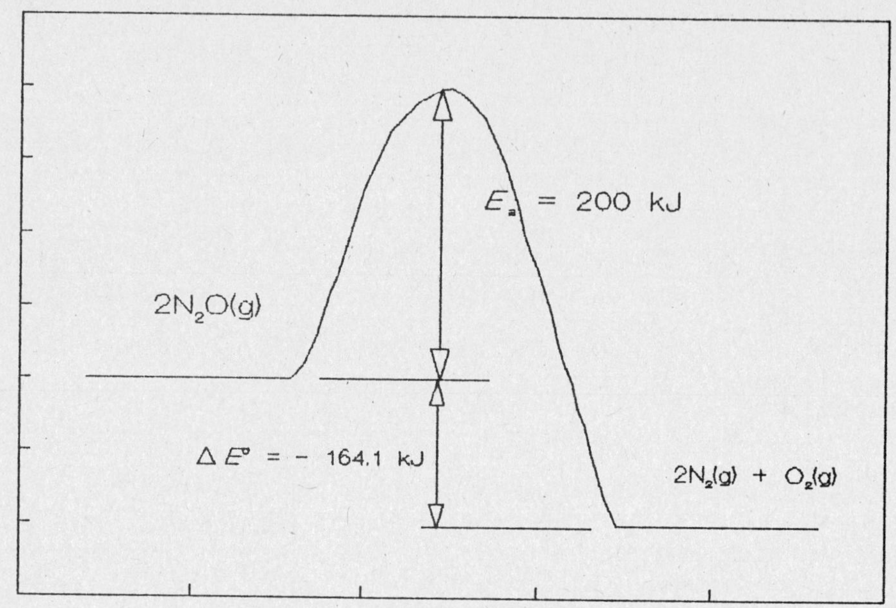

18.37 T_1 = 361 °C = 634 K T_2 = 371 °C = 644 K

$$E_a = (2.303)R\left[\frac{T_1T_2}{T_2 - T_1}\right]\log\left[\frac{k_2}{k_1}\right]$$

$$= (2.303)(8.314 \text{ J/K mol})\left[\frac{(634 \text{ K})(644 \text{ K})}{644 \text{ K} - 634 \text{ K}}\right]\left[\frac{1 \text{ kJ}}{1000 \text{ J}}\right]\log\left[\frac{7.2 \times 10^{-4} \text{ s}^{-1}}{4.6 \times 10^{-4} \text{ s}^{-1}}\right]$$

$$= 150 \text{ kJ/mol}$$

18.39 T_1 = 300. °C = 573 K T_2 = 330. °C = 603 K

$$E_a = (2.303)R\left[\frac{T_1T_2}{T_2 - T_1}\right]\log\left[\frac{k_2}{k_1}\right]$$

$$= (2.303)(8.314 \text{ J/K mol})\left[\frac{(573 \text{ K})(603 \text{ K})}{603 \text{ K} - 573 \text{ K}}\right]\log\left[\frac{2.1 \times 10^{-10} \text{ s}^{-1}}{2.6 \times 10^{-11} \text{ s}^{-1}}\right]$$

$$= 2.0 \times 10^{5} \text{ J/mol}$$

See Fig. 18.4.

Fig. 18–5

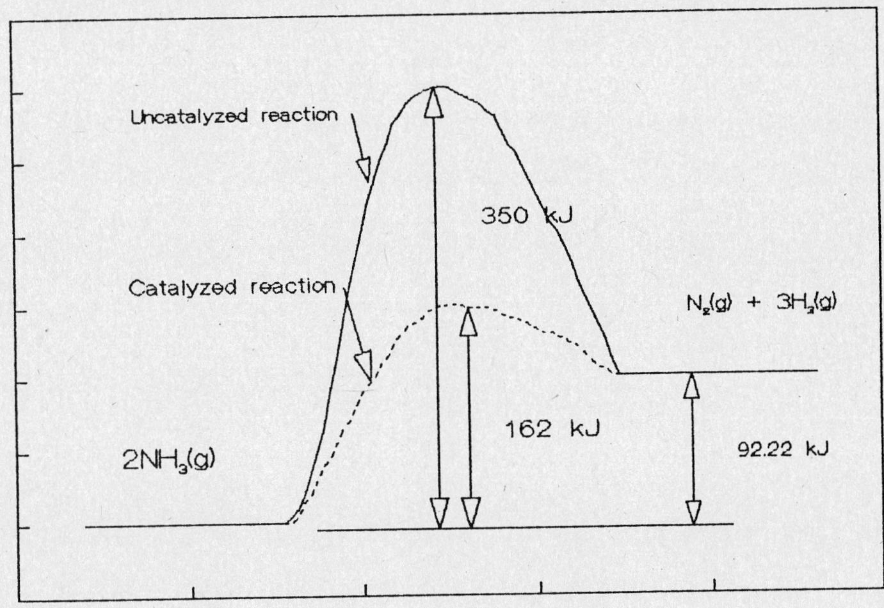

18.41 T_1 = 0 °C = 273 K T_2 = 25 °C = 298 K

$$\log\left[\frac{k_2}{k_1}\right] = \frac{E_a}{(2.303)R}\left[\frac{T_2 - T_1}{T_1T_2}\right] = \frac{(65\ \text{kJ/mol})(1000\ \text{J/1 kJ})}{(2.303)(8.314\ \text{J/K mol})}\left[\frac{298\ \text{K} - 273\ \text{K}}{(298\ \text{K})(273\ \text{K})}\right]$$

$$= 1.0$$

$$\frac{k_2}{k_1} = 10 \text{ times faster}$$

18.43 See Fig. 18-5.

18.45 For the reaction of A + 2B, the reaction rate will

(a) double if [A] is doubled

(b) double if [B] is doubled

(c) increase if more catalyst is added

(d) slow down at a lower rate than if D were not removed

(e) quadruple if [A] and [B] are both doubled

(f) increase if the temperature is increased

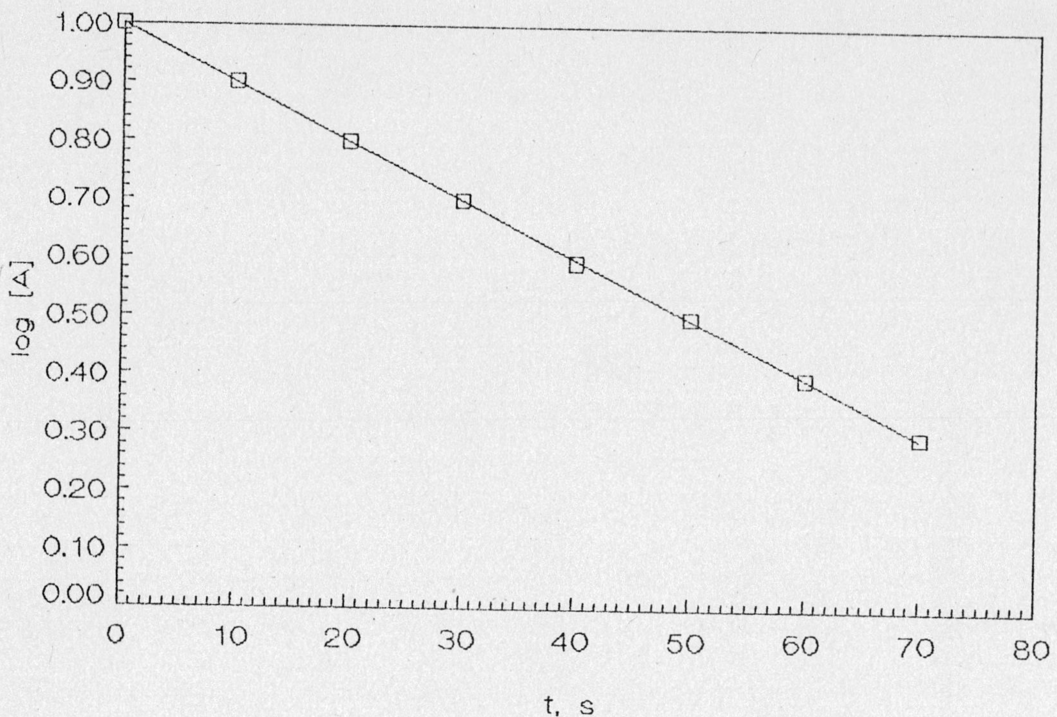

18.47 (a)

t,s	[A], mol/L	log[A]	$\frac{1}{[A]}$, L/mol
0	10.00	1.0000	0.100
10	7.95	0.900	0.126
20	6.31	0.800	0.158
30	5.00	0.699	0.200
40	3.92	0.593	0.255
50	3.16	0.500	0.316
60	2.50	0.398	0.400
70	1.99	0.299	0.503

(b) The plot of log [A] against time is linear (see Fig. 18-6); therefore, the reaction is first order.

(c) $k = -(2.303)(\text{slope}) = -(2.303)(-1.00 \times 10^{-2}\ s^{-1}) = 2.30 \times 10^{-2}\ s^{-1}$

(d) (1) Rate = $k[A]^2[B]$, no

(2) Rate = $k[A][B] \approx k'[A]$ in large excess of [B], ok

(3) Rate = $k[C][A][B] = k(k'[A])[A][B] = k'[A]^2[B]$, no

(4) Rate = $k[A]$, ok

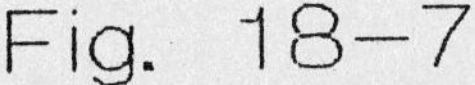

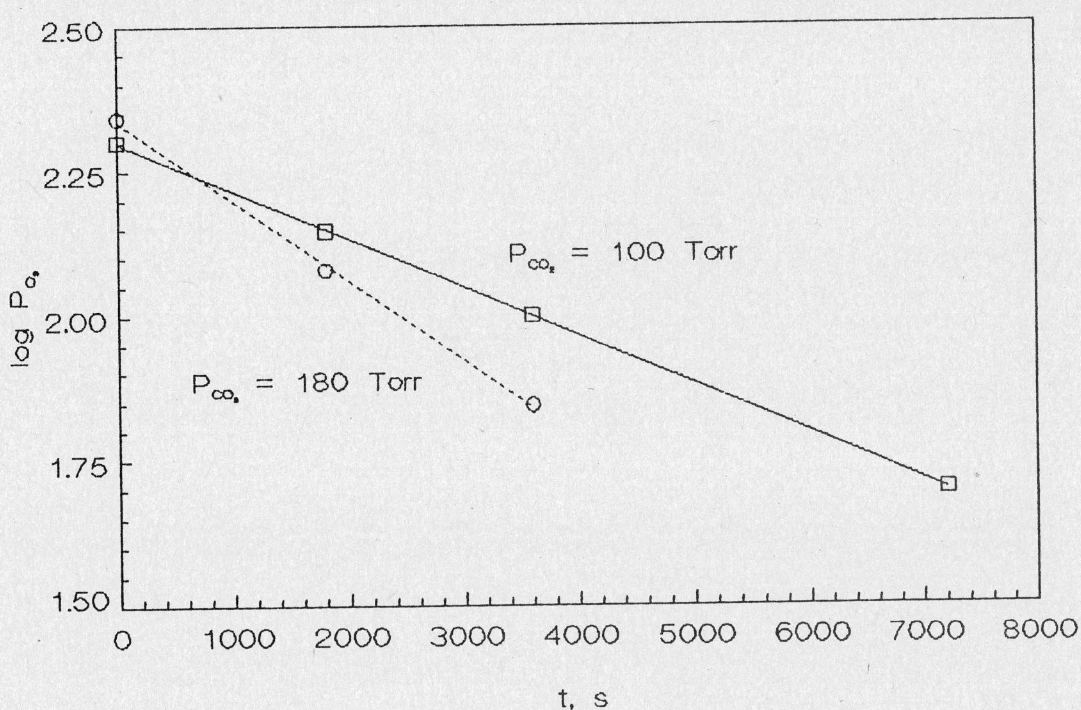

18.49 (a) Rate = $k\, P^x_{O_3}\, P^y_{O_2}$

Experiments (1) and (2) show that the rate is halved for P_{O_2} doubling and P_{O_3} constant. Thus the reaction is negative first order with respect to O_2 and $x = -1$. Experiments (2) and (3) show that the rate quadruples for P_{O_3} doubling and P_{O_2} constant. Thus the reaction is second order with respect to O_3 and $y = 2$.

(b) P_{CO_2} = 100 Torr

t, s	P_{O_3}, Torr	log P_{O_3}
0	200	2.30
1800	140	2.15
3600	100	2.00
7200	50	1.70

P_{CO_2} = 180 Torr

t, s	P_{O_3}, Torr	log P_{O_3}
0	220	2.34
1800	120	2.08
3600	70	1.85

Because each plot of log P_{O_3} versus t (see Fig. 18-7) is linear, the order of reaction is 1 with respect to O_3 for each experiment.

$$k = -(2.303)(\text{slope})$$

$$k_1 = -(2.303)\left[\frac{2.13 - 1.80}{2000\ s - 6000\ s}\right] = 1.9 \times 10^{-4}\ s^{-1}$$

$$k_2 = -(2.303)\left[\frac{2.20 - 1.93}{1000\ s - 3000\ s}\right] = 3.1 \times 10^{-4}\ s^{-1}$$

(c) The catalyzed reaction does have a different mechanism than the uncatalyzed reaction because the order of reaction with respect to O_3 is different for each reaction.

(d) $$k' = \frac{k}{P_{CO_2}}$$

$$k_1' = \frac{1.9 \times 10^{-4}\ s^{-1}}{100\ \text{Torr}} = 1.9 \times 10^{-6}\ s^{-1}\ \text{Torr}^{-1}$$

$$k_2' = \frac{3.1 \times 10^{-4}\ s^{-1}}{180\ \text{Torr}} = 1.7 \times 10^{-6}\ s^{-1}\ \text{Torr}^{-1}$$

$$\text{average } k' = \frac{1.9 \times 10^{-6}\ s^{-1}\ \text{Torr}^{-1} + 1.7 \times 10^{-6}\ s^{-1}\ \text{Torr}^{-1}}{2}$$

$$= 1.8 \times 10^{-6}\ s^{-1}\ \text{Torr}^{-1}$$

18.51

T, °C	T, K	k, s^{-1}	log k	$1/T$, K^{-1}
0	273	7.36×10^{-7}	-6.133	3.66×10^{-3}
25	298	3.33×10^{-5}	-4.478	3.36×10^{-3}
35	308	1.29×10^{-4}	-3.889	3.25×10^{-3}
45	318	4.58×10^{-4}	-3.339	3.14×10^{-3}
55	328	1.51×10^{-3}	-2.821	3.05×10^{-3}
65	338	4.64×10^{-3}	-2.333	2.96×10^{-3}

The data are plotted in Fig. 18-8. The slope is -5.40×10^3 K giving

$$E_a = -(2.303)R(\text{slope})$$

$$= -(2.303)(8.314\ \text{J/K mol})(-5.4 \times 10^3\ \text{K}) = 1.0 \times 10^5\ \text{J/mol}$$

Fig. 18-8

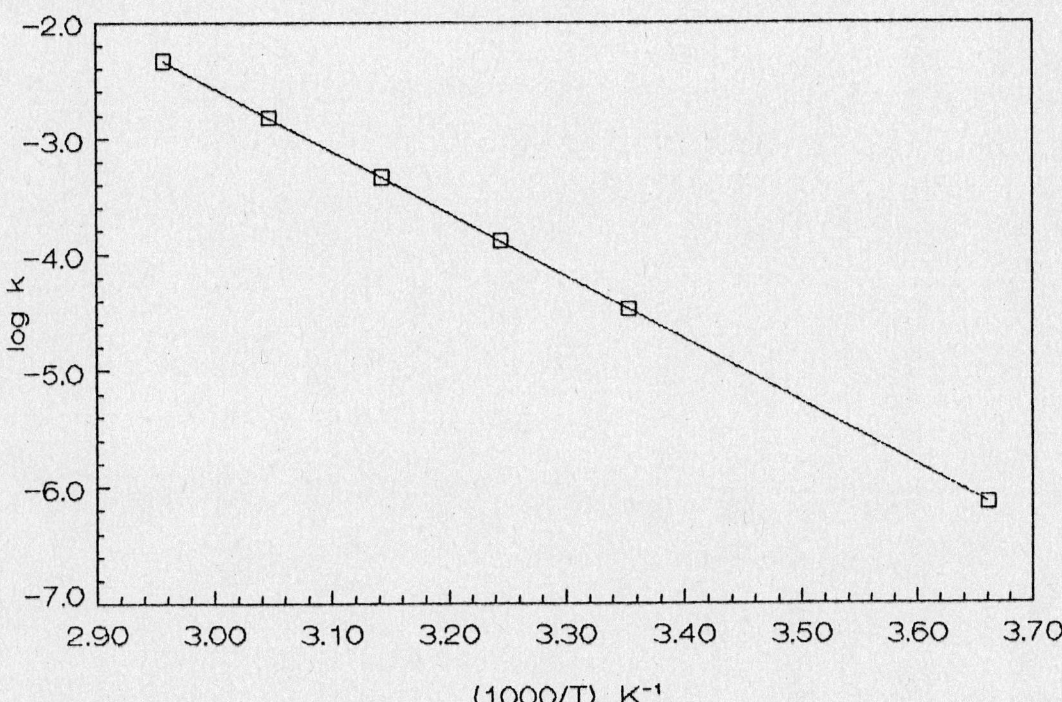
-2.0
-3.0
-4.0
-5.0
-6.0
-7.0
log k
2.90
3.00
3.10
3.20
3.30
3.40
3.50
3.60
3.70
(1000/T), K⁻¹

CHAPTER 19

CHEMICAL EQUILIBRIUM

Solutions to Exercises

19.1 The concentration equilibrium constant expressions for each of the gas-phase reactions are written as

(a) $K_c = \frac{[NO_2]^2}{[NO]^2[O_2]}$ (b) $K_c = \frac{[H_2][F_2]}{[HF]^2}$

19.2 The rate of the forward reaction would be less than the rate of the reverse reaction because there would be fewer chlorine molecules to react. The reverse reaction rate will slow down as some of the PCl_5 molecules decompose and the forward reaction rate will increase as the concentrations of PCl_3 and Cl_2 molecules increase. Equilibrium will be established once the rates are equal. Although the concentrations of the molecules will be different under the new equilibrium conditions, the value of $[PCl_5]/[PCl_3][Cl_2]$ will be the same as before.

19.3 Table 19.2 shows that the values of K_c for the hydrocyanic acid reaction is smaller than that for the acetic acid. Thus as H^+ ion is added, the CN^- ion will react in preference to the CH_3COO^- ion.

19.4 $\Delta n = [(1) + (5)] - [2] = 4$

$K_p = K_c(RT)^{\Delta n} = (2.6 \times 10^{-9})[(0.08314)(1500.)]^4 = 0.63$

19.5 The equilibrium constant expressions for the reaction of an acid on Mg are

$$K_c = \frac{[Mg^{2+}][H_2]}{[H^+]^2} \quad \text{and} \quad K_p = \frac{[Mg^{2+}]P_{H_2}}{[H^+]^2}$$

19.6 $K_p = \frac{1}{P_{F_2}} = 1.24 \times 10^{184}$

$$P_{F_2} = \frac{1}{1.24 \times 10^{184}} = 8.06 \times 10^{-185} \text{ bar}$$

19.7 $Cl_2 + F_2 \rightleftharpoons 2ClF$ $K_p = (2.09 \times 10^{18})$

$2ClF_3 \rightleftharpoons Cl_2 + 3F_2$ $K_p = (4.57 \times 10^{41})^{-1}$

$2ClF_3 \rightleftharpoons 2ClF + 2F_2$

$K_p = (2.09 \times 10^{18})(4.57 \times 10^{41})^{-1} = 4.57 \times 10^{-24}$

$ClF_3 \rightleftharpoons ClF + F_2$

$K_p = (4.57 \times 10^{-24})^{1/2} = 2.14 \times 10^{-12}$

19.8 $$Q = \frac{P_{BrF}\,P_{F_2}}{P_{BrF_3}} = \frac{(0.01)(1.36)}{(0.52)} = 0.03$$

Because $Q < K_p$, some of the BrF_3 will decompose.

19.9 (a) As P_{BrF_5} (or $[BrF_5]$) is increased, some of it decomposes and additional Br_2 and F_2 are formed.

(b) As P_{F_2} (or $[F_2]$) is decreased, additional Br_2 is formed by the further decomposition of BrF_5.

(c) As P_{Br_2} and P_{F_2} (or $[Br_2]$ and $[F_2]$) are both increased, they react, so additional BrF_5 is formed.

(d) No changes are observed because a catalyst does not affect the equilibrium that a system attains.

19.10 (a) Because the reaction is endothermic, we can write it as

$$2BrF_5(g) + 858\text{ kJ} \rightleftharpoons Br_2(g) + 5F_2(g)$$

An increase in temperature shifts the equilibrium toward the products side, so the value of K_p at 1500 K would be larger than at 1000 K.

(b) If the total pressure of the system were increased, additional BrF_5 would form because there are fewer moles of reactant than of products.

19.12 Let $x = [Br_2]$.

	$2BrF_5(g)$ $\rightleftharpoons$	$Br_2(g)$ +	$5F_2(g)$
initial	1.00	0	0
change	$-2x$	$+x$	$+5x$
equilibrium	$1.00 - 2x$	x	$5x$

$$K_c = \frac{[Br_2][F_2]^5}{[BrF_5]^2} = \frac{(x)(5x)^5}{(1.00 - 2x)^2} = 2.6 \times 10^{-9}$$

Assume that $1.00 - 2x \approx 1.00$.

$$\frac{(x)(5x)^5}{(1.00)^2} = 2.6 \times 10^{-9}$$

$$x^6 = \frac{2.6 \times 10^{-9}}{(5)^5} = 8.3 \times 10^{-13}$$

$$x = 9.7 \times 10^{-3}$$

The approximation $1.00 - 2x \approx 1.00$ is valid.

$[Br_2] = 9.7 \times 10^{-3}$ mol/L

19.13 Let $x = [Cl_2]$ that reacts.

	Cl_2	+	F_2	$\rightleftharpoons$	$2ClF$
initial	1.00		2.00		0
change	$-x$		$-x$		$+2x$
equilibrium	$1.00 - x$		$2.00 - x$		$2x$

$$K_c = \frac{[ClF]^2}{[Cl_2][F_2]} = \frac{(2x)^2}{(1.00 - x)(2.00 - x)} = 4.78 \times 10^{-19}$$

Assume $1.00 - x \approx 1.00$ and $2.00 - x = 2.00$

$$\frac{(2x)^2}{(1.00)(2.00)} = 4.78 \times 10^{-19}$$

$x = 4.89 \times 10^{-10}$ mol/L

The approximations are valid, so

$[ClF] = 2x = 9.78 \times 10^{-10}$ mol/L

19.14 Let $x = [ClF_3]$ that reacts.

	$ClF_3(g)$	$\rightarrow$	$ClF(g)$	+	$F_2(g)$
initial	17.25		1.3×10^{-6}		4.62×10^{-7}
change	$-x$		$+x$		$+x$
equilibrium	$17.25 - x$		$1.3 \times 10^{-6} + x$		$4.62 \times 10^{-7} + x$

$$K_c = \frac{[ClF][F_2]}{[ClF_3]} = \frac{(1.3 \times 10^{-6} + x)(4.62 \times 10^{-7} + x)}{(17.25 - x)} = 8.77 \times 10^{-14}$$

Assume $17.25 - x = 17.25$

$6.0 \times 10^{-13} + (1.8 \times 10^{-6})x + x^2 = 1.51 \times 10^{-12}$

$x^2 + (1.8 \times 10^{-6})x - 9.1 \times 10^{-13} = 0$

$x = 4 \times 10^{-7}$ mol/L

The approximation is valid, so

$[ClF_3] = 17.25 - x = 17.25$ mol/L

$[ClF] = 1.3 \times 10^{-6} + x = 1.7 \times 10^{-7}$ mol/L

$[F_2] = 4.62 \times 10^{-7} + x = 9 \times 10^{-7}$ mol/L

Solutions to Odd-Numbered Questions and Problems

19.1 A dynamic equilibrium is established in a chemical reaction. Equilibria are reached when the rates of the forward and reverse reactions become equal. Consider the preparation of an aqueous solution of acetic acid--as liquid CH_3COOH is mixed with water, initially the rate of formation of H^+ and CH_3COO^- ions is greater than the reaction between the ions to form CH_3COOH molecules. Once equilibrium is reached, the rates of reaction become the same.

19.3 (a) $K_c = \dfrac{[SO_2][H_2O]}{[SO_3][H_2]}$ (b) $K_c = \dfrac{[NO]^4[H_2O]^6}{[NH_3]^4[O_2]^5}$

(c) $K_c = \dfrac{[CO_2]^3[H_2O]^4}{[C_3H_8][O_2]^5}$

19.5 $K_c = [H^+][H_2PO_4^-]/[H_3PO_4]$

The addition of H^+ ions results in an increase in the rate of formation of H_3PO_4 (the reverse reaction) because of the increased number of collisions possible between the ions. But, as more H_3PO_4 molecules are formed, the number ionizing to give the products and hence the rate of ionization (the forward reaction) also increases. The amount of reactants and products will continue to adjust until equilibrium is reached once again. Although the concentration of the species at the new equilibrium conditions will be different than the concentration of the old equilibrium condition, the value of $[H^+][H_2PO_4^-]/[H_3PO_4]$ will be unchanged.

19.7 Products are favored in reactions having large equilibrium constants. Reactions (b) and (d) favor the products.

19.9 The formation of $BaCrO_4$ is more highly favored because of the larger equilibrium constant value.

19.11 $K_c = \dfrac{[NO]^2}{[N_2][O_2]} = \dfrac{(1.1 \times 10^{-5})^2}{(6.4 \times 10^{-3})(1.7 \times 10^{-3})} = 1.1 \times 10^{-5}$

19.13 $K_c = \dfrac{[PCl_5]}{[PCl_3][Cl_2]} = \dfrac{(12)}{(10)(9)} = 0.1$

19.15 $[SO_3] = 0.040$ mol/L

$[SO_2] = 0.060 - 0.040 = 0.020$ mol/L

$[O_2] = 0.050 - (\frac{1}{2})(0.040) = 0.030$ mol/L

$$K_c = \frac{[SO_3]^2}{[SO_2]^2[O_2]} = \frac{(0.040)^2}{(0.020)^2(0.030)} = 130$$

19.17 (a) $$K_p = \frac{P_{SO_2}\ P_{H_2O}}{P_{SO_3}\ P_{H_2}}$$ (b) $$K_p = \frac{P^4_{NO}\ P^6_{H_2O}}{P^4_{NH_3}\ P^5_{O_2}}$$

(c) $$K_p = \frac{P^3_{CO_2}\ P^4_{H_2O}}{P_{C_3H_8}\ P^5_{O_2}}$$

19.19 $$K_p = \frac{P_{H_2}\ P_{I_2}}{P^2_{HI}} = \frac{(0.06443)(0.06540)}{(0.4821)^2} = 0.01813$$

19.21 $K_p = K_c(RT)^{\Delta n} = (193)[(0.08314)(2500.)]^0 = 193$

19.23 (a) $K_c = [Ag^+]^2[S^{2-}]$ (b) $K_p = P_{CO_2}$ $K_c = [CO_2]$

(c) $K_p = P^2_{CO}$ $K_c = [CO]^2$ (d) $K_c = [Br^-]^2/[I^-]^2[Br_2]$

19.25 $$K_c = \frac{K_p}{(RT)^{\Delta n}} = \frac{6.41 \times 10^{50}}{[(0.08314)(1100.)]^{-1/2}} = 6.13 \times 10^{51}$$

19.27 Combining the equations to give the desired equation

$$CO(g) \rightleftharpoons C(s) + \tfrac{1}{2}O_2(g) \qquad K_p = \frac{1}{2.9 \times 10^{10}} = 3.4 \times 10^{-11}$$

$$C(s) + O_2(g) \rightleftharpoons CO_2(g) \qquad K_p = 4.8 \times 10^{20}$$

$$CO(g) + \tfrac{1}{2}O_2(g) \rightleftharpoons CO_2(g) \qquad K_p = (3.4 \times 10^{-11})(4.8 \times 10^{20})$$

$$= 1.6 \times 10^{10}$$

19.29

$$2[NH_3(aq) + H_2O(l) \rightleftharpoons NH_4^+ + OH^-] \qquad K_a^2$$

$$CO_2(g) + H_2O(l) \rightleftharpoons H_2CO_3(aq) \qquad K_b$$

$$H_2CO_3(aq) \rightleftharpoons 2H^+ + CO_3^{2-} \qquad K_c$$

$$2[H^+ + OH^- \rightleftharpoons H_2O(l)] \qquad K_d^{-2}$$

$$2NH_3(aq) + CO_2(g) + H_2O(l) \rightleftharpoons 2NH_4^+ + CO_3^{2-} \qquad K = \frac{K_a^2\ K_b\ K_c}{K_d^2}$$

19.31 The form of the reaction quotient is the same as that for the equilibrium constant. The reaction quotient is used for reactions not at equilibrium.

19.33 $Q_c = \dfrac{[ClF]^2}{[Cl_2][F_2]} = \dfrac{(3.65)^2}{(0.2)(0.1)} = 700$

Because $Q > K_c$, ClF will decompose.

19.35 (a) $Q_c = \dfrac{1}{[Ba^{2+}][CO_3{}^{2-}]} = \dfrac{1}{(0.50 \times 10^{-3})(1.0 \times 10^{-6})} = 2.0 \times 10^9$

(b) Because $Q_c > K_c$, the reverse reaction is favored, so no precipitate will form.

19.37 Le Chatelier's principle states that if a system at equilibrium is subjected to a stress, the system will react in a way that tends to relieve the stress. The factors usually considered to have an effect on equilibrium are concentration changes, temperature changes, and total pressure changes. The presence of a catalyst or an inhibitor has no net effect on a system at chemical equilibrium because the rates of forward and reverse reactions are changed equally.

19.39 If the total pressure of the system were decreased, additional BrF_5 would decompose because the number of moles of products is greater than the number of reactants.

19.41 The pink color of a weather indicator indicates moist air. The water vapor in the moist air reacts with blue $[Co(H_2O)_4]Cl_2$ to form the pink $[Co(H_2O)_6]Cl_2$.

19.43 The concentration of NO_2 at equilibrium will

(a) (i) increase if additional O_2 is introduced.

(b) (i) increase if additional NO is introduced.

(c) (i) increase if the total pressure is increased.

(d) (i) increase if the temperature is decreased.

19.45 Many times the expression can be simplified by neglecting x in terms of the form $(C \pm x)$ provided the value of C is larger compared to the value of K. The approximations should always be checked for validity--commonly using the rule that up to a 5 % error is permitted.

19.47 $K_p = \dfrac{P^2_{IF}}{P_{I_2}\, P_{F_2}}$

Solving the K_p expression for P_{I_2} and substituting the numerical data gives

$$P_{I_2} = \frac{P^2_{IF}}{K_p\, P_{I_2}} = \frac{(0.32)^2}{(1.15 \times 10^7)(1.9 \times 10^{-2})} = 4.7 \times 10^{-7} \text{ bar}$$

19.49 Let x = $[O_2]$ found.

	$2CO_2$	⇄	$2CO$	+	O_2
initial	1.00		0		0
change	$-2x$		$+2x$		$+x$
equilibrium	$1.00 - 2x$		$2x$		x

$$K_c = \frac{[CO]^2[O_2]}{[CO_2]} = \frac{(2x)^2(x)}{(1.00 - 2x)^2} = 2.96 \times 10^{-92}$$

Assume $1.00 - 2x \approx 1.00$, then

$4x^3 = (2.96 \times 10^{-92})(1.00) = 2.96 \times 10^{-92}$

$x = 1.95 \times 10^{-31}$ mol/L

The approximation is valid, so $[CO] = 2x = 2(1.95 \times 10^{-31}$ mol/L$) = 3.90 \times 10^{-31}$ mol/L.

19.51 Let x = $[C_6H_5COOH_2]$ that reacts.

	$(C_6H_5COOH)_2$(in C_6H_6)	⇄	$2C_6H_5COOH$(in C_6H_6)
initial	0.45		0
change	$-x$		$+2x$
equilibrium	$0.45 - x$		$2x$

$$K_c = \frac{[C_6H_5COOH]^2}{[(C_6H_5COOH)_2]} = \frac{(2x)^2}{(0.45 - x)} = 3.7 \times 10^{-3}$$

Assume that $0.45 - x \approx 0.45$.

$4x^2 = (3.7 \times 10^{-3})(0.45) = 1.7 \times 10^{-3}$

$x = 0.021$ mol/L

The approximation is valid because $[(0.021)/(0.45)](100) = 4.7$ %.

Thus $[C_6H_5COOH] = 2x = 0.042$ mol/L.

19.53 Let x = $[H^+]$.

	H^+	+	BrO^-	$\rightleftharpoons$	HBrO
initial	0		0		0.32
change	$+x$		$+x$		$-x$
equilibrium	x		x		$0.32 - x$

$$K_c = \frac{[HBrO]}{[H^+][BrO^-]} = \frac{(0.32 - x)}{(x)(x)} = 4.5 \times 10^8$$

Assume $0.32 - x \approx 0.32$

$x^2 = (0.32)/(4.5 \times 10^8) = 7.1 \times 10^{-10}$

$x = 2.7 \times 10^{-5}$ mol/L

The approximation is valid, so $[H^+] = [BrO^-] = x = 2.7 \times 10^{-5}$ mol/L.

19.55 Let x = $[N_2]$ that reacts.

	N_2	+	C_2H_2	$\rightleftharpoons$	2HCN
initial	5.0		2.0		0
change	$-x$		$-x$		$+2x$
equilibrium	$5.0 - x$		$2.0 - x$		$2x$

$$K_c = \frac{[HCN]^2}{[N_2][C_2H_2]} = \frac{(2x)^2}{(5.0 - x)(2.0 - x)} = 2.3 \times 10^{-4}$$

Assume $5.0 - x \approx 5.0$ and $2.0 - x \approx 2.0$

$4x^2 = (2.3 \times 10^{-4})(5.0)(2.0) = 2.3 \times 10^{-3}$

$x = 0.024$ mol/L

The approximations are valid, so [HCN] = $2x$ = 2(0.024 mol/L) = 0.048 mol/L.

19.57 Let x = $[Br_2]$ that reacts.

	$Br_2(g)$	+	$F_2(g)$	$\rightleftharpoons$	2BrF(g)
initial	0.125		0.125		0
change	$-x$		$-x$		$+2x$
equilibrium	$0.125 - x$		$1.25 - x$		$2x$

$$K_c = \frac{[BrF]^2}{[Br_2][F_2]} = \frac{(2x)^2}{(0.125 - x)(1.25 - x)} = 54.7$$

$4x^2 = (54.7)(0.125 - x)(1.25 - x) = 8.55 - 75.5x + 54.7x^2$

$50.7x^2 - 75.5x + 8.55 = 0$

$$x = \frac{75.5 \pm \sqrt{(75.5)^2 - (4)(50.7)(8.55)}}{2(50.7)} = \frac{75.5 \pm 63.0}{2(50.7)} = 1.37 \text{ or } 0.123 \text{ mol/L}$$

$[Br_2]$ = 0.125 - 0.123 = 0.002 mol/L

$[F_2]$ = 1.25 - 0.123 = 1.13 mol/L

[BrF] = (2)(0.123) = 0.246 mol/L

19.59 Let $x = [Fe^{3+}]$.

	Fe^{3+}	$+ 3C_2O_4^{2-}$	$\rightleftharpoons$	$[Fe(C_2O_4)_3]^{3-}$
initial	0.00050	0.5000		0
Fe^{3+} reacted	0	0.4985		0.00050
change	$+x$	$+3x$		$-x$
equilibrium	x	$0.4985 + 3x$		$0.00050 - x$

$$K_c = \frac{[Fe(C_2O_4)_3{}^{3-}]}{[Fe^{3+}][C_2O_4{}^{2-}]^3} = \frac{(0.00050 - x)}{(x)(0.4985 + 3x)^3} = 1.67 \times 10^{20}$$

Assume that $0.00050 - x \approx 0.00050$ and $0.4985 + 3x \approx 0.4985$.

$$x = \frac{(0.00050)}{(1.67 \times 10^{20})(0.4985)^3} = 2.4 \times 10^{-23}$$

The approximations are valid because $0.00050 - (2.4 \times 10^{-23}) \approx 0.00050$ and $0.4985 - (2.4 \times 10^{-23}) \approx 0.4985$. Thus $[Fe^{3+}] = 2.4 \times 10^{-23}$ mol/L.

CHAPTER 20

ACIDS AND BASES

Solutions to Exercises

20.1 (a) In reaction (i) HS^- is acting as an acid (a proton donor).
(b) In reaction (ii) HS^- is acting as a base (a proton acceptor).

20.2 The equation for the equilibrium between hydrofluoric acid and sulfate ion is $HF(aq) + SO_4^{2-} \rightleftharpoons HSO_4^- + F^-$. The species acting as acids are HF and HSO_4^-. The stronger base is F^-.

20.3 The following oxoacids and their respective oxoanions are named as (a) H_2SO_3, sulfurous acid, sulfite ion; (b) H_2SO_4, sulfuric acid, sulfate ion; and (c) $H_2S_2O_7$, pyrosulfuric acid, pyrosulfate ion.

20.4 HClO is stronger than HBrO because acid strength increases with increasing electronegativity of the nonmetal and $HClO_2$ is stronger than HClO because acid strength increases with increasing numbers of oxygen atoms bonded to the central atom.

20.5 Each mole of H_2SO_3 donates two moles of H^+--2 equiv/mol--and so the equivalent mass is one half of the molar mass. Each mole of NaOH donates one mole of OH^---1 equiv/mol--and so the equivalent mass is the same as the molar mass.

20.6 Each mole of lithium hydroxide donates one mole of OH^---1 equiv/mol.

$$\left[\frac{0.5 \text{ equiv}}{1 \text{ L}}\right]\left[\frac{1 \text{ mol}}{1 \text{ equiv}}\right] = 0.5 \text{ mol/L} = 0.5 \text{ M}$$

20.7

$$[H^+] = \frac{K_w}{[OH^-]} = \frac{1.00 \times 10^{-14}}{0.5} = 2 \times 10^{-14} \text{ mol/L}$$

20.8 $pH = -\log [H^+] = -\log (1.3 \times 10^{-5}) = 4.89$
This black coffee is acidic because its pH is less than 7.

20.9 log [H^+] = -pH = -7.4

[H^+] = 4 x 10^{-8} mol/L

Because the [H^+] is less than 1.00 x 10^{-7} mol/L, blood is slightly alkaline.

20.10 (a) pOH = 14.00 - pH = 14.00 - 8.36 = 5.64

(b) log [OH^-] = -pOH = -5.64

[OH^-] = 2.3 x 10^{-6} mol/L

(c) log [H^+] = -pH = -8.36

[H^+] = 4.4 x 10^{-9} mol/L

20.11 (a) pH = -log [H^+] = -log(0.0025) = 2.60

(b) pOH = -log [OH^-] = -log(0.0025) = 2.60

pH = 14.00 - pOH = 14.00 - 2.60 = 11.40

(c) pH = 3.65

(d) pH = 14.00 - 9.26 = 4.74

The most alkaline solution is (b) because it has the highest pH.

20.12 $C_6H_5COOH(aq) + H_2O(l) \rightleftharpoons H_3O^+ + C_6H_5COO^-$

[H^+] = [$C_6H_5COO^-$]

$$K_a = \frac{[H^+][C_6H_5COO^-]}{[C_6H_5COOH]} = \frac{(2.6 \times 10^{-3})(2.6 \times 10^{-3})}{(0.100)} = 6.8 \times 10^{-5}$$

20.13 $HCOOH(aq) + H_2O(l) \rightleftharpoons H_3O^+ + HCOO^-$

$$K_a = \frac{[H^+][HCOO^-]}{[HCOOH]} = \frac{(0.15)[HCOO^-]}{(0.037)} = 1.77 \times 10^{-4}$$

[$HCOO^-$] = 4.4 x 10^{-5} mol/L

20.14 Let x = [H^+] = [ClO^-].

	$HClO(aq) + H_2O(l)$	$\rightleftharpoons$ H_3O^+	+ ClO^-
initial	0.0108	0	0
change	$-x$	$+x$	$+x$
equilibrium	0.0108 - x	x	x

$$K_a = \frac{[H^+][ClO^-]}{[HClO]} = \frac{(x)(x)}{(0.0108 - x)} = 2.90 \times 10^{-8}$$

Assume that 0.0108 - $x \approx$ 0.0108.

x^2 = (2.90 x 10^{-8})(0.0108) = 3.13 x 10^{-10}

x = 1.77 x 10^{-5}

The approximation is valid because $0.0108 - 0.0000177 \approx 0.0108$.

$[H^+] = [ClO^-] = 1.77 \times 10^{-5}$ mol/L

$[HClO] = 0.0108$ mol/L

20.15 Let $x = [H^+] = [Br^-]$ at equilibrium.

	$HBrO(aq) + H_2O(l)$	$\rightleftharpoons H_3O^+$	$+ BrO^-$
initial	0.020	0	0
change	$-x$	$+x$	$+x$
equilibrium	$0.020 - x$	x	x

$$K_a = \frac{[H^+][BrO^-]}{[HBrO]} = \frac{(x)(x)}{(0.020 - x)} = 2.2 \times 10^{-9}$$

Assume that $0.020 - x \approx 0.020$.

$x^2 = (0.020)(2.2 \times 10^{-9}) = 4.4 \times 10^{-11}$

$x = 6.6 \times 10^{-6}$

The approximation is valid because $0.020 - (6.6 \times 10^{-6}) \approx 0.020$.

$[H^+] = 6.6 \times 10^{-6}$ mol/L

$$\% \text{ ionization} = \left[\frac{[H^+]_{\text{equilibrium}}}{[HBrO]_{\text{initial}}}\right] \times 100 = \left[\frac{6.6 \times 10^{-6}}{0.020}\right] \times 100 = 0.033\ \%$$

$pH = -\log [H^+] = -\log(6.6 \times 10^{-6}) = 5.18$

20.16 Let $x = [OH^-] = [NH_4^+]$.

	$NH_3(aq) + H_2O(l)$	$\rightleftharpoons NH_4^+$	$+ OH^-$
initial	0.015	0	0
change	$-x$	$+x$	$+x$
equilibrium	$0.015 - x$	x	x

$$K_b = \frac{[NH_4^+][OH^-]}{[NH_3]} = \frac{(x)(x)}{(0.015 - x)} = 1.6 \times 10^{-5}$$

Assume that $0.015 - x \approx 0.015$.

$x^2 = (0.015)(1.6 \times 10^{-5}) = 2.4 \times 10^{-7}$

$x = 4.9 \times 10^{-4}$

The approximation is valid because $\% \text{ error} = \left[\frac{4.9 \times 10^{-4}}{0.015}\right] \times 100 = 3.3\ \%$.

$[OH^-] = 4.9 \times 10^{-4}$ mol/L

$pOH = -\log [OH^-] = -\log(4.9 \times 10^{-4}) = 3.31$

$pH = 14.00 - pOH = 14.00 - 3.31 = 10.69$

$$\% \text{ ionization} = \left[\frac{[OH^-]_{equilibrium}}{[NH_3]_{initial}}\right] \times 100 = \left[\frac{4.9 \times 10^{-4}}{0.015}\right] \times 100 = 3.3\ \%.$$

There is a larger percentage of ionization in the more dilute solution.

20.17 $NH_3(aq) + H_2O(l) \rightleftarrows NH_4^+ + OH^-$

$$K_a = \frac{K_w}{K_b} = \frac{1.00 \times 10^{-14}}{1.6 \times 10^{-5}} = 6.3 \times 10^{-10}$$

20.18 In each of the Lewis acid-base reactions, the respective acid is (a) SO_3, attraction of electron pairs by an atom in a multiple bond; (b) SiF_4, occupation of empty *d* orbitals; and (c) S, octet completion.

Solutions to Odd-Numbered Questions and Problems

20.1 A Brønsted-Lowry acid is any molecule or ion that can act as a proton donor. All water-ion acids are considered to be Brønsted-Lowry acids because they are proton donors.

acid	→	proton	+	conjugate base
(a) $HNO_3(aq)$	→	H^+	+	NO_3^-
(b) HSO_4^-	→	H^+	+	SO_4^{2-}
(c) $H_2SO_4(aq)$	→	H^+	+	HSO_4^-
(d) $HCl(aq)$	→	H^+	+	Cl^-
(e) $H_2O(l)$	→	H^+	+	OH^-

20.3 (a) $HCO_3^- + H_2O(l) \rightleftarrows H_3O^+ + CO_3^{2-}$

$HCO_3^- + H_3O^+ \rightleftarrows H_2CO_3(aq) + H_2O(l)$

(b) CO_3^{2-} is the conjugate base of HCO_3^- and H_2CO_3 is the conjugate acid of HCO_3^-.

20.5 (a) The Brønsted-Lowry acids include (iv) NH_4^+, (v) NH_3, (vi) H_2O, and (vii) HBr.

(b) The Brønsted-Lowry bases include (i) SO_3^{2-}, (iii) Cl^-, (v) NH_3, and (vi) H_2O.

20.7 A Brønsted-Lowry acid-base reaction is the transfer of a proton from a proton donor to a proton acceptor.

acid₁	+	base₂	→	conjugate base₁	+	conjugate acid₂
(a) $H_2O(l)$	+	$NH_3(aq)$	→	OH^-	+	NH_4^+
(b) H_3O^+	+	OH^-	→	$H_2O(l)$	+	$H_2O(l)$
(c) NH_4^+	+	CO_3^{2-}	→	$NH_3(aq)$	+	HCO_3^-
(d) $HNO_3(aq)$	+	$H_2O(l)$	→	NO_3^-	+	H_3O^+

20.9 (a) $HF(aq) + Cl^- \rightarrow$ no reaction

(b) $H_2S(aq) + CN^- \rightarrow HS^- + HCN(aq)$

(c) $HCOOH(aq) + Cl^- \rightarrow$ no reaction

20.11 The names of the following binary acids are (a) HF, hydrofluoric acid; (b) HBr, hydrobromic acid; (c) H_2S, hydrosulfuric acid; and (d) H_2Se, hydroselenic acid.

20.13 The names are (a) phosphoric acid, phosphorous acid, and hypophosphorous acid; (b) arsenic and arsenous acid, and (c) antimonic acid and antimonous acid.

20.15 An ortho oxoacid is the acid with the largest number of OH groups. The loss of one or more water molecules yields a meta acid.

$$H_3PO_4 \rightarrow H_2O + HPO_3$$

20.17 The names of the following oxoanions are (a) BO_3^{3-}, borate ion; (b) AsO_3^{3-}, arsenite ion; (c) NO_2^-, nitrite ion; (d) BiO_3^{3-}, bismuthate ion; (e) SeO_3^{2-}, selenite ion; (f) IO^-, hypoiodite ion; and (g) ClO_4^-, perchlorate ion.

20.19 The names of the following salts of oxoacids are (a) $Ga_2(SeO_4)_3$, gallium(III) selenate; (b) $Co_3(PO_4)_2$, cobalt(II) phosphate; (c) $Zn(BrO_3)_2$, zinc bromate; (d) $TlNO_3$, thallium(I) nitrate; and (e) $Th(SO_4)_2$, thorium(IV) sulfate.

20.21 The formulas of the following salts of oxoanions are (a) lead selenate, $PbSeO_4$; (b) cobalt(II) sulfite, $CoSO_3$; (c) cadmium iodate, $Cd(IO_3)_2$; (d) cesium perchlorate, $CsClO_4$; and (e) copper(I) sulfate, Cu_2SO_4.

20.23 The nonmetal radius is more important than electronegativity for determining the acid strength of binary acids that fall into the same family of the periodic table. Thus, H_2Te is a stronger acid than H_2Se.

20.25 The stronger acid of each pair is (a) NH_3, (b) HI, (c) HBr, and (d) H_2S.

20.27 Acid strength of a series of oxoacids of a given element increases with increasing oxidation state of the central atom. Thus, H_2SO_4 is a stronger acid than H_2SO_3.

20.29 The stronger acid of each pair is (a) HIO_3, (b) $Sb(OH)_3$, and (c) HNO_3.

20.31 The equivalent mass of an acid or a base is the molar mass divided by the number of equivalents of H^+ or OH^- ions that the compound supplies per mole in an acid-base reaction. The equivalent mass of many substances can vary from reaction to reaction, for example in the reactions $H_2SO_4(aq) + NaOH(aq) \rightarrow NaHSO_4(aq) + H_2O(l)$ and $H_2SO_4(aq) + 2NaOH(aq) \rightarrow NaSO_4(aq) + 2H_2O(l)$, the equivalent mass of H_2SO_4 changes from 98 g/equiv to 49 g/equiv.

20.33 The relationship between the molar mass and the equivalent mass in acid-base reactions for (a) HBr is 1 equiv/1 mol; (b) H_3PO_4 is 3 equiv/1 mol; and (c) $Ca(OH)_2$ is 2 equiv/1 mol.

20.35 (a) $$\left[\frac{0.13 \text{ equiv } H_2SO_4}{1 \text{ L soln}}\right]\left[\frac{1 \text{ mol } H_2SO_4}{2 \text{ equiv } H_2SO_4}\right] = 0.065 \text{ mol } H_2SO_4/L$$

(b) $$\left[\frac{0.13 \text{ equiv KOH}}{1 \text{ L soln}}\right]\left[\frac{1 \text{ mol KOH}}{1 \text{ equiv KOH}}\right] = 0.13 \text{ mol KOH/L}$$

(c) $$\left[\frac{0.13 \text{ equiv } HNO_3}{1 \text{ L soln}}\right]\left[\frac{1 \text{ mol } HNO_3}{1 \text{ equiv } HNO_3}\right] = 0.13 \text{ mol } HNO_3/L$$

20.37 The chemical equation showing the ionization of water is

$$H_2O(l) + H_2O(l) \rightleftharpoons H_3O^+ + OH^-$$

The equilibrium constant expression is $K_w = [H^+][OH^-]$. The special symbol used for this equilibrium constant is K_w. In pure water, $[H^+] = [OH^-]$. From this relationship, acidic means $[H^+] > [OH^-]$ and alkaline means $[OH^-] > [H^+]$.

20.39 (a) $$[OH^-] = \frac{K_w}{[H^+]} = \frac{1.00 \times 10^{-14}}{1.3 \times 10^{-4}} = 7.7 \times 10^{-11} \text{ mol/L}$$

(b) $$[OH^-] = \frac{K_w}{[H^+]} = \frac{1.00 \times 10^{-14}}{7 \times 10^{-9}} = 1 \times 10^{-6} \text{ mol/L}$$

20.41 (a) $pH = -\log [H^+] = -\log(9.5 \times 10^{-6}) = 5.02$

(b) $pH = -\log(0.0001472) = 3.8321$

(c) $pH = -\log(0.00203) = 2.693$

(d) $pH = -\log(12 \times 10^{-12}) = 10.92$

(e) pH = -log(1.0 x 10^1) = -1.00

The acidic solutions are (a), (b), (c), and (e).

20.43 (a) pOH = -log $[OH^-]$ = -log(0.00035) = 3.46

pH = 14.00 - pOH = 14.00 - 3.46 = 10.54

(b) pOH = -log(11 x 10^{-13}) = 11.96

pH = 14.00 - 11.96 = 2.04

(c) pOH = -log(4.5 x 10^{-7}) = 6.35

pH = 14.00 - 6.35 = 7.65

(d) pOH = -log(10) = -1.0

pH = 14.0 - (-1.0) = 15.0

The alkaline solutions are (a), (c), and (d).

20.45 (a) pOH = -log $[OH^-]$ = -log(0.2374) = 0.6245

pH = 14.0000 - pOH = 14.0000 - 0.6245 = 13.3755

(b) pH = -log $[H^+]$ = -log(0.0365) = 1.438

(c) pH = -log $[H^+]$ = -log(4.2 x 10^{-5}) = 4.38

(d) pOH = -log $[OH^-]$ = -log(0.0826) = 1.083

pH = 14.000 - pOH = 14.000 - 1.083 = 12.917

20.47 $HA(aq) + H_2O(l) \rightleftharpoons H_3O^+ + A^-$ $\qquad K_a = \frac{[H^+][A^-]}{[HA]}$

The special symbol used for this equilibrium constant is K_a.

20.49 $HIO(aq) + H_2O(l) \rightleftharpoons H_3O^+ + IO^-$ $\qquad K_a = \frac{[H^+][IO^-]}{[HIO]} = 2.3 \times 10^{-11}$

$$[IO^-] = \frac{K_a[HIO]}{[H^+]} = \frac{(2.3 \times 10^{-11})(0.427)}{(0.035)} = 2.8 \times 10^{-10} \text{ mol/L}$$

20.51 Let $x = [H^+]$ at equilibrium.

	$HF(aq) + H_2O(l)$ $\rightleftharpoons$	H_3O^+	+ F^-
initial	0.025	0	0
change	$-x$	$+x$	$+x$
equilibrium	$0.025 - x$	x	x

$$K_a = \frac{[H^+][F^-]}{[HF]} = \frac{(x)(x)}{(0.025 - x)} = 6.5 \times 10^{-4}$$

Assume that $0.025 - x \approx 0.025$.

$x^2 = (6.5 \times 10^{-4})(0.025) = 1.6 \times 10^{-5}$

$x = 4.0 \times 10^{-3}$ mol/L

The approximation is not valid because % error = $\frac{4.0 \times 10^{-3}}{0.025} \times 100 = 16\ \%$.

$x^2 = 1.6 \times 10^{-5} - (6.5 \times 10^{-4})x$

$x^2 + (6.5 \times 10^{-4})x - 1.6 \times 10^{-5} = 0$

$$x = \frac{-(6.5 \times 10^{-4}) \pm \sqrt{(6.5 \times 10^{-4})^2 - (4)(1)(-1.6 \times 10^{-5})}}{(2)(1)}$$

$$= \frac{-(6.5 \times 10^{-4}) \pm (8.0 \times 10^{-3})}{2} = 3.7 \times 10^{-3}\ \text{mol/L}$$

$[H^+] = [F^-] = 3.7 \times 10^{-3}$ mol/L

$[HF] = 0.025 - x = 0.025 - 0.0037 = 0.021$ mol/L

$pH = -\log [H^+] = -\log(3.7 \times 10^{-3}) = 2.43$

20.53 The concentration of acetic acid that ionizes is $(0.019)(0.050\ \text{mol/L}) = 9.5 \times 10^{-4}$ mol/L

	$CH_3COOH(aq)$	$\rightleftharpoons$	CH_3COO^-	+ H^+
initial	0.050		0	0
change	-0.00095		+0.00095	+0.00095
equilibrium	0.049		0.00095	0.00095

The value of K_a will be

$$K_a = \frac{[CH_3COO^-][H^+]}{[CH_3COOH]} = \frac{(9.5 \times 10^{-4})(9.5 \times 10^{-4})}{(0.049)} = 1.8 \times 10^{-5}$$

20.55 Let $x = [H^+]$.

	$CH_3COOH(aq) + H_2O(l)$	$\rightleftharpoons$	H_3O^+	+ CH_3COO^-
initial	c		0	0
change	$-x$		$+x$	$+x$
equilibrium	$c - x$		x	x

$$K_a = \frac{[H^+][CH_3COO^-]}{[CH_3COOH]} = \frac{(x)(x)}{(c - x)} = 1.754 \times 10^{-5}$$

(a) $c = 0.100$ mol/L

Assume that $0.100 - x \approx 0.100$.

$x^2 = (1.754 \times 10^{-5})(0.100) = 1.75 \times 10^{-6}$

$x = 1.32 \times 10^{-3}$ mol/L

The approximation is valid because % error = $\frac{(0.00132)}{(0.100)} \times 100 = 1.32\ \%$.

$$\% \text{ ionization} = \left[\frac{[H^+]_{\text{equilibrium}}}{[CH_3COOH]_{\text{initial}}}\right] \times 100 = \frac{(0.00132)}{(0.100)} \times 100 = 1.32\ \%$$

(b) c = 0.0100 mol/L

Assume that 0.0100 - $x \approx$ 0.0100.

x^2 = (1.754 x 10^{-5})(0.0100) = 1.75 x 10^{-7}

x = 4.18 x 10^{-4} mol/L mol/L

The approximation is valid because % error = $\frac{(4.18 \times 10^{-4})}{(0.0100)}$ x 100 = 4.18 %.

$[H^+]$ = 4.18 x 10^{-4} mol/L

$$\% \text{ ionization} = \frac{(4.18 \times 10^{-4})}{(0.0100)} \times 100 = 4.18\ \%$$

20.57 (a) $MOH(aq) \rightleftharpoons M^+ + OH^-$ $\qquad K_b = \frac{[M^+][OH^-]}{[MOH]}$

(b) $B(aq) + H_2O(l) \rightleftharpoons BH^+ + OH^-$ $\qquad K_b = \frac{[BH^+][OH^-]}{[B]}$

The special symbol used for these equilibrium constants is K_b.

20.59 $K_b = \frac{[CH_3NH_2^+][OH^-]}{[CH_3NH_2]} = \frac{(1.8 \times 10^{-3})^2}{(0.0082)} = 4.0 \times 10^{-4}$

20.61 Let x = $[OH^-]$.

	$NH_3(aq) + H_2O(l) \rightleftharpoons$	NH_4^+	+ OH^-
initial	0.10	0	0
change	$-x$	$+x$	$+x$
equilibrium	0.10 - x	x	x

$$K_b = \frac{[NH_4^+][OH^-]}{[NH_3]} = \frac{(x)(x)}{(0.10 - x)} = 1.6 \times 10^{-5}$$

Assume that 0.10 - $x \approx$ 0.10.

x^2 = (1.6 x 10^{-5})(0.10) = 1.6 x 10^{-6}

x = 1.3 x 10^{-3} mol/L

The approximation is valid, so $[NH_4^+]$ = $[OH^-]$ = 1.3 x 10^{-3} mol/L and $[NH_3]$ = 0.10 - x = 0.10 mol/L.

20.63 $CN^- + H_2O(l) \rightleftharpoons HCN(aq) + OH^-$

$$K_b = \frac{[HCN][OH^-]}{[CN^-]} = \frac{(1.3 \times 10^{-3})(4.6 \times 10^{-3})}{(0.37)} = 1.6 \times 10^{-5}$$

20.65 Let x = $[OH^-]$ at equilibrium.

	CN^-	+ $H_2O(l)$ ⇌	$HCN(aq)$	+ OH^-
initial	0.13		0	0
change	$-x$		$+x$	$+x$
equilibrium	$0.13 - x$		x	x

$$K_b = \frac{[HCN][OH^-]}{[CN^-]} = \frac{(x)(x)}{(0.13 - x)} = 1.6 \times 10^{-5}$$

Assume that $0.13 - x \approx 0.13$.

$x^2 = (1.6 \times 10^{-5})(0.13) = 2.1 \times 10^{-6}$

$x = 1.4 \times 10^{-3}$ mol/L

This approximation is valid because % error = $\frac{0.0014}{0.13} \times 100 = 1.4$ %.

$[OH^-] = 1.4 \times 10^{-3}$ mol/L

$pOH = -\log [OH^-] = -\log(1.4 \times 10^{-3}) = 2.85$

$pH = 14.00 - pOH = 14.00 - 2.85 = 11.15$

20.67

	B + $H_2O(l)$ ⇌	BH^+	+ OH^-
initial	0.100	0	0
change	-(0.032)(0.100)	+(0.032)(0.100)	+(0.032)(0.100)
equilibrium	0.097	0.0032	0.0032

$$K_b = \frac{[BH^+][OH^-]}{[B]} = \frac{(0.0032)(0.0032)}{(0.097)} = 1.1 \times 10^{-4}$$

20.69 The relationship between the ionization constant for an acid and the ionization constant for its conjugate base is $K_a K_b = K_w$.

20.71

$$K_b = \frac{K_w}{K_a} = \frac{1.00 \times 10^{-14}}{2.2 \times 10^{-9}} = 4.5 \times 10^{-6}$$

20.73 $HA(aq) + H_2O(l) \rightleftharpoons H_3O^+ + A^-$

for Cl^- $K_b = \frac{K_w}{K_a} = \frac{1.00 \times 10^{-14}}{3 \times 10^{8}} = 3 \times 10^{-23}$

for CN^- $K_b = \frac{1.00 \times 10^{-14}}{6.2 \times 10^{-10}} = 1.6 \times 10^{-5}$

The CN^- ion has the larger value of K_b.

20.75 A Lewis acid is a molecule or ion that can accept one or more electron pairs. See Fig. 20-1.

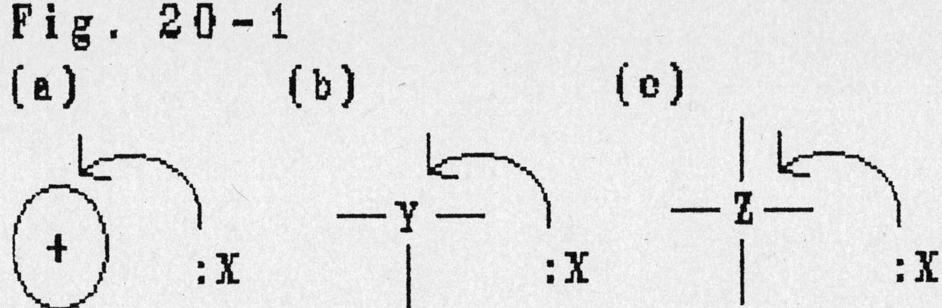

20.77 A Lewis acid-base reaction is the donation of an electron pair from one atom to a covalent bond formed with another atom. All of the given reactions are Lewis acid-base reactions. The respective Lewis acid in each reaction is (a) Ag^+, (b) I_2, (c) H_3O^+, and (d) H_2O.

20.79 $H_2O(l) + H_2O(l) \rightleftharpoons H_3O^+ + OH^-$

$[H^+] = [OH^-]$

$K_w = [H^+][OH^-] = [H^+]^2 = 9.614 \times 10^{-14}$

$[H^+] = 3.101 \times 10^{-7}$ mol/L

$pH = -\log [H^+] = -\log(3.101 \times 10^{-7}) = 6.5085$

Even though $[H^+] > 1.00 \times 10^{-7}$, it is equal to $[OH^-]$, so the solution is neutral.

20.81 $HX(aq) \rightarrow H^+ + X^-$

$\Delta H° = [(1\ mol)\Delta H_f°(H^+) + (1\ mol)\Delta H_f°(X^-)] - [(1\ mol)\Delta H_f°(HX)]$

for HF

$\Delta H°_H = [(1\ mol)(0) + (1\ mol)(-332.63\ kJ/mol)] - [(1\ mol)(-320.08\ kJ/mol)]$

$= -12.55$ kJ

The $\Delta H°$ for ionization of HF(aq) is -12.55 kJ/mol.

for HCl

$\Delta H°_{HCl} = [(1\ mol)(0) + (1\ mol)(-246.0\ kJ/mol)] - [(1\ mol)(-167.16\ kJ/mol)]$

$= -78.8$ kJ

The $\Delta H°$ for ionization of HCl(aq) is -78.8 kJ/mol.

The reaction between OH^- and H^+ will generate the same amount of energy for both reactions, so the controlling factor is $\Delta H°$ for the ionization of the acid. Thus, HCl would yield more heat upon reaction with NaOH(aq).

20.83 (a) $\log K = -pK = -14.869$

$K = 1.35 \times 10^{-15}$

(b) For the ionization of D_2O, $[D_3O^+] = [OD^-]$.

$K = [D_3O^+][OD^-] = [D_3O^+]^2 = 1.35 \times 10^{-15}$

$[D_3O^+] = [OD^-] = 3.67 \times 10^{-8}$ mol/L

(c) $pD = -\log [D^+] = -\log(3.67 \times 10^{-8}) = 7.435$

(d) $pD + pOD = 14.869$

(e) An acidic solution in a solvent system is one in which $pD < 7.435$.

The substances acting as

(f) water-ion acids are CH_3COOH and HD_2O^+.

(g) classical bases include none of them.

(h) Brønsted-Lowry acids are CH_3COOH and HD_2O^+.

(i) Brønsted-Lowry bases are D_2O and CH_3COO^-.

(j) Let $x = [D_2O]$ that reacts.

	$D_2O(l)$ +	$H_2O(l)$ $\rightleftharpoons$	$2HDO(l)$
initial	27.7	27.7	0
change	$-x$	$-x$	$+2x$
equilibrium	$27.7 - x$	$27.7 - x$	$2x$

$$K = \frac{[HDO]^2}{[D_2O][H_2O]} = \frac{(2x)^2}{(27.7 - x)(27.7 - x)} = 3.56$$

$(4)x^2 = 2730 - (197)x + (3.56)x^2$

$(0.44)x^2 + (197)x - 2730 = 0$

$$x = \frac{-(197) \pm \sqrt{(197)^2 - (4)(0.44)(-2730)}}{(2)(0.44)} = \frac{-(197) \pm (209)}{0.88} = 14 \text{ mol/L}$$

$[HDO] = 2x = 2(14) = 28$ mol/L

CHAPTER 21

IONS AND IONIC EQUILIBRIA: ACIDS AND BASES

Solutions to Exercises

21.1 (a) The CH_3COO^- ion is the conjugate base of a weak acid, CH_3COOH, and would be expected to react as a base, to give an alkaline solution.

$$CH_3COO^- + H_2O(l) \rightleftharpoons CH_3COOH(aq) + OH^-$$

(b) The NO_3^- ion is the conjugate base of a strong acid, HNO_3, and, as a weaker base than water, will simply be hydrated in aqueous solution.

(c) The Ba^{2+} ion has a low polarizing ability (low charge-to-size ratio) and is simply hydrated in aqueous solution.

(d) The Zn^{2+} ion has a high enough charge-to-size ratio so that the hydrated Zn^{2+} ion acts as a stronger acid than water and gives an acidic solution.

$$Zn^{2+} + 2H_2O(l) \rightleftharpoons [Zn(OH)]^+ + H_3O^+$$

21.2 (a) Iron(III) nitrate, $Fe(NO_3)_3$, is the salt of a cation that will react with water and an anion that will not, so the aqueous solution will be acidic.

(b) Ammonium cyanide, NH_4CN, is the salt of a cation and an anion both of which react with water, so the aqueous solution will be close to neutral.

(c) Potassium iodide, KI, is the salt of a cation and anion neither of which reacts with water, so the aqueous solution will be close to neutral.

(d) Lithium acetate, $LiCH_3COO$, is the salt of a cation that will not react with water and an anion that will, so the aqueous solution will be alkaline.

21.3 The only ion of NH_4Cl that reacts with water is the NH_4^+ ion, which reacts to form an acidic solution. Let $x = [H^+]$ at equilibrium.

	$NH_4^+ + H_2O(l)$	⇌	H_3O^+	+ $NH_3(aq)$
initial	0.10		0	0
change	$-x$		$+x$	$+x$
equilibrium	$0.10 - x$		x	x

$$K_a = \frac{[H^+][NH_3]}{[NH_4^+]} = \frac{K_w}{K_b} = \frac{1.00 \times 10^{-14}}{1.6 \times 10^{-5}} = 6.3 \times 10^{-10} = \frac{(x)(x)}{(0.10 - x)}$$

Assume that $0.10 - x \approx 0.10$.

$x^2 = (0.10)(6.3 \times 10^{-10}) = 6.3 \times 10^{-11}$

$x = 7.9 \times 10^{-6}$ mol/L

The approximation is valid because $0.10 - (7.9 \times 10^{-6}) \approx 0.10$.

$[H^+] = 7.9 \times 10^{-6}$ mol/L

$pH = -\log [H^+] = -\log(7.9 \times 10^{-6}) = 5.10$

21.4 Let $x = [H^+]$ at equilibrium.

	$HClO(aq) + H_2O(l)$	⇌	H_3O^+	+ ClO^-
initial	0.050		0	0.100
change	$-x$		$+x$	$+x$
equilibrium	$0.050 - x$		x	$0.100 + x$

$$K_a = \frac{[H^+][ClO^-]}{[HClO]} = \frac{(x)(0.100 + x)}{(0.050 - x)} = 2.90 \times 10^{-8}$$

Assume that $0.100 + x \approx 0.100$ and $0.050 - x \approx 0.050$.

$(0.100)x = 1.5 \times 10^{-9}$

$x = 1.5 \times 10^{-8}$ mol/L

The approximations are valid because $0.100 - (1.5 \times 10^{-8}) \approx 0.100$ and $0.050 - (1.5 \times 10^{-8}) \approx 0.050$.

$[H^+] = 1.5 \times 10^{-8}$ mol/L

$pH = -\log [H^+] = -\log(1.5 \times 10^{-8}) = 7.82$

21.5 $\log [H^+] = -pH = -4.00$

$[H^+] = 1.00 \times 10^{-4}$ mol/L

Let $x = [F^-]$.

	$HF(aq) + H_2O(l)$	⇌	H_3O^+	+	F^-
initial	0.10		0		x
change	-1.00×10^{-4}		$+1.00 \times 10^{-4}$		$+1.00 \times 10^{-4}$
equilibrium	0.10		1.00×10^{-4}		$x + 1.00 \times 10^{-4}$

$$K_a = \frac{[H^+][F^-]}{[HF]} = \frac{(1.00 \times 10^{-4})(x + 1.00 \times 10^{-4})}{(0.10)} = 6.5 \times 10^{-4}$$

Assume that $x + 1.00 \times 10^{-4} \approx x$.

$$x = \frac{(0.10)(6.5 \times 10^{-4})}{(1.00 \times 10^{-4})} = 0.65 \text{ mol/L}$$

The approximation is valid because $0.65 + 0.0001 \approx 0.65$. Thus $[F^-] = 0.65$ mol/L.

21.6 The amount of base added is

$$\left[\frac{0.10 \text{ mol NaOH}}{1 \text{ L}}\right](10.0 \text{ mL})\left[\frac{1 \text{ L}}{1000 \text{ mL}}\right] = 0.0010 \text{ mol NaOH}$$

NH_4^+	+	OH^-	→	$NH_3(aq)$	+	$H_2O(l)$
0.010 mol		0.0010 mol		0.0010 mol		

$$[NH_4^+] = \frac{(0.010 \text{ mol} - 0.0010 \text{ mol})}{0.1100 \text{ L}} = 0.08 \text{ mol/L}$$

$$[NH_3] = \frac{(0.010 \text{ mol} + 0.001 \text{ mol})}{0.1100 \text{ L}} = 0.10 \text{ mol/L}$$

$$[OH^-] = \frac{[NH_3]}{[NH_4^+]} K_b = \frac{(0.10)(1.6 \times 10^{-5})}{(0.08)} = 2 \times 10^{-5} \text{ mol/L}$$

$pOH = -\log [OH^-] = -\log(2 \times 10^{-5}) = 4.7$

$pH = 14.0 - pOH = 14.0 - 4.7 = 9.3$

21.7 Because $H_2SO_4(aq)$ is a strong acid in the first ionization step, $[H_2SO_4] \approx 0$ and $[H^+] = [HSO_4^-] = 0.0100$ mol/L before the second ionization step occurs. For the second step, letting $x = [SO_4^{2-}]$ at equilibrium gives

	$HSO_4^- + H_2O(l)$	⇌	H_3O^+	+	SO_4^{2-}
initial	0.0100		0.0100		0
change	$-x$		$+x$		$+x$
equilibrium	$0.0100 - x$		$0.0100 + x$		x

$$K_{a_2} = \frac{[H^+][SO_4^{2-}]}{[HSO_4^-]} = \frac{(0.0100 + x)(x)}{(0.0100 - x)} = 1.0 \times 10^{-2}$$

$(0.0100)x + x^2 = 1.0 \times 10^{-4} - (0.010)x$

$x^2 + (0.020)x - 1.0 \times 10^{-4} = 0$

$$x = \frac{-(0.020) \pm \sqrt{(0.020)^2 - (4)(1)(-1.0 \times 10^{-4})}}{(2)(1)} = \frac{-(0.020) \pm (0.028)}{2}$$

$= 0.004$ mol/L

$[SO_4^{2-}] = 0.004$ mol/L

$[H^+] = 0.0100 + 0.004 = 0.014$ mol/L

$[HSO_4^-] = 0.0100 - 0.004 = 0.006$ mol/L

21.8 For the first ionization let $x = [H^+] = [HS^-]$ at equilibrium.

	$H_2S(aq) + H_2O(l)$ $\rightleftharpoons$	H_3O^+	+ HS^-
initial	0.010	0	0
change	$-x$	$+x$	$+x$
equilibrium	$0.010 - x$	x	x

$$K_{a_1} = \frac{[H^+][HS^-]}{[H_2S]} = \frac{(x)(x)}{(0.010 - x)} = 1.0 \times 10^{-7}$$

Assume that $0.010 - x \approx 0.010$.

$x^2 = (1.0 \times 10^{-7})(0.010) = 1.0 \times 10^{-9}$

$x = 3.2 \times 10^{-5}$ mol/L

The approximation is valid because $0.010 - 3.2 \times 10^{-5} \approx 0.010$.

$[H^+] = [HS^-] = 3.2 \times 10^{-5}$ mol/L

$[H_2S] = 0.010$ mol/L

For the second ionization let $x = [S^{2-}]$ at equilibrium.

	$HS^- + H_2O(l)$ $\rightleftharpoons$	H_3O^+	+ S^{2-}
initial	3.2×10^{-5}	3.2×10^{-5}	0
change	$-x$	$+x$	$+x$
equilibrium	$3.2 \times 10^{-5} - x$	$3.2 \times 10^{-5} + x$	x

$$K_{a_2} = \frac{[H^+][S^{2-}]}{[HS^-]} = \frac{(3.2 \times 10^{-5} + x)(x)}{(3.2 \times 10^{-5} - x)} = 3 \times 10^{-13}$$

Assume that $3.2 \times 10^{-5} + x \approx 3.2 \times 10^{-5}$ and $3.2 \times 10^{-5} - x \approx 3.2 \times 10^{-5}$.

$x = 3 \times 10^{-13}$ mol/L

The approximations are valid because $3.2 \times 10^{-5} \pm 3 \times 10^{-13} \approx 3.2 \times 10^{-5}$.

$[S^{2-}] = 3 \times 10^{-13}$ mol/L

21.9 $$\frac{[H^+]^2[S^{2-}]}{[H_2S]} = 3 \times 10^{-20}$$

$$\frac{(x)^2(1 \times 10^{-9})}{(0.01)} = 3 \times 10^{-20}$$

$x^2 = (3 \times 10^{-20})(0.01)/(1 \times 10^{-9}) = 3 \times 10^{-13}$

$x = 5 \times 10^{-7}$ mol/L

$[H^+] = 5 \times 10^{-7}$ mol/L

$pH = -\log [H^+] = -\log(5 \times 10^{-7}) = 6.3$

21.10 $HClO_4(aq) + NaOH(aq) \rightarrow NaClO_4(aq) + H_2O(l)$

$$(42.6 \text{ mL soln})\left[\frac{0.0972 \text{ mol NaOH}}{1 \text{ L soln}}\right]\left[\frac{1 \text{ L}}{1000 \text{ mL}}\right] = 0.00414 \text{ mol NaOH}$$

$$(0.00414 \text{ mol NaOH})\left[\frac{1 \text{ mol } HClO_4}{1 \text{ mol NaOH}}\right] = 0.00414 \text{ mol } HClO_4$$

$$\frac{(0.00414 \text{ mol } HClO_4)}{(50.0 \text{ mL soln})}\left[\frac{1000 \text{ mL}}{1 \text{ L}}\right] = 0.0828 \text{ mol } HClO_4/\text{L} = 0.0828 \text{ M } HClO_4$$

21.11 $NaOH(aq) + KHP(aq) \rightarrow KNaP(aq) + H_2O(l)$

$$(0.4963 \text{ g KHP})\left[\frac{1 \text{ mol KHP}}{204.23 \text{ g KHP}}\right] = 0.002430 \text{ mol KHP}$$

$$(0.002430 \text{ mol KHP})\left[\frac{1 \text{ mol NaOH}}{1 \text{ mol KHP}}\right] = 0.002430 \text{ mol NaOH}$$

$$\frac{(0.002430 \text{ mol NaOH})}{(19.61 \text{ mL soln})}\left[\frac{1000 \text{ mL}}{1 \text{ L}}\right] = 0.1239 \text{ mol NaOH/L} = 0.1239 \text{ M NaOH}$$

Solutions to Odd-Numbered Questions and Problems

21.1 The cation that undergoes no significant reaction with water upon dissolution is a weaker acid than water. Dissolution of these cations will have no effect on the pH of the solution.

21.3 The cations that will react with water to form H^+ are (b) NH_4^+, (c) $[Al(H_2O)_6]^{3+}$, and (d) $[Fe(H_2O)_6]^{3+}$. The chemical equations for the reactions are

(b) $NH_4^+ + H_2O(l) \rightleftharpoons NH_3(aq) + H_3O^+$

(c) $[Al(H_2O)_6]^{3+} + H_2O(l) \rightleftharpoons [Al(OH)(H_2O)_5]^{2+} + H_3O^+$

(d) $[Fe(H_2O)_6]^{3+} + H_2O(l) \rightleftharpoons [Fe(OH)(H_2O)_5]^{2+} + H_3O^+$

21.5 $NH_4^+ + H_2O(l) \rightleftharpoons NH_3(aq) + H_3O^+$

$$K_a = \frac{K_w}{K_b} = \frac{1.00 \times 10^{-14}}{1.6 \times 10^{-5}} = 6.3 \times 10^{-10}$$

21.7 An anion that undergoes no significant reaction with water upon dissolution is a weaker base than water. Dissolution of these anions will have no effect on the pH of the solution.

21.9 The anions that will react with water to form OH^- are (a) S^{2-}, (b) HS^-, (c) F^-, and (e) NO_2^-. The chemical equations for the hydrolysis reactions are

(a) $S^{2-} + H_2O(l) \rightleftarrows HS^- + OH^-$

(b) $HS^- + H_2O(l) \rightleftarrows H_2S(aq) + OH^-$

(c) $F^- + H_2O(l) \rightleftarrows HF(aq) + OH^-$

(e) $NO_2^- + H_2O(l) \rightleftarrows HNO_2(aq) + OH^-$.

21.11 $F^- + H_2O(l) \rightleftarrows HF(aq) + OH^-$

$$K_b = \frac{K_w}{K_a} = \frac{1.00 \times 10^{-14}}{6.5 \times 10^{-4}} = 1.5 \times 10^{-11}$$

21.13 The ions of the salt that do not react with water (hydrolyze) will not change the pH of the water. If one or both of the ions of the salt react with water, then the solution will no longer be neutral.

21.15 Aqueous solutions of the following salts that are

(a) acidic include (i) $(NH_4)HSO_4$, (ii) $(NH_4)_2SO_4$, and (v) $Al(NO_3)_3$

(b) alkaline include (iv) LiCN

(c) essentially neutral include (iii) KCl

21.17 Let $x = [OH^-]$ at equilibrium.

	$IO^- + H_2O(l)$	$\rightleftarrows$ $HIO(aq)$	$+ OH^-$
initial	0.10	0	0
change	$-x$	$+x$	$+x$
equilibrium	$0.10 - x$	x	x

$$K_b = \frac{[HIO][OH^-]}{[IO^-]} = \frac{(x)(x)}{(0.10 - x)} = 4.3 \times 10^{-4}$$

$$x^2 = 4.3 \times 10^{-5} - (4.3 \times 10^{-4})x$$

$$x^2 + (4.3 \times 10^{-4})x - 4.3 \times 10^{-5} = 0$$

$$x = \frac{-(4.3 \times 10^{-4}) \pm \sqrt{(4.3 \times 10^{-4})^2 - (4)(1)(-4.3 \times 10^{-5})}}{(2)(1)}$$

$$= \frac{-(4.3 \times 10^{-4}) \pm (1.3 \times 10^{-2})}{2} = 6.3 \times 10^{-3} \text{ mol/L}$$

$[OH^-] = 6.3 \times 10^{-3}$ mol/L

$pOH = -\log [OH^-] = -\log(6.3 \times 10^{-3}) = 2.20$

$pH = 14.00 - pOH = 14.00 - 2.20 = 11.80$

21.19 The equilibrium will be driven toward the reactants side of the equation by the additional formate ions, thus a smaller fraction of formic acid molecules will undergo ionization.

21.21 $CH_3CH_2COOH(aq) + H_2O(l) \rightleftharpoons CH_3CH_2COO^- + H_3O^+$

$$K_a = \frac{[CH_3CH_2COO^-][H_3O^+]}{[CH_3CH_2COOH]} = \frac{[CH_3CH_2COO^-](0.10)}{(0.010)} = 1.3 \times 10^{-5}$$

$[CH_3CH_2COO^-] = 1.3 \times 10^{-6}$ mol/L

21.23 The concentration of each acid is 0.010 mol/1.00 L = 0.010 mol/L. The common ion is H^+ from the HCl.

$$HClO(aq) \rightleftharpoons H^+ + ClO^-$$

	$HClO(aq) + H_2O(l)$	$\rightleftharpoons$	H^+	+	ClO^-
initial	0.010		0.010		0
change	$-x$		$+x$		$+x$
equilibrium	$0.010 - x$		$0.010 + x$		x

$$K_a = \frac{[H^+][ClO^-]}{[HClO]} = \frac{(0.010 + x)(x)}{(0.010 - x)} = 2.90 \times 10^{-8}$$

Assume that $(0.010 + x) \approx (0.010 - x) \approx 0.010$.

$$\frac{(0.010)(x)}{(0.010)} = 2.9 \times 10^{-8}$$

$x = [ClO^-] = 2.9 \times 10^{-8}$ mol/L

The approximations are valid, so the concentration of ClO^- is 2.9×10^{-8} mol/L.

21.25 Because $HClO_4$ is a strong acid, $[H^+] = 0.10$ mol/L giving

$pH = -\log [H^+] = -\log(0.10) = 1.00$

This solution is not a buffer solution because the acid-conjugate base pair is not that of a weak acid.

21.27 Let $x = [H^+]$ at equilibrium.

	$CH_3COOH(aq) + H_2O(l)$	$\rightleftharpoons$	CH_3COO^-	+	H_3O^+
initial	0.010		0.050		0
change	$-x$		$+x$		$+x$
equilibrium	$0.010 - x$		$0.050 + x$		x

$$K_a = \frac{[CH_3COO^-][H^+]}{[CH_3COOH]} = \frac{(0.050 + x)(x)}{(0.010 - x)} = 1.754 \times 10^{-5}$$

Assume that $0.050 - x \approx 0.050$ and $0.010 - x \approx 0.010$.

$x = (1.754 \times 10^{-5})(0.010)/(0.050) = 3.5 \times 10^{-6}$ mol/L

The approximations are valid because $0.050 - (3.5 \times 10^{-6}) \approx 0.050$ and $0.010 - (3.5 \times 10^{-6}) \approx 0.010$.

$[H^+] = 3.5 \times 10^{-6}$ mol/L

$pH = -\log [H^+] = -\log(3.5 \times 10^{-6}) = 5.46$

21.29 $NO_2^- + H_3O^+ \rightleftarrows HNO_2(aq) + H_2O(l)$

$$(1.00 \text{ L soln})\left[\frac{0.100 \text{ mol } NO_2^-}{1 \text{ L soln}}\right] = 0.100 \text{ mol } NO_2^-$$

$$(0.050 \text{ mol HCl})\left[\frac{1 \text{ mol } HNO_2}{1 \text{ mol HCl}}\right] = 0.050 \text{ mol } HNO_2 \text{ formed}$$

$$(0.050 \text{ mol HCl})\left[\frac{1 \text{ mol } NO_2^-}{1 \text{ mol HCl}}\right] = 0.050 \text{ mol } NO_2^- \text{ reacted}$$

0.100 mol - 0.050 mol = 0.050 mol NO_2^- remaining

$$[HNO_2] = \frac{0.050 \text{ mol } HNO_2}{1 \text{ L}} = 0.050 \text{ mol/L}$$

$$[NO_2^-] = \frac{0.050 \text{ mol } NO_2^-}{1 \text{ L}} = 0.050 \text{ mol/L}$$

Let $x = [H^+]$ at equilibrium.

	$HNO_2(aq) + H_2O(l)$	$\rightleftarrows$ H_3O^+	+ NO_2^-
initial	0.050	0	0.050
change	$-x$	$+x$	$+x$
equilibrium	$0.050 - x$	x	$0.050 + x$

$$K_a = \frac{[H^+][NO_2^-]}{[HNO_2]} = \frac{(x)(0.050 + x)}{(0.050 - x)} = 7.2 \times 10^{-4}$$

Assume that $0.050 + x \approx 0.050$ and $0.050 - x \approx 0.050$.

$x = (7.2 \times 10^{-4})(0.050)/(0.050) = 7.2 \times 10^{-4}$ mol/L

The approximations are valid because % error $= \dfrac{0.00072}{0.050} \times 100 = 1.4$ %.

$[H^+] = 7.2 \times 10^{-4}$ mol/L

$pH = -\log [H^+] = -\log(7.2 \times 10^{-4}) = 3.14$

For twofold dilution

$$[HNO_2] = \frac{0.050 \text{ mol } HNO_2}{2 \text{ L}} = 0.025 \text{ mol/L}$$

$$[NO_2^-] = \frac{0.050 \text{ mol } NO_2^-}{2 \text{ L}} = 0.025 \text{ mol/L}$$

Let x = $[H^+]$ at equilibrium.

	$HNO_2(aq) + H_3O(l)$	$\rightleftharpoons$ H_3O^+	+ NO_2^-
initial	0.025	0	0.025
change	$-x$	$+x$	$+x$
equilibrium	$0.025 - x$	x	$0.025 + x$

$$K_a = \frac{[H^+][NO_2^-]}{[HNO_2]} = \frac{(x)(0.025 + x)}{(0.025 - x)} = 7.2 \times 10^{-4}$$

Assume that $0.025 + x \approx 0.025$ and $0.025 - x \approx 0.025$.

$x = (7.2 \times 10^{-4})(0.025)/(0.025) = 7.2 \times 10^{-4}$ mol/L

The approximations are valid because % error = $\frac{0.00072}{0.025} \times 100 = 2.9$ %.

$[H^+] = 7.2 \times 10^{-4}$ mol/L

$pH = -\log [H^+] = -\log(7.2 \times 10^{-4}) = 3.14$

21.31 $pH = -\log [H^+] = 5.00$

$[H^+] = 1.0 \times 10^{-5}$ mol/L

Let x = $[C_6H_5COO^-]$ initially present.

	$C_6H_5COOH(aq) + H_2O(l)$	$\rightleftharpoons$ H_3O^+	+ $C_6H_5COO^-$
initial	0.010	0	x
change	-1.0×10^{-5}	$+1.0 \times 10^{-5}$	$+1.0 \times 10^{-5}$
equilibrium	0.010	1.0×10^{-5}	$x + 1.0 \times 10^{-5}$

$$K_a = \frac{[H^+][C_6H_5COO^-]}{[C_6H_5COOH]} = \frac{(1.0 \times 10^{-5})(x + 1.0 \times 10^{-5})}{(0.010)} = 6.6 \times 10^{-5}$$

Assume that $x + 1.0 \times 10^{-5} \approx x$.

$x = (6.6 \times 10^{-5})(0.010)/(1.0 \times 10^{-5}) = 0.066$ mol/L

The approximation is valid because $0.066 + 1.0 \times 10^{-5} \approx 0.066$.

$[C_6H_5COO^-] = 0.066$ mol/L

21.33 $pOH = 14.00 - pH = 14.00 - 8.80 = 5.20$

$\log [OH^-] = -pOH = -5.20$

$[OH^-] = 6.3 \times 10^{-6}$ mol/L

Let x = $[NH_4^+]$ initially present.

	$NH_3(aq) + H_2O(l)$	⇌	NH_4^+	+	OH^-
initial	0.050		x		0
change	-6.3×10^{-6}		$+6.3 \times 10^{-6}$		$+6.3 \times 10^{-6}$
equilibrium	0.050		$x + 6.3 \times 10^{-6}$		6.3×10^{-6}

$$K_b = \frac{[NH_4^+][OH^-]}{[NH_3]} = \frac{(x + 6.3 \times 10^{-6})(6.3 \times 10^{-6})}{(0.050)} = 1.6 \times 10^{-5}$$

Assume that $x + 6.3 \times 10^{-6} \approx x$.

$x = (1.6 \times 10^{-5})(0.050)/(6.3 \times 10^{-6}) = 0.13$ mol/L

The approximation is valid because $0.13 + 6.3 \times 10^{-6} \approx 0.13$.

$[NH_4^+] = 0.13$ mol/L

21.35 (a) Let $x = [H_3O^+]$.

	$CH_3COOH(aq) + H_2O(l)$	⇌	CH_3COO^-	+	H_3O^+
initial	0.500		0.500		
change	$-x$		$+x$		$+x$
equilibrium	$0.500 - x$		$0.500 + x$		x

$$K_a = \frac{[CH_3COO^-][H_3O^+]}{[CH_3COOH]} = \frac{(0.500 + x)(x)}{0.500 - x} = 1.754 \times 10^{-5}$$

Assume that $0.500 - x \approx 0.500 + x \approx 0.500$

$x = [H_3O^+] = 1.75 \times 10^{-5}$ mol/L

The approximation is valid, so

pH = $-\log [H_3O^+] = -\log(1.75 \times 10^{-5}) = 4.757$

(b) $\left[\frac{0.500 \text{ mol}}{1 \text{ L}}\right](1.000 \text{ L}) = 0.500$ mol CH_3COO^- present

$\left[\frac{0.500 \text{ mol}}{1 \text{ L}}\right](1.000 \text{ L}) = 0.500$ mol CH_3COOH present

$CH_3COO^- + H^+ \rightarrow CH_3COOH$

$(0.001 \text{ mol HCl})\left[\frac{1 \text{ mol } CH_3COOH}{1 \text{ mol HCl}}\right] = 0.001$ mol CH_3COOH formed

$(0.001 \text{ mol HCl})\left[\frac{1 \text{ mol } CH_3COO^-}{1 \text{ mol HCl}}\right] = 0.001$ mol CH_3COO^- reacted

(0.500 mol) - (0.001 mol) = 0.499 mol CH_3COO^- remaining

(0.500 mol) + (0.001 mol) = 0.501 mol CH_3COOH present

$[CH_3COO^-] = \frac{0.499 \text{ mol}}{1.000 \text{ L}} = 0.499$ mol/L

$[CH_3COOH] = \dfrac{0.501\ mol}{1.000\ L} = 0.501\ mol/L$

Let $x = [H_3O^+]$.

	$CH_3COOH + H_2O(l)$	$\rightleftharpoons$	CH_3COO^-	+	H_3O^+
initial	0.501		0.499		0
change	$-x$		$+x$		$+x$
equilibrium	$0.501 - x$		$0.499 + x$		x

$$K_a = \frac{[H_3O^+][CH_3COO^-]}{[CH_3COOH]} = \frac{(x)(0.499 + x)}{(0.501 - x)} \approx \frac{x(0.499)}{(0.501)} = 1.754 \times 10^{-5}$$

$x = 1.76 \times 10^{-5}$ mol/L = $[H^+]$

pH = $-\log(1.76 \times 10^{-5}) = 4.754$

(c) $CH_3COOH + OH^- \rightarrow CH_3COO^- + H_2O$

$$(0.001\ mol\ NaOH)\left[\frac{1\ mol\ CH_3COOH}{1\ mol\ NaOH}\right] = 0.001\ mol\ CH_3COOH\ reacted$$

$$(0.001\ mol\ NaOH)\left[\frac{1\ mol\ CH_3COO^-}{1\ mol\ NaOH}\right] = 0.001\ mol\ CH_3COO^-\ produced$$

(0.500 mol) - (0.001 mol) = 0.499 mol CH_3COOH remaining

(0.500 mol) + (0.001 mol) = 0.501 mol CH_3COO^- present

$[CH_3COOH] = \dfrac{0.499\ mol}{1.000\ L} = 0.499\ mol/L$

$[CH_3COO^-] = \dfrac{0.501\ mol}{1.000\ L} = 0.501\ mol/L$

Let $x - [H_3O^+]$.

	$CH_3COOH + H_2O(l)$	$\rightleftharpoons$	CH_3COO^-	+	H_3O^+
initial	0.499		0.501		0
change	$-x$		$+x$		$+x$
equilibrium	$0.499 - x$		$0.501 + x$		x

$$K_a = \frac{[H_3O^+][CH_3COO^-]}{[CH_3COOH]} = \frac{(x)(0.501 + x)}{(0.499 - x)} \approx \frac{x(0.501)}{(0.499)} = 1.754 \times 10^{-5}$$

$x = 1.75 \times 10^{-5}$ mol/L = $[H^+]$

pH = $-\log(1.75 \times 10^{-5}) = 4.757$

(d) $[CH_3COO^-] = \dfrac{0.500\ mol}{1.100\ L} = 0.455\ M$

$[CH_3COOH] = \frac{0.500 \text{ mol}}{1.100 \text{ L}} = 0.455 \text{ M}$

Let $x = [H_3O^+]$.

	$CH_3COOH + H_2O(l)$	⇌	CH_3COO^-	+	H_3O^+
initial	0.455		0.455		0
change	$-x$		$+x$		$+x$
equilibrium	$0.455 - x$		$0.455 + x$		x

$$K_a = \frac{[H_3O^+][CH_3COO^-]}{[CH_3COOH]} = \frac{(0.455 + x)(x)}{(0.455 - x)} \approx \frac{(0.455)x}{(0.455)} = 1.754 \times 10^{-5}$$

$x = 1.75 \times 10^{-5}$ mol/L = $[H^+]$

pH = $-\log(1.75 \times 10^{-5}) = 4.757$

21.37 For the original acid solution, letting $x = [H_3O^+]$

	$CH_3CHOHCOOH(aq) + H_2O(l)$	⇌	$CH_3CHOHCOO^-$	+	H_3O^+
initial	0.200		0		0
change	$-x$		$+x$		$+x$
equilibrium	$0.200 - x$		x		x

$$K_a = \frac{[CH_3CHOHCOO^-][H_3O^+]}{[CH_3CHOHCOOH]} = \frac{(x)(x)}{0.200 - x} = 1.36 \times 10^{-4}$$

Assume $0.200 - x \approx 0.200$. The approximate solution is

$x^2 = (1.36 \times 10^{-4})(0.200) = 2.72 \times 10^{-5}$

$x = 5.22 \times 10^{-3}$ mol/L

The approximation is valid, so the pH of the original acid solution is

pH = $-\log [H_3O^+] = -\log(5.22 \times 10^{-3}) = 2.282$

For the original buffer solution, letting $x = [H_3O^+]$

	$CH_3CHOHCOOH(aq) + H_2O(l)$	⇌	$CH_3CHOHCOO^-$	+	H_3O^+
initial	0.200		0.200		0
change	$-x$		$+x$		$+x$
equilibrium	$0.200 - x$		$0.200 + x$		x

$$K_a = \frac{(0.200 + x)(x)}{(0.200 - x)} = 1.36 \times 10^{-4}$$

Assuming $(0.200 + x) \approx (0.200 - x) \approx 0.0100$, the approximate solution for x is

$$\frac{(0.200)(x)}{(0.200)} = 1.36 \times 10^{-4}$$

$x = 1.36 \times 10^{-4}$ mol/L

The approximation is valid giving the pH of the original buffer solution as

pH = $-\log(1.36 \times 10^{-4}) = 3.866$

After the addition of 0.005 mol OH^- to the acid solution, the equilibrium table letting $x = [H_3O^+]$ is

	$CH_3CHOHCOOH(aq) + H_2O(l)$	$\rightleftharpoons$ $CH_3CHOHCOO^-$	+ H_3O^+
initial	0.195	0.005	0
change	$-x$	$+x$	$+x$
equilibrium	$0.195 - x$	$0.005 + x$	x

$$K_a = \frac{(0.005 + x)(x)}{(0.195 - x)} = 1.36 \times 10^{-4}$$

Assuming $0.005 + x \approx 0.005$ and $0.195 - x \approx 0.195$, the solution for x is

$$x = \frac{(1.36 \times 10^{-4})(0.195)}{0.005} = 5 \times 10^{-3} \text{ mol/L}$$

The approximation is not valid, so

$(0.005 + x)(x) = (1.36 \times 10^{-4})(0.195 - x)$

$(0.005)x + x^2 = 2.65 \times 10^{-5} - (1.36 \times 10^{-4})x$

$x^2 + (0.005)x - 2.65 \times 10^{-5} = 0$

$$x = \frac{-(0.005) \pm \sqrt{(0.005)^2 - (4)(1)(-2.65 \times 10^{-5})}}{2}$$

$= -8 \times 10^{-3}$ or 3×10^{-3} mol/L

Choosing the positive value,

pH = $-\log(3 \times 10^{-3}) = 2.5$

After the addition of 0.005 mol OH^- to the buffer solution, the equilibrium table letting $x = [H_3O^+]$ is

	$CH_3CHOHCOOH(aq) + H_2O(l)$	$\rightleftharpoons$ $CH_3CHOHCOO^-$	+ H_3O^+
initial	0.195	0.205	0
change	$-x$	$+x$	$+x$
equilibrium	$0.195 - x$	$0.205 + x$	x

$$K_a = \frac{(0.205 + x)(x)}{(0.195 - x)} = 1.36 \times 10^{-4}$$

Assuming $(0.195 - x) \approx 0.195$ and $0.205 + x \approx 0.205$, the solution for x is

$$x = \frac{(1.36 \times 10^{-4})(0.195)}{(0.205)} = 1.29 \times 10^{-4}$$

and the corresponding pH is

pH = $-\log(1.29 \times 10^{-4}) = 3.889$

21.39 $H_2SO_4(aq) + H_2O(l) \rightarrow HSO_4^- + H_3O^+$

$HSO_4^- + H_2O(l) \rightleftharpoons SO_4^{2-} + H_3O^+$

The HSO_4^- ion will react with water to form H_3O^+, thus lowering the pH, and the SO_4^{2-} ion will react with water to form OH^-, thus raising the pH.

21.41 Let x = $[H^+]$ at equilibrium.

	$H_2SO_3(aq) + H_2O(l)$	$\rightleftharpoons$ H_3O^+	+ HSO_3^-
initial	0.100	0	0
change	$-x$	$+x$	$+x$
equilibrium	$0.100 - x$	x	x

$$K_{a1} = \frac{[H^+][HSO_3^-]}{[H_2SO_3]} = \frac{(x)(x)}{(0.100 - x)} = 1.43 \times 10^{-2}$$

$x^2 = 1.43 \times 10^{-3} - (1.43 \times 10^{-2})x$

$x^2 + (1.43 \times 10^{-2})x - 1.43 \times 10^{-3} = 0$

$$x = \frac{-(1.43 \times 10^{-2}) \pm \sqrt{(1.43 \times 10^{-2})^2 - (4)(1)(-1.43 \times 10^{-3})}}{(2)(1)}$$

$$= \frac{-(1.43 \times 10^{-2}) \pm (7.69 \times 10^{-2})}{2} = 0.0313 \text{ mol/L}$$

$[H_2SO_3] = 0.100 - 0.0313 = 0.069$ mol/L

Let y = $[SO_3^{2-}]$ at equilibrium.

	$HSO_3^- + H_2O(l)$	$\rightleftharpoons$ H_3O^+	+ SO_3^{2-}
initial	0.0313	0.0313	0
change	$-y$	$+y$	$+y$
equilibrium	$0.0313 - y$	$0.0313 + y$	y

$$K_{a2} = \frac{[H^+][SO_3^{2-}]}{[HSO_3^-]} = \frac{(0.0313 + y)(y)}{(0.0313 - y)} = 5.0 \times 10^{-8}$$

Assume that $0.0313 \pm y \approx 0.0313$.

$y = (5.0 \times 10^{-8})(0.0313)/(0.0313) = 5.0 \times 10^{-8}$ mol/L

The approximations are valid because $0.0313 \pm 5.0 \times 10^{-8} \approx 0.0313$.

$[SO_3^{2-}] = 5.0 \times 10^{-8}$ mol/L

$[H^+] = 0.0313 + 5.0 \times 10^{-8} = 0.0313$ mol/L

$[HSO_3^-] = 0.0313 - 5.0 \times 10^{-8} = 0.0313$ mol/L

21.43 Let $x = [HC_2O_4^-]$ after first ionization.

	$H_2C_2O_4(aq) + H_2O(l)$	$\rightleftharpoons$	H_3O^+	$+ HC_2O_4^-$
initial	0.25		0	0
change	$-x$		$+x$	$+x$
equilibrium	$0.25 - x$		x	x

$$K_{a1} = \frac{[H^+][HC_2O_4^-]}{[H_2C_2O_4]} = \frac{(x)(x)}{(0.25 - x)} = 5.60 \times 10^{-2}$$

$x^2 = 1.4 \times 10^{-2} - (5.6 \times 10^{-2})x$

$x^2 + (5.6 \times 10^{-2})x - 1.4 \times 10^{-2} = 0$

$$x = \frac{-(0.056) \pm \sqrt{(0.056)^2 - (4)(1)(-0.014)}}{(2)(1)} = \frac{-(0.056) \pm (0.24)}{2}$$

$= 0.092$ mol/L

Let $y = [C_2O_4^{2-}]$ at equilibrium.

	$HC_2O_4^-(aq) + H_2O(l)$	$\rightleftharpoons$	H_3O^+	$+ C_2O_4^{2-}$
initial	0.092		0.092	0
change	$-y$		$+y$	$+y$
equilibrium	$0.092 - y$		$0.092 + y$	y

$$K_{a2} = \frac{[H^+][C_2O_4^{2-}]}{[HC_2O_4^-]} = \frac{(0.092 + y)(y)}{(0.092 - y)} = 6.2 \times 10^{-5}$$

Assume that $0.092 \pm y \approx 0.092$.

$y = (6.2 \times 10^{-5})(0.092)/(0.092) = 6.2 \times 10^{-5}$

The approximations are valid because $0.092 \pm 6.2 \times 10^{-5} \approx 0.092$.

$[H^+] = 0.092 + 6.2 \times 10^{-5} \approx 0.092$ mol/L

$pH = -\log [H^+] = -\log(9.2 \times 10^{-2}) = 1.04$

21.45 $H^+ + OH^- \rightleftharpoons H_2O(l)$

The net ionic equation for all these reactions is the same and so the enthalpy change is the same.

21.47 $CH_3COOH(aq) + NaOH(aq) \rightarrow NaCH_3COO(aq) + H_2O(l)$

$$(25.00 \text{ mL acid solution})\left[\frac{1 \text{ L}}{1000 \text{ mL}}\right]\left[\frac{0.263 \text{ mol } CH_3COOH}{1 \text{ L acid solution}}\right]\left[\frac{1 \text{ mol NaOH}}{1 \text{ mol } CH_3COOH}\right]$$

$$\times \left[\frac{1 \text{ L base solution}}{0.106 \text{ mol NaOH}}\right] = 0.0620 \text{ L base solution} = 62.0 \text{ mL base solution}$$

21.49 $Na_2CO_3(s) + 2HCl(aq) \rightarrow 2NaCl(aq) + H_2O(l) + CO_2(g)$

$$(2.1734\text{ g }Na_2CO_3)\left[\frac{1\text{ mol }Na_2CO_3}{105.99\text{ g }Na_2CO_3}\right] = 0.020506\text{ mol }Na_2CO_3$$

$$(0.020506\text{ mol }Na_2CO_3)\left[\frac{2\text{ mol HCl}}{1\text{ mol }Na_2CO_3}\right] = 0.041012\text{ mol HCl}$$

$$\left[\frac{0.041012\text{ mol HCl}}{23.69\text{ mL}}\right]\left[\frac{1000\text{ mL}}{1\text{ L}}\right] = 1.731\text{ mol HCl/L}$$

21.51 $Na_2CO_3(s) + 2HCl(aq) \rightarrow 2NaCl(aq) + H_2O(l) + CO_2(g)$

$$(13.72\text{ mL soln})\left[\frac{1\text{ L}}{1000\text{ mL}}\right]\left[\frac{0.107\text{ mol HCl}}{1\text{ L soln}}\right] = 0.00147\text{ mol HCl}$$

$$(0.00147\text{ mol HCl})\left[\frac{1\text{ mol }Na_2CO_3}{2\text{ mol HCl}}\right] = 0.000735\text{ mol }Na_2CO_3$$

$$(0.000735\text{ mol }Na_2CO_3)\left[\frac{105.99\text{ g }Na_2CO_3}{1\text{ mol }Na_2CO_3}\right] = 0.0779\text{ g }Na_2CO_3$$

$$\frac{0.0779\text{ g }Na_2CO_3}{0.253\text{ g sample}} \times 100 = 30.8\ \%$$

21.53

$$(67.43\text{ mL soln})\left[\frac{1\text{ L}}{1000\text{ mL}}\right]\left[\frac{0.103\text{ mol NaOH}}{1\text{ L soln}}\right] = 0.00695\text{ mol NaOH}$$

$$(0.00695\text{ mol NaOH})\left[\frac{1\text{ mol }CH_3COOH}{1\text{ mol NaOH}}\right] = 0.00695\text{ mol }CH_3COOH$$

$$(0.00695\text{ mol }CH_3COOH)\left[\frac{60.06\text{ g }CH_3COOH}{1\text{ mol }CH_3COOH}\right] = 0.417\text{ g }CH_3COOH$$

$$\frac{0.417\text{ g }CH_3COOH}{10.13\text{ g sample}} \times 100 = 4.12\ \%$$

Yes, the vinegar met the minimum specifications of 4 mass % acetic acid.

21.55 $CH_3COOH(aq) + NaOH(aq) \rightarrow Na^+ + CH_3COO^- + H_2O(l)$

$$n_{CH_3COOH} = 0.125\text{ mol }CH_3COOH - (0.100\text{ mol NaOH})\left[\frac{1\text{ mol }CH_3COOH}{1\text{ mol NaOH}}\right]$$

$$= 0.025\text{ mol }CH_3COOH\text{ remaining}$$

$$n_{CH_3COO^-} = (0.100\text{ mol NaOH})\left[\frac{1\text{ mol }CH_3COO^-}{1\text{ mol NaOH}}\right] = 0.100\text{ mol }CH_3COO^-\text{ formed}$$

Let $x = [H_3O^+]$.

Fig. 21-1

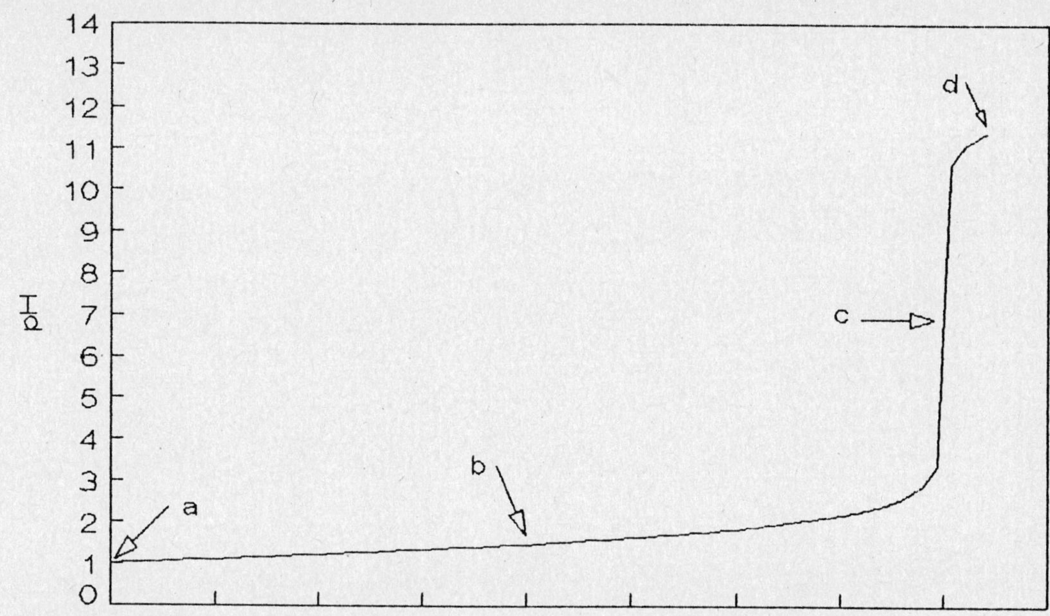

	$CH_3COOH(aq)$ + $H_2O(l)$ ⇌	CH_3COO^-	+ H_3O^+
initial	0.025	0.100	0
change	$-x$	$+x$	$+x$
equilibrium	$0.025 - x$	$0.100 + x$	x

$$K_a = \frac{[CH_3COO^-][H_3O^+]}{[CH_3COOH]} = \frac{(0.100 + x)(x)}{0.025 - x} = 1.754 \times 10^{-5}$$

Assuming $0.100 + x \approx 0.100$ and $0.025 - x \approx 0.025$, the approximate solution for x is

$$x = \frac{(0.025)(1.754 \times 10^{-5})}{(0.100)} = 4.4 \times 10^{-6} \text{ mol/L} = [H_3O^+]$$

The approximation is valid, so

$pH = -\log [H_3O^+] = -\log(4.4 \times 10^{-6}) = 5.36$

21.57 See Fig. 21-1 for the sketch. That which determines the pH of the solution at point

(a) is the concentration of the acid.

Fig. 21-2

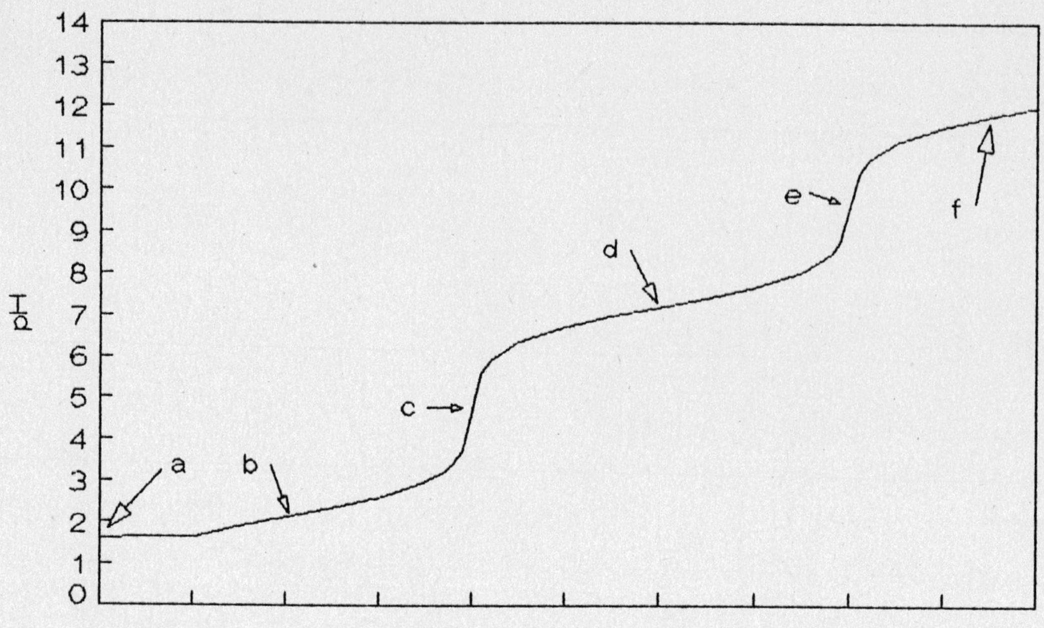

(b) is the concentration of the unreacted acid.

(c) is water itself.

(d) is the concentration of excess base.

21.59 See Fig. 21-2 for the sketch. That which determines the pH of the solution at point

(a) is the concentration and K_{a_1} of the acid.

(b) is K_{a_1}, the concentration of acid, and the buffer effect.

(c) is $(pK_{a_1} + pK_{a_2})/2$.

(d) is K_{a_2}, the concentration of anion, and the buffer effect.

(e) is the hydrolysis of the anion.

(f) is the concentration of excess base.

21.61 $HNO_3(aq) + NaOH(aq) \rightarrow NaNO_3(aq) + H_2O(l)$

$$(25.0 \text{ mL soln})\left[\frac{1 \text{ L}}{1000 \text{ mL}}\right]\left[\frac{0.250 \text{ mol } HNO_3}{1 \text{ L soln}}\right] = 0.00625 \text{ mol } HNO_3$$

(a) $[H^+] = 0.250$ mol/L

$pH = -\log [H^+] = -\log(0.250) = 0.602$

(b) $(10.0\ \text{mL soln})\left[\frac{1\ \text{L}}{1000\ \text{mL}}\right]\left[\frac{0.100\ \text{mol NaOH}}{1\ \text{L soln}}\right] = 0.00100\ \text{mol NaOH}$

$(0.00100\ \text{mol NaOH})\left[\frac{1\ \text{mol}\ HNO_3}{1\ \text{mol NaOH}}\right] = 0.00100\ \text{mol}\ HNO_3$ reacted

0.00625 mol - 0.00100 mol = 0.00525 mol HNO_3 remaining

$[H^+] = \left[\frac{0.00525\ \text{mol}}{35.0\ \text{mL}}\right]\left[\frac{1000\ \text{mL}}{1\ \text{L}}\right] = 0.150\ \text{mol/L}$

pH = -log(0.150) = 0.824

(c) $(25.0\ \text{mL soln})\left[\frac{1\ \text{L}}{1000\ \text{mL}}\right]\left[\frac{0.100\ \text{mol NaOH}}{1\ \text{L soln}}\right] = 0.00250\ \text{mol NaOH}$

$(0.00250\ \text{mol NaOH})\left[\frac{1\ \text{mol}\ HNO_3}{1\ \text{mol NaOH}}\right] = 0.00250\ \text{mol}\ HNO_3$ reacted

0.00625 mol - 0.00250 mol = 0.00375 mol HNO_3 remaining

$[H^+] = \left[\frac{0.00375\ \text{mol}\ HNO_3}{50.0\ \text{mL}}\right]\left[\frac{1000\ \text{mL}}{1\ \text{L}}\right] = 0.0750\ \text{mol/L}$

pH = -log(0.0750) = 1.125

(d) $(50.0\ \text{mL soln})\left[\frac{1\ \text{L}}{1000\ \text{mL}}\right]\left[\frac{0.100\ \text{mol NaOH}}{1\ \text{L soln}}\right] = 0.00500\ \text{mol NaOH}$

$(0.00500\ \text{mol NaOH})\left[\frac{1\ \text{mol}\ HNO_3}{1\ \text{mol NaOH}}\right] = 0.00500\ \text{mol}\ HNO_3$ reacted

0.00625 mol - 0.00500 mol = 0.00125 mol HNO_3 remaining

$[H^+] = \left[\frac{0.00125\ \text{mol}}{75.0\ \text{mL}}\right]\left[\frac{1000\ \text{mL}}{1\ \text{L}}\right] = 0.0167\ \text{mol/L}$

pH = -log(0.0167) = 1.777

(e) $(62.5\ \text{mL soln})\left[\frac{1\ \text{L}}{1000\ \text{mL}}\right]\left[\frac{0.100\ \text{mol NaOH}}{1\ \text{L soln}}\right] = 0.00625\ \text{mol NaOH}$

$(0.00625\ \text{mol NaOH})\left[\frac{1\ \text{mol}\ HNO_3}{1\ \text{mol NaOH}}\right] = 0.00625\ \text{mol}\ HNO_3$ reacted

0.00625 mol - 0.00625 mol = 0.00000 mol HNO_3 remaining

$NaNO_3$ does not hydrolyze, so the solution is neutral.

pH = 7.000

(f) $(75.0\ \text{mL soln})\left[\frac{1\ \text{L}}{1000\ \text{mL}}\right]\left[\frac{0.100\ \text{mol NaOH}}{1\ \text{L soln}}\right] = 0.00750\ \text{mol NaOH}$

$$(0.00625 \text{ mol } HNO_3)\left[\frac{1 \text{ mol NaOH}}{1 \text{ mol } HNO_3}\right] = 0.00625 \text{ mol NaOH reacted}$$

0.00750 mol - 0.00625 mol = 0.00125 mol NaOH remaining

$$[OH^-] = \left[\frac{0.00125 \text{ mol}}{100.0 \text{ mL}}\right]\left[\frac{1000 \text{ mL}}{1 \text{ L}}\right] = 0.0125 \text{ mol/L}$$

pOH = -log $[OH^-]$ = -log(0.0125) = 1.903

pH = 14.000 - pOH = 14.000 - 1.903 = 12.097

21.63 $HCOOH(aq) + NaOH(aq) \rightleftharpoons Na(HCOO)(aq) + H_2O(l)$

$$(25.00 \text{ mL soln})\left[\frac{1 \text{ L}}{1000 \text{ mL}}\right]\left[\frac{0.250 \text{ mol HCOOH}}{1 \text{ L soln}}\right] = 0.00625 \text{ mol HCOOH}$$

(a) Let $x = [H^+]$ at equilibrium.

	$HCOOH(aq) + H_2O(l)$ $\rightleftharpoons$	H_3O^+	+ $HCOO^-$
initial	0.250	0	0
change	$-x$	$+x$	$+x$
equilibrium	$0.250 - x$	x	x

$$K_a = \frac{[H^+][HCOO^-]}{[HCOOH]} = \frac{(x)(x)}{(0.250 - x)} = 1.772 \times 10^{-4}$$

Assume that $0.250 - x \approx 0.250$.

$x^2 = (1.772 \times 10^{-4})(0.250) = 4.43 \times 10^{-5}$

$x = 6.66 \times 10^{-3}$ mol/L

The approximation is valid because % error = $\dfrac{0.00666}{0.250} \times 100 = 2.66$ %.

$[H^+] = 6.66 \times 10^{-3}$ mol/L

pH = -log $[H^+]$ = -log(6.66×10^{-3}) = 2.177

(b) $$(10.0 \text{ mL soln})\left[\frac{1 \text{ L}}{1000 \text{ mL}}\right]\left[\frac{0.100 \text{ mol NaOH}}{1 \text{ L soln}}\right] = 0.00100 \text{ mol NaOH}$$

$$(0.00100 \text{ mol NaOH})\left[\frac{1 \text{ mol HCOOH}}{1 \text{ mol NaOH}}\right] = 0.00100 \text{ mol HCOOH reacted}$$

0.00625 mol - 0.00100 mol = 0.00525 mol HCOOH remaining

$$[HCOOH] = \left[\frac{0.00525 \text{ mol HCOOH}}{35.00 \text{ mL}}\right]\left[\frac{1000 \text{ mL}}{1 \text{ L}}\right] = 0.150 \text{ mol/L}$$

$$(0.00100 \text{ mol NaOH})\left[\frac{1 \text{ mol } HCOO^-}{1 \text{ mol NaOH}}\right] = 0.00100 \text{ mol } HCOO^- \text{ formed}$$

$$[HCOO^-] = \left[\frac{0.00100 \text{ mol } HCOO^-}{35.00 \text{ mL}}\right]\left[\frac{1000 \text{ mL}}{1 \text{ L}}\right] = 0.0286 \text{ mol/L}$$

Let x = [H^+] at equilibrium.

	HCOOH(aq) + H_2O(l)	⇄ H_3O^+	+ $HCOO^-$
initial	0.150	0	0.0286
change	$-x$	$+x$	$+x$
equilibrium	0.150 - x	x	0.0286 + x

$$K_a = \frac{[H^+][HCOO^-]}{[HCOOH]} = \frac{(x)(0.0286 + x)}{(0.150 - x)} = 1.772 \times 10^{-4}$$

Assume that $0.0286 + x \approx 0.0286$ and $0.150 - x \approx 0.150$.

$x = (1.772 \times 10^{-4})(0.150)/(0.0286) = 9.29 \times 10^{-4}$ mol/L

The approximations are valid because % error = $\frac{0.000929}{0.0286} \times 100 =$ 3.25 % and $\frac{0.000929}{0.150} \times 100 = 0.619$ %.

[H^+] = 9.29×10^{-4} mol/L

pH = $-\log(9.29 \times 10^{-4}) = 3.032$

(c) $$(25.00 \text{ mL soln})\left[\frac{1 \text{ L}}{1000 \text{ mL}}\right]\left[\frac{0.100 \text{ mol NaOH}}{1 \text{ L soln}}\right] = 0.00250 \text{ mol NaOH}$$

$$(0.00250 \text{ mol NaOH})\left[\frac{1 \text{ mol HCOOH}}{1 \text{ mol NaOH}}\right] = 0.00250 \text{ mol HCOOH reacted}$$

0.00625 mol - 0.00250 mol = 0.00375 mol HCOOH remaining

$$[HCOOH] = \left[\frac{0.00375 \text{ mol}}{50.0 \text{ mL}}\right]\left[\frac{1000 \text{ mL}}{1 \text{ L}}\right] = 0.0750 \text{ mol/L}$$

$$(0.00250 \text{ mol NaOH})\left[\frac{1 \text{ mol } HCOO^-}{1 \text{ mol NaOH}}\right] = 0.00250 \text{ mol } HCOO^-$$

$$[HCOO^-] = \left[\frac{0.00250 \text{ mol } HCOO^-}{50.0 \text{ mL}}\right]\left[\frac{1000 \text{ mL}}{1 \text{ L}}\right] = 0.0500 \text{ mol/L formed}$$

Let x = [H^+] at equilibrium.

	HCOOH(aq) + H_2O(l)	⇄ H_3O^+	+ $HCOO^-$
initial	0.0750	0	0.0500
change	$-x$	$+x$	$+x$
equilibrium	0.0750 - x	x	0.0500 + x

$$K_a = \frac{[H^+][HCOO^-]}{[HCOOH]} = \frac{(x)(0.0500 + x)}{(0.0750 - x)} = 1.772 \times 10^{-4}$$

Assume that $0.0500 \pm x \approx 0.0500$ and $0.0750 - x \approx 0.0750$.

$x = (1.772 \times 10^{-4})(0.0750)/(0.0500) = 2.66 \times 10^{-4}$ mol/L

The approximations are valid because % error = $\frac{0.000266}{0.0500} \times 100 =$ 0.532 % and $\frac{0.000266}{0.0750} \times 100 = 0.355$ %.

$[H^+] = 2.66 \times 10^{-4}$ mol/L

pH = $-\log(2.66 \times 10^{-4}) = 3.575$

(d) $(50.0 \text{ mL soln})\left[\frac{1 \text{ L}}{1000 \text{ mL}}\right]\left[\frac{0.100 \text{ mol NaOH}}{1 \text{ L soln}}\right] = 0.00500$ mol NaOH

$(0.00500 \text{ mol NaOH})\left[\frac{1 \text{ mol HCOOH}}{1 \text{ mol NaOH}}\right] = 0.00500$ mol HCOOH reacted

0.00625 mol - 0.00500 mol = 0.00125 mol HCOOH remaining

$[HCOOH] = \left[\frac{0.00125 \text{ mol HCOOH}}{75.0 \text{ mL}}\right]\left[\frac{1000 \text{ mL}}{1 \text{ L}}\right] = 0.0167$ mol/L

$(0.00500 \text{ mol NaOH})\left[\frac{1 \text{ mol } HCOO^-}{1 \text{ mol NaOH}}\right] = 0.00500$ mol $HCOO^-$

$[HCOO^-] = \left[\frac{0.00500 \text{ mol } HCOO^-}{75.0 \text{ mL}}\right]\left[\frac{1000 \text{ mL}}{1 \text{ L}}\right] = 0.0667$ mol/L

Let $x = [H^+]$ at equilibrium.

	$HCOOH(aq) + H_2O(l)$	$\rightleftharpoons$ H_3O^+	+ $HCOO^-$
initial	0.0167	0	0.0667
change	$-x$	$+x$	$+x$
equilibrium	$0.0167 - x$	x	$0.0667 + x$

$$K_a = \frac{[H^+][HCOO^-]}{[HCOOH]} = \frac{(x)(0.0667 + x)}{(0.0167 - x)} = 1.772 \times 10^{-4}$$

Assume that $0.0667 + x \approx 0.0667$ and $0.0167 - x \approx 0.0167$.

$x = (1.772 \times 10^{-4})(0.0167)/(0.0667) = 4.44 \times 10^{-5}$ mol/L

The assumptions are valid because $0.0667 + 0.0000444 \approx 0.0667$ and $0.0167 - 0.0000444 \approx 0.0167$.

$[H^+] = 4.44 \times 10^{-5}$ mol/L

pH = $-\log(4.44 \times 10^{-5}) = 4.353$

(e) $(62.5 \text{ mL soln})\left[\frac{1 \text{ L}}{1000 \text{ mL}}\right]\left[\frac{0.100 \text{ mol NaOH}}{1 \text{ L soln}}\right] = 0.00625$ mol NaOH

$(0.00625 \text{ mol NaOH})\left[\frac{1 \text{ mol HCOOH}}{1 \text{ mol NaOH}}\right] = 0.00625$ mol HCOOH reacted

0.00625 mol - 0.00625 mol = 0.00000 mol HCOOH remaining

$$(0.00625 \text{ mol NaOH})\left[\frac{1 \text{ mol } HCOO^-}{1 \text{ mol NaOH}}\right] = 0.00625 \text{ mol } HCOO^- \text{ formed}$$

$$[HCOO^-] = \left[\frac{0.00625 \text{ mol } HCOO^-}{87.5 \text{ mL}}\right]\left[\frac{1000 \text{ mL}}{1 \text{ L}}\right] = 0.0714 \text{ mol/L}$$

Let $x = [OH^-]$ at equilibrium.

	$HCOO^-$ + $H_2O(aq)$	⇄ $HCOOH(aq)$	+ OH^-
initial	0.0714	0	0
change	$-x$	$+x$	$+x$
equilibrium	$0.0714 - x$	x	x

$$K_b = \frac{[HCOOH][OH^-]}{[HCOO^-]} = \frac{(x)(x)}{(0.0714 - x)} = \frac{K_w}{K_a} = \frac{1.000 \times 10^{-14}}{1.772 \times 10^{-4}} = 5.643 \times 10^{-11}$$

Assume that $0.0714 - x \approx 0.0714$.

$x^2 = (5.643 \times 10^{-11})(0.0714) = 4.03 \times 10^{-12}$

$x = 2.01 \times 10^{-6}$ mol/L

The approximation is valid because $0.0714 - 2.01 \times 10^{-6} \approx 0.0714$.

$[OH^-] = 2.01 \times 10^{-6}$ mol/L

$pOH = -\log [OH^-] = -\log(2.01 \times 10^{-6}) = 5.697$

$pH = 14.000 - pOH = 14.000 - 5.697 = 8.303$

(f) $$(75.0 \text{ mol soln})\left[\frac{1 \text{ L}}{1000 \text{ mL}}\right]\left[\frac{0.100 \text{ mol NaOH}}{1 \text{ L soln}}\right] = 0.00750 \text{ mol NaOH}$$

$$(0.00625 \text{ mol HCOOH})\left[\frac{1 \text{ mol NaOH}}{1 \text{ mol HCOOH}}\right] = 0.00625 \text{ mol NaOH reacted}$$

0.00725 mol - 0.00625 mol = 0.00125 mol NaOH remaining

$$[OH^-] = \left[\frac{0.00125 \text{ mol}}{100.0 \text{ mL}}\right]\left[\frac{1000 \text{ mL}}{1 \text{ L}}\right] = 0.0125 \text{ mol/L}$$

$pOH = -\log(0.0125) = 1.903$

$pH = 14.000 - 1.903 = 12.097$

21.65 For a mixture of methyl red with thymolphthalein, a solution would be red below a pH of 4.4, yellow from 6.2 to 9.3, and green (yellow + blue) over 10.3. Thus the color observed at a pH value of (a) 3 would be red, (b) 7 would be yellow, and (c) 11 would be green.

21.67 The approximate concentration of acetate ion at equilibrium will be (0.01 mol/L)/2 = 0.005 mol/L.

Let x = [OH⁻].

	$CH_3COO^- + H_2O(l)$	⇌ CH_3COOH	$+ OH^-$
initial	0.005	0	0
change	$-x$	$+x$	$+x$
equilibrium	$0.005 - x$	x	x

$$K_b = \frac{[CH_3COOH][OH^-]}{[CH_3COO^-]} = \frac{(x)(x)}{(0.005 - x)} \approx \frac{x^2}{0.005} = 5.701 \times 10^{-10}$$

$x = 2 \times 10^{-6}$ mol/L

pOH = $-\log(2 \times 10^{-6}) = 5.7$

pH = 14.0 - pOH = 14.0 - 5.7 = 8.3

Table 2.15 shows that *m*-cresol purple or thymol blue would be a satisfactory indicator for this titration.

21.69 K_b for the hydrolysis reaction involving the anion of barbital is

$$K_b = \frac{K_w}{K_a} = \frac{1.00 \times 10^{-14}}{3.7 \times 10^{-8}} = 2.7 \times 10^{-7}$$

Setting up the usual table with x = [OH⁻]

	$C_8H_{11}N_2O_3^- + H_2O(l)$	⇌ $C_8H_{12}N_2O_3$	$+ OH^-$
initial	0.001	0	0
change	$-x$	$+x$	$+x$
equilibrium	$0.001 - x$	x	x

$$K_b = \frac{[C_8H_{12}N_2O_3][OH^-]}{[C_8H_{11}N_2O_3^-]} = \frac{(x)(x)}{(0.001 - x)} \approx \frac{x^2}{0.001} = 2.7 \times 10^{-7}$$

$x = [OH^-] = 2 \times 10^{-5}$ mol/L

The approximation $(0.001) - x \approx 0.001$ is valid, so

pOH = $-\log [OH^-] = 4.7$

pH = 14.00 - pOH = 9.3

The equivalence point should be near pH 9.3. Table 21.5 shows that a suitable indicator would be either thymol blue or phenolphthalein.

CHAPTER 22

IONS AND IONIC EQUILIBRIA: COMPLEX IONS AND IONIC SOLIDS

Solutions to Exercises

22.1 $[Cu(NH_3)_4]^{2+} \rightleftharpoons Cu^{2+} + 4NH_3(aq)$ $\quad K_d = \dfrac{[Cu^{2+}][NH_3]^4}{[Cu(NH_3)_4{}^{2+}]}$

$$[Cu^{2+}] = \frac{K_d[Cu(NH_3)_4{}^{2+}]}{[NH_3]^4} = \frac{(1 \times 10^{-13})(0.001)}{(6)^4} = 8 \times 10^{-20}\ mol/L$$

22.2 Let x = molar solubility of $Mg(OH)_2$.

$$\begin{array}{lcl} Mg(OH)_2(s) & \rightleftharpoons & Mg^{2+} + 2OH^- \\ & & \quad x \qquad\ 2x \end{array}$$

$$x = \left[\frac{0.0070\ g\ Mg(OH)_2}{1\ L\ soln}\right]\left[\frac{1\ mol\ Mg(OH)_2}{58.33\ g\ Mg(OH)_2}\right] = 1.2 \times 10^{-4}\ mol\ Mg(OH)_2/L$$

$$K_{sp} = [Mg^{2+}][OH^-]^2 = (x)(2x)^2 = 4x^3 = 4(1.2 \times 10^{-4})^3 = 6.9 \times 10^{-12}$$

22.3 Let x = molar solubility of $SrSO_4$.

$$\begin{array}{lcl} SrSO_4(s) & \rightleftharpoons & Sr^{2+} + SO_4{}^{2-} \\ & & \ \ x \qquad\ \ x \end{array}$$

$K_{sp} = [Sr^{2+}][SO_4{}^{2-}] = (x)(x) = 3.5 \times 10^{-7}$

$x^2 = 3.5 \times 10^{-7}$

$x = 5.9 \times 10^{-4}\ mol/L$

$$\left[\frac{5.9 \times 10^{-4}\ mol\ SrSO_4}{1\ L\ soln}\right]\left[\frac{183.68\ g\ SrSO_4}{1\ mol\ SrSO_4}\right] = 0.11\ g\ SrSO_4/L$$

22.4 Let x = molar solubility of $SrSO_4$.

$$\begin{array}{lcl} SrSO_4(s) & \rightleftharpoons & Sr^{2+} \quad + \quad SO_4{}^{2-} \\ & & 0.10 + x \qquad x \end{array}$$

$K_{sp} = [Sr^{2+}][SO_4{}^{2-}] = (0.10 + x)(x) = 3.5 \times 10^{-7}$

Assume that $0.10 + x \approx 0.10$.

$0.10x = 3.5 \times 10^{-7}$

$x = 3.5 \times 10^{-6}\ mol/L$

The approximation is valid because $0.10 + 3.5 \times 10^{-6} \approx 0.10$.

$$\left[\frac{3.5 \times 10^{-6} \text{ mol SrSO}_4}{1 \text{ L soln}}\right]\left[\frac{183.68 \text{ g SrSO}_4}{1 \text{ mol SrSO}_4}\right] = 0.00064 \text{ g SrSO}_4\text{/L}$$

$$\frac{0.00064 \text{ g}}{0.11 \text{ g}} \times 100 = 0.58 \text{ \%}$$

The solubility of $SrSO_4$ is reduced to about 0.6 % of the original value with no extra Sr^{2+} present.

22.5 $AgCl(s) \rightleftharpoons Ag^+ + Cl^-$

$Q_i = [Ag^+][Cl^-] = (1 \times 10^{-6})(0.010) = 1 \times 10^{-8}$

Because $Q_i > K_{sp}$, precipitation will occur.

22.6 $[OH^-] = (0.010)(0.001/1.001) = 1 \times 10^{-5}$ mol/L

$[Mg^{2+}] = (0.010)(1.000/1.001) = 1.0 \times 10^{-2}$ mol/L

$Q_i = [Mg^{2+}][OH^-]^2 = (1.0 \times 10^{-2})(1.0 \times 10^{-5})^2 = 1 \times 10^{-12}$

Because $Q_i < K_{sp}$, no precipitation will occur.

22.7 $Fe(OH)_2(s) \rightleftharpoons Fe^{2+} + 2OH^-$ $\qquad K_{sp} = [Fe^{2+}][OH^-]^2 = 8 \times 10^{-16}$

$$[OH^-] = \sqrt{\frac{K_{sp}}{[Fe^{2+}]}} = \sqrt{\frac{8 \times 10^{-16}}{0.010}} = 3 \times 10^{-7} \text{ mol/L}$$

$pOH = -\log [OH^-] = -\log(3 \times 10^{-7}) = 6.5$

$pH = 14.0 - pOH = 14.0 - 6.5 = 7.5$

22.8 For the simultaneous reactions

$Ag(CH_3COO)(s) \rightleftharpoons Ag^+ + CH_3COO^-$ $\qquad K = K_{sp}$

$CH_3COO^- + H^+ \rightleftharpoons CH_3COOH$ $\qquad K = \frac{1}{K_a}$

$\underset{}{Ag(CH_3COO)(s) + H^+ \rightleftharpoons \underset{s}{Ag^+} + \underset{s}{CH_3COOH}}$ $\qquad K = \frac{K_{sp}}{K_a} = \frac{4.4 \times 10^{-3}}{1.754 \times 10^{-5}} = 2.5 \times 10^2$

$$K = \frac{[Ag^+][CH_3COOH]}{[H^+]} = \frac{(s)(s)}{[H^+]} = 2.5 \times 10^2$$

$s^2 = (2.5 \times 10^2)(0.010) = 2.5$

$s = 1.6$ mol/L

22.9 At low concentrations of Cl^- ion, the solubility of silver ion will be determined by the K_{sp} equilibrium expression

$AgCl(s) \rightleftharpoons Ag^+ + Cl^-$ $\qquad K_{sp} = [Ag^+][Cl^-] = 1.8 \times 10^{-10}$

At $[Ag^+] = 1.0 \times 10^{-4}$ mol/L

$$[Cl^-] = \frac{K_{sp}}{[Ag^+]} = \frac{1.8 \times 10^{-10}}{1 \times 10^{-4}} = 2 \times 10^{-6} \text{ mol/L}$$

At higher Cl^- ion concentrations, the solubility of silver ion will be described by the equilibrium between the precipitate and the complex ion

$AgCl(aq) \rightleftharpoons Ag^+ + Cl^-$ $\quad K_{sp}$

$Ag^+ + 2Cl^- \rightleftharpoons [AgCl_2]^-$ $\quad 1/K_d$

$AgCl(s) + Cl^- \rightleftharpoons [AgCl_2]^-$ $\quad K = K_{sp}/K_d = \frac{[AgCl_2^-]}{[Cl^-]}$

At $[AgCl_2^-] = 1 \times 10^{-4}$ mol/L

$$[Cl^-] = \frac{[AgCl_2^-]}{K_{sp}/K_d} = \frac{1 \times 10^{-4}}{(1.8 \times 10^{-10})/(9 \times 10^{-6})} = 5 \text{ mol/L}$$

22.10 $BaSO_4(s) \rightleftharpoons Ba^{2+} + SO_4^{2-}$ $\quad K_{sp} = [Ba^{2+}][SO_4^{2-}] = 1.7 \times 10^{-10}$

$SrSO_4(s) \rightleftharpoons Sr^{2+} + SO_4^{2-}$ $\quad K_{sp} = [Sr^{2+}][SO_4^{2-}] = 3.5 \times 10^{-7}$

(a) As the solid Na_2SO_4 is added to the solution, it will dissolve until the $[SO_4^{2-}] = 4.7 \times 10^{-7}$ mol/L, at which point the ion product for $SrSO_4$ $[(0.75)(4.7 \times 10^{-7}) = 3.5 \times 10^{-7}]$ is equal to K_{sp} and $SrSO_4$ begins to precipitate.

(b) As more Na_2SO_4 dissolves, $SrSO_4$ continues to precipitate until $[SO_4^{2-}] = 1.7 \times 10^{-6}$ mol/L, at which point the ion product for $BaSO_4$ $[(0.00010)(1.7 \times 10^{-6}) = 1.7 \times 10^{-10}]$ is equal to K_{sp} and $BaSO_4$ begins to precipitate. The concentrations of Sr^{2+} ion remaining at the point at which $BaSO_4$ begins to precipitate is

$$[Sr^{2+}] = \frac{3.5 \times 10^{-7}}{1.7 \times 10^{-6}} = 0.21 \text{ mol/L}$$

Thus the Sr^{2+} ion remaining in solution is

$$\frac{0.21 \text{ M}}{0.75 \text{ M}} \times 100 = 28 \%$$

22.11 (a) $Zn(OH)_2(s) \rightleftharpoons Zn^{2+} + 2OH^-$ $\quad K_{sp} = [Zn^{2+}][OH^-]^2 = 1.2 \times 10^{-17}$

$Fe(OH)_3(s) \rightleftharpoons Fe^{3+} + 3OH^-$ $\quad K_{sp} = [Fe^{3+}][OH^-]^3 = 3 \times 10^{-39}$

The $[OH^-]$ concentrations needed for the beginning of each precipitation are

$$[OH^-] = \sqrt{\frac{K_{sp}}{[Zn^{2+}]}} = \sqrt{\frac{1.2 \times 10^{-17}}{1 \times 10^{-4}}} = 3 \times 10^{-7} \text{ mol/L}$$

$$[OH^-] = \sqrt[3]{\frac{K_{sp}}{[Fe^{3+}]}} = \sqrt[3]{\frac{3 \times 10^{-39}}{1 \times 10^{-4}}} = 3 \times 10^{-12}\ mol/L$$

The $Fe(OH)_3$ will precipitate once the OH^- ion concentration reaches 3×10^{-12} mol/L and the $Zn(OH)_2$ will not precipitate until the OH^- ion concentration reaches 3×10^{-7} mol/L.

(b) The maximum amount of $Fe(OH)_3$ formed before the $Zn(OH)_2$ begins to coprecipitate will be obtained at a concentration of OH^- ion that is slightly less than 3×10^{-7} mol/L or

$$pOH = -\log [OH^-] = -\log(3 \times 10^{-7}) = 6.5$$
$$pH = 14.0 - pOH = 14.0 - 6.5 = 7.5$$

Solutions to Odd-Numbered Questions and Problems

22.1 $[Fe(C_2O_4)_3]^{3-} \rightleftharpoons [Fe(C_2O_4)_2]^- + C_2O_4^{2-}$ $\quad K_{d_1} = \frac{[Fe(C_2O_4)_2^-][C_2O_4^{2-}]}{[Fe(C_2O_4)_3^{3-}]}$

$[Fe(C_2O_4)_2]^- \rightleftharpoons [Fe(C_2O_4)]^+ + C_2O_4^{2-}$ $\quad K_{d_2} = \frac{[Fe(C_2O_4)^+][C_2O_4^{2-}]}{[Fe(C_2O_4)_2^-]}$

$[Fe(C_2O_4)]^+ \rightleftharpoons Fe^{3+} + C_2O_4^{2-}$ $\quad K_{d_3} = \frac{[Fe^{3+}][C_2O_4^{2-}]}{[Fe(C_2O_4)^+]}$

$K_d = K_{d_1} K_{d_2} K_{d_3}$

22.3 $[Ag(CN)_2]^- \rightleftharpoons Ag^{2+} + 2CN^-$

$$K_d = \frac{[Ag^+][CN^-]^2}{[Ag(CN)_2^-]} = \frac{[Ag^+](0.25)^2}{(0.10)} = 1 \times 10^{-22}$$

$[Ag^+] = (1 \times 10^{-22})(0.10)/(0.25)^2 = 2 \times 10^{-22}$ mol/L

22.5 $[Ag(NH_3)_2]^+ \rightleftharpoons Ag^+ + 2NH_3(aq)$ $\quad K_d = \frac{[Ag^+][NH_3]^2}{[Ag(NH_3)_2^+]}$

Because of the very small value of K_d, practically all of the Ag^+ will be in the form of the complex. Therefore we assume $[Ag(NH_3)_2^+] = 0.10$ mol/L and the remaining concentration of NH_3 is

$$[NH_3] = \left[\frac{1.3\ mol}{L}\right] - \left[\frac{0.10\ mol\ complex}{L}\right]\left[\frac{2\ mol\ NH_3}{1\ mol\ complex}\right] = 1.1\ mol/L$$

$$[Ag^+] = \frac{(K_d)[Ag(NH_3)_2^+]}{[NH_3]^2} = \frac{(6.2 \times 10^{-8})(0.10)}{(1.1)^2} = 5.1 \times 10^{-9}\ mol/L$$

22.7 The expression for the solubility product of the following compounds is

(a) Ag_2SO_4, $K_{sp} = [Ag^+]^2[SO_4^{2-}]$

(b) Fe_2S_3, $K_{sp} = [Fe^{3+}]^2[S^{2-}]^3$

(c) $SrCrO_4$, $K_{sp} = [Sr^{2+}][CrO_4^{2-}]$

22.9 (a) $AgCl(s) \rightleftharpoons Ag^+ + Cl^-$
$\qquad\qquad\qquad x \qquad x$

$$x^2 = \left[\frac{1.9 \times 10^{-3}\ g\ AgCl}{1\ L}\right]\left[\frac{1\ mol\ AgCl}{143.32\ g\ AgCl}\right] = 1.3 \times 10^{-5}\ mol/L$$

$$K_{sp} = [Ag^+][Cl^-] = x^2 = (1.3 \times 10^{-5})^2 = 1.7 \times 10^{-10}$$

(b) $PbBr_2(s) \rightleftharpoons Pb^{2+} + 2Br^-$
$\qquad\qquad\qquad x \qquad 2x$

$$x = \left[\frac{7.8\ g\ PbBr_2}{1\ L}\right]\left[\frac{1\ mol\ PbBr_2}{367.0\ g\ PbBr_2}\right] = 0.021\ mol/L$$

$$K_{sp} = [Pb^{2+}][Br^-]^2 = (x)(2x)^2 = 4x^3 = 4(0.021)^3 = 3.7 \times 10^{-5}$$

(c) $Co(OH)_2(s) \rightleftharpoons Co^{2+} + 2OH^-$
$\qquad\qquad\qquad x \qquad 2x$

$$x = \left[\frac{3.4 \times 10^{-4}\ g\ Co(OH)_2}{1\ L}\right]\left[\frac{1\ mol\ Co(OH)_2}{92.95\ g\ Co(OH)_2}\right] = 3.7 \times 10^{-6}\ mol/L$$

$$K_{sp} = [Co^{2+}][OH^-]^2 = (x)(2x)^2 = 4x^3 = 4(3.7 \times 10^{-6})^3 = 2.0 \times 10^{-16}$$

22.11 $Mn(OH)_2(s) \rightleftharpoons Mn^{2+} + 2OH^-$
$\qquad\qquad\qquad x \qquad 2x$

$K_{sp} = [Mn^{2+}][OH^-]^2 = (x)(2x)^2 = 4x^3$

$pOH = 14.00 - pH = 14.00 - 9.57 = 4.43$

$\log [OH^-] = -pOH = -4.43$

$[OH^-] = 3.7 \times 10^{-5}\ mol/L$

$K_{sp} = 4(3.7 \times 10^{-5})^3 = 2.0 \times 10^{-13}$

22.13 (a) $Ba(OH)_2(s) \rightleftharpoons Ba^{2+} + 2OH^-$
$\qquad\qquad\qquad x \qquad 2x$

$K_{sp} = [Ba^{2+}][OH^-]^2 = (x)(2x)^2 = 4x^3 = 1.3 \times 10^{-2}$

$x^3 = 3.3 \times 10^{-3}$

$x = 0.15\ mol/L$

$$\left[\frac{0.15\ mol\ Ba(OH)_2}{1\ L}\right]\left[\frac{171.35\ g\ Ba(OH)_2}{1\ mol\ Ba(OH)_2}\right] = 26\ g/L$$

(b) $PbI_2(s) \rightleftharpoons Pb^{2+} + 2I^-$
$\qquad\qquad\qquad x \qquad 2x$

$K_{sp} = [Pb^{2+}][I^-]^2 = (x)(2x)^2 = 4x^3 = 7.1 \times 10^{-9}$

$x^3 = 1.8 \times 10^{-9}$

$x = 1.2 \times 10^{-3}\ mol/L$

$$\left[\frac{1.2 \times 10^{-3} \text{ mol } PbI_2}{1 \text{ L}}\right]\left[\frac{461.0 \text{ g } PbI_2}{1 \text{ mol } PbI_2}\right] = 0.55 \text{ g/L}$$

22.15 Let $x = [Ag^+]$.

	$AgBr(s)$ ⇄	Ag^+ +	Br^-
initial		0	0.010
change		$+x$	$+x$
equilibrium		x	$0.010 + x$

$K_{sp} = [Ag^+][Br^-] = (x)(0.010 + x) = 4.9 \times 10^{-13}$

Assume that $0.010 + x \approx 0.010$.

$x = (4.9 \times 10^{-13})/(0.010) = 4.9 \times 10^{-11}$ mol/L

The approximation is valid because $0.010 + 4.9 \times 10^{-11} \approx 0.010$.

$[Ag^+] = 4.9 \times 10^{-11}$ mol/L

22.17 (a) Let $x = [CrO_4^{2-}]$ at equilibrium.

$Ag_2CrO_4(s) \rightleftarrows 2Ag^+ + CrO_4^{2-}$
$\quad 2x \quad x$

$K_{sp} = [Ag^+]^2[CrO_4^{2-}] = (2x)^2(x) = 4x^3 = 2.5 \times 10^{-12}$

$x = 8.5 \times 10^{-5}$ mol/L

(b) $Ag_2CrO_4(s) \rightleftarrows 2Ag^+ + CrO_4^{2-}$
$\quad 0.010 + 2x \quad x$

$K_{sp} = (0.010 + 2x)^2(x) = 2.5 \times 10^{-12}$

Assuming $0.010 + 2x \approx 0.010$,

$(0.010)^2(x) = 2.5 \times 10^{-12}$

$x = 2.5 \times 10^{-8}$ mol/L

The approximation is valid, so the solubility is 2.5×10^{-8} mol/L.

(c) $Ag_2CrO_4(s) \rightleftarrows 2Ag^+ + CrO_4^{2-}$
$\quad 2x \quad 0.010 + x$

$K_{sp} = (2x)^2(0.010 + x) = 2.5 \times 10^{-12}$

Assuming $0.010 + x \approx 0.010$,

$(2x)^2(0.010) = 2.5 \times 10^{-12}$

$x = 7.9 \times 10^{-6}$ mol/L

The approximation is valid, so the solubility is 7.9×10^{-6} mol/L.

22.19 $[F^-] = \left[\frac{1 \text{ mg } F^-}{1 \text{ L}}\right]\left[\frac{1 \text{ g}}{1000 \text{ mg}}\right]\left[\frac{1 \text{ mol } F^-}{19 \text{ g } F^-}\right] = 5 \times 10^{-5}$ M

Let $x = [Ca^{2+}]$.

$CaF_2(s) \rightleftharpoons Ca^{2+} + 2F^-$
$\quad\quad\quad\quad\quad x \quad\quad 5 \times 10^{-5}$

$K_{sp} = [Ca^{2+}][F^-]^2 = (x)(5 \times 10^{-5})^2 = 2.7 \times 10^{-11}$

$x = 1 \times 10^{-2}$ mol/L $= [Ca^{2+}]$

$$\left[\frac{1 \times 10^{-2} \text{ mol } Ca^{2+}}{1 \text{ L}}\right]\left[\frac{40 \text{ g } Ca^{2+}}{1 \text{ mol } Ca^{2+}}\right] = 0.4 \text{ g } Ca^{2+}/\text{L}$$

22.21 $$[Ag^+] = \left[\frac{1.0 \text{ g } AgNO_3}{50 \text{ mL}}\right]\left[\frac{1000 \text{ mL}}{1 \text{ L}}\right]\left[\frac{1 \text{ mol } AgNO_3}{169.88 \text{ g } AgNO_3}\right] = 0.1 \text{ mol/L}$$

$[Cl^-] = 0.050$ mol/L

$AgCl(s) \rightleftharpoons Ag^+ + Cl^-$

$Q_i = [Ag^+][Cl^-] = (0.1)(0.050) = 0.005$

Because $Q_i > K_{sp}$, precipitation will occur.

22.23 $PbBr_2(s) \rightleftharpoons Pb^{2+} + 2Br^-$

$$[Pb^{2+}] = \left[\frac{0.332 \text{ g } Pb(NO_3)_2}{1 \text{ L}}\right]\left[\frac{1 \text{ mol } Pb(NO_3)_2}{331.2 \text{ g } Pb(NO_3)_2}\right]\left[\frac{1 \text{ mol } Pb^{2+}}{1 \text{ mol } Pb(NO_3)_2}\right]$$
$$= 1.00 \times 10^{-3} \text{ mol/L}$$

$$[Br^-] = \left[\frac{1.03 \text{ g NaBr}}{1 \text{ L}}\right]\left[\frac{1 \text{ mol NaBr}}{102.89 \text{ g NaBr}}\right]\left[\frac{1 \text{ mol } Br^-}{1 \text{ mol NaBr}}\right] = 1.00 \times 10^{-2} \text{ mol/L}$$

$Q_i = [Pb^{2+}][Br^-]^2 = (1.00 \times 10^{-3})(1.00 \times 10^{-2})^2 = 1.00 \times 10^{-7}$

Because $Q_i < K_{sp}$, no precipitation will occur.

22.25 The solubility would be greater because the concentration of the sulfide ion would be less than expected as a result of the reaction with H^+ ion. Because NO_3^- ion is the weak conjugate base of a strong acid, no reaction occurs with the H^+ ion and so there is no effect on the solubility.

22.27 Most water-insoluble metal hydroxides can be dissolved in concentrated strong acids.

$Fe(OH)_3(s) + 3H_3O^+ \rightarrow Fe^{3+} + 6H_2O(l)$

22.29 $M(OH)_2(s) \rightleftharpoons M^{2+} + 2OH^-$

$K_{sp} = [M^{2+}][OH^-]^2$

$$[OH^-] = \sqrt{\frac{K_{sp}}{[M^{2+}]}}$$

for $Fe(OH)_2$: $$[OH^-] = \sqrt{\frac{8 \times 10^{-16}}{0.010}} = 3 \times 10^{-7} \text{ mol/L}$$

$$pOH = -\log [OH^-] = -\log(3 \times 10^{-7}) = 6.5$$
$$pH = 14.0 - pOH = 14.0 - 6.5 = 7.5$$

for $Cu(OH)_2$: $[OH^-] = \sqrt{\dfrac{1.3 \times 10^{-20}}{0.010}} = 1.1 \times 10^{-9}$

$$pOH = -\log(1.1 \times 10^{-9}) = 8.96$$
$$pH = 14.0 - 8.96 = 5.04$$

22.31 (a) $[Cl^-] = \dfrac{K_{sp}}{Ag^+} = \dfrac{1.8 \times 10^{-10}}{0.01} = 2 \times 10^{-8}$ mol/L

$$[Cl^-] = \sqrt{\frac{K_{sp}}{Pb^{2+}}} = \sqrt{\frac{2 \times 10^{-5}}{0.01}} = 4 \times 10^{-2} \text{ mol/L}$$

Therefore AgCl will precipitate first.

(b) $[Ag^+] = \dfrac{1.8 \times 10^{-10}}{4 \times 10^{-2}} = 5 \times 10^{-9}$ mol/L

$$\frac{5 \times 10^{-9} \text{ mol/L}}{0.01 \text{ mol/L}} = 5 \times 10^{-7} \text{ remains}$$

22.33 Most of the F^- ion from the CaF_2 will be present in the form of HF and the two equilibrium expressions can be combined

$$CaF_2(s) \rightleftarrows Ca^{2+} + 2F^- \qquad K = K_{sp}$$
$$2F^- + 2H^+ \rightleftarrows 2HF(aq) \qquad K = \left[\frac{1}{K_a}\right]^2$$

$$CaF_2(s) + 2H^+ \rightleftarrows Ca^{2+} + 2HF(aq) \qquad K = \frac{K_{sp}}{K_a^2} = \frac{2.7 \times 10^{-11}}{(6.5 \times 10^{-4})^2} = 6.4 \times 10^{-5}$$

Letting x represent the solubility of CaF_2, $[Ca^{2+}] = x$ and $[HF] = 2x$ giving

$$K = \frac{[Ca^{2+}][HF]^2}{[H^+]^2} = \frac{(x)(2x)^2}{(0.010)^2} = 6.4 \times 10^{-5}$$

$$4x^3 = (6.4 \times 10^{-5})(0.010)^2$$
$$x = 1.2 \times 10^{-3} \text{ mol/L}$$

The solubility of CaF_2 in 0.010 M HCl is 1.2×10^{-3} mol/L.

22.35 As the pH increases from acidic to neutral values, the formation of $Cr(OH)_3$ is important

$$Cr(OH)_3(s) \rightleftarrows Cr^{3+} + 3OH^- \qquad K_{sp} = [Cr^{3+}][OH^-]^3$$

The OH^- ion concentration in equilibrium with $[Cr^{3+}] = 1 \times 10^{-4}$ mol/L is

$$[OH^-]^3 = \frac{K_{sp}}{[Cr^{3+}]} = \frac{6 \times 10^{-31}}{1 \times 10^{-4}} = 6 \times 10^{-27}$$

$$[OH^-] = 2 \times 10^{-9} \text{ mol/L}$$

$$pOH = -\log [OH^-] = -\log(2 \times 10^{-9}) = 8.7$$

$$pH = 14.0 - pOH = 14.0 - 8.7 = 5.3$$

As the pH increases to more alkaline values, the formation of $[Cr(OH)_4]^-$ becomes important

$$Cr(OH)_3(s) \rightleftharpoons Cr^{3+} + 3OH^- \qquad K_{sp}$$

$$Cr^{3+} + 4OH^- \rightleftharpoons [Cr(OH)_4]^- \qquad 1/K_d$$

$$Cr(OH)_3(s) + OH^- \rightleftharpoons [Cr(OH)_4]^- \qquad K = K_{sp}/K_d = \frac{[Cr(OH)_4^-]}{[OH^-]}$$

The OH^- ion concentration in equilibrium with $[Cr(OH)_4^-] = 1 \times 10^{-4}$ mol/L is

$$[OH^-] = \frac{[Cr(OH)_4^-]}{K_{sp}/K_d} = \frac{1 \times 10^{-4}}{(6 \times 10^{-31})/(1 \times 10^{-28})} = 2 \times 10^{-2} \text{ mol/L}$$

$$pOH = -\log(2 \times 10^{-2}) = 1.7$$

$$pH = 14.0 - 1.7 = 12.3$$

The pH range over which $[Cr(III)] < 1 \times 10^{-4}$ mol/L is 5.3 to 12.3.

12.37 $[Ag(SCN)_4]^{3-} \rightleftharpoons Ag^+ + 4SCN^-$

$\qquad\qquad x \qquad 4x + 0.10$

$$K_d = \frac{[Ag^+][SCN^-]^4}{[Ag(SCN)_4{}^{3-}]} = \frac{(x)(4x + 0.10)^4}{(0.10)} = 2.1 \times 10^{-10}$$

Assuming $4x + 0.10 \approx 0.10$ gives

$$\frac{(x)(0.10)^4}{(0.10)} = 2.1 \times 10^{-10}$$

$$x = 2.1 \times 10^{-7} \text{ mol/L}$$

The assumption is valid, so $[Ag^+] = 2.1 \times 10^{-7}$ mol/L. The ion product is

$$Ag_2SO_4(s) \rightleftharpoons 2Ag^+ + SO_4{}^{2-}$$

$$Q_i = [Ag^+]^2[SO_4{}^{2-}] = (2.1 \times 10^{-7})^2(0.1) = 4 \times 10^{-15}$$

Because $Q_i < K_{sp}$, no precipitate will form.

22.39 $MS(s) \rightleftharpoons M^{2+} + S^{2-}$

$$K_{sp} = [M^{2+}][S^{2-}]$$

$$[S^{2-}] = \frac{K_{sp}}{[M^{2+}]}$$

$$[H^+] = \sqrt{\frac{(3 \times 10^{-21})}{[S^{2-}]}} = \sqrt{\frac{(3 \times 10^{-21})[M^{2+}]}{K_{sp}}}$$

for MnS: $$[H^+] = \sqrt{\frac{(3 \times 10^{-21})(0.010)}{(2.3 \times 10^{-13})}} = 1 \times 10^{-5} \text{ mol/L}$$

$$pH = -\log [H^+] = -\log(1 \times 10^{-5}) = 5.0$$

for ZnS: $$[H^+] = \sqrt{\frac{(3 \times 10^{-21})(0.010)}{(2 \times 10^{-24})}} = 4 \text{ mol/L}$$

$$pH = -\log(4) = -0.6$$

To selectively precipitate ZnS, use $-0.6 < pH < 5.0$.

CHAPTER 23

THERMODYNAMICS

Solutions to Exercises

23.1 For the system in which occurs the

(a) formation of a raindrop in a cloud, a decrease in entropy is predicted.

(b) crystallization of the metal alloy from the molten state in molding a bookend, a decrease in entropy is predicted.

(c) the beating of an egg for an omelet, an increase in entropy is predicted.

23.2 $\Delta S° = \frac{\Delta H°}{T} = \frac{(-68.28\ \text{kJ/mol})(1000\ \text{J/1 kJ})}{(690. + 273)\ \text{K}} = -70.9\ \text{J/K mol}$

This is a decrease in entropy for the cesium.

23.3 There are six contributions to the value of $S°$: (1) $S_0° = 0$, (2) heating the solid from 0 K to 55.19 K, (3) fusion at 55.19 K, (4) heating the liquid from 55.19 K to 85.23 K, (5) vaporization at 85.23 K, and (6) heating the gas from 85.23 K to 298 K.

23.4 (a) One mole of solid and two and one half moles of gas react to form one mole of liquid, so the entropy change is (i) large and negative.

(b) Three moles of gas and one mole of liquid react to form two moles of molecules in solution and one mole of gas, so the entropy change is (i) large and negative.

(c) One mole of gas and one mole of liquid form two moles of ions in solution, so the entropy change is (i) large and negative.

(d) Two moles of solid and one mole of gas react to form two moles of gas, so the entropy change is (ii) large and positive.

23.5

$$\begin{array}{lcll} 2O_3(g) & \rightarrow & 3O_2(g) & \Delta S^\circ = 137.56 \text{ J/K} \\ \underline{O_2(g) + O(g)} & \underline{\rightarrow} & \underline{O_3(g)} & \underline{\Delta S^\circ = -127.27 \text{ J/K}} \\ O_3(g) + O(g) & \rightarrow & 2O_2(g) & \Delta S^\circ = 10.29 \text{ J/K} \end{array}$$

23.6 $\Delta G^\circ = \Delta H^\circ - T\,\Delta S^\circ = (-391.9 \text{ kJ}) - (298.15 \text{ K})(10.29 \text{ J/K})(1 \text{ kJ}/1000 \text{ J})$

$= -395.0 \text{ kJ}$

The reaction is spontaneous at this temperature.

23.7 $\Delta G^\circ = [(1 \text{ mol})\Delta G_f^\circ(BrF) + (1 \text{ mol})\Delta G_f^\circ(BrF_5)] - [(2 \text{ mol})\Delta G_f^\circ(BrF_3)]$

$= [(1 \text{ mol})(-78.58 \text{ kJ/mol}) + (1 \text{ mol})(-144.85 \text{ kJ/mol})]$

$- [(2 \text{ mol})(-139.13 \text{ kJ/mol})]$

$= 54.83 \text{ kJ}$

The reaction is not spontaneous under these conditions.

23.8

$$\begin{array}{rcll} (3)[Fe_2O_3(s) & \rightarrow & 2FeO(s) + \frac{1}{2}O_2(g) & \Delta G^\circ = 252.7 \text{ kJ}] \\ (2)[3FeO(s) + \frac{1}{2}O_2(g) & \rightarrow & Fe_3O_4(s) & \Delta G^\circ = -281.1 \text{ kJ}] \\ \hline 3Fe_2O_3(s) & \rightarrow & \frac{1}{2}O_2(g) + 2Fe_3O_4(s) & \Delta G^\circ = 195.9 \text{ kJ} \end{array}$$

The reactant is favored over the products as shown by the positive value of ΔG°. Fe_2O_3 is quite stable towards thermal decomposition.

23.9

$$Q = \frac{P_{BrF}\,P_{BrF_5}}{P^2_{BrF_3}} = \frac{(0.0010)(0.0010)}{(5.2)^2} = 3.7 \times 10^{-8}$$

$\Delta G = \Delta G^\circ + (2.303)RT \log Q$

$$= (54.83 \text{ kJ}) + (2.303)(8.314 \text{ J/K mol})(1000.\text{ K})\left[\frac{1 \text{ kJ}}{1000 \text{ J}}\right] \log(3.7 \times 10^{-8})$$

$= 54.83 \text{ kJ} - 142 \text{ kJ} = -87 \text{ kJ}$

This reaction is more favorable and under these conditions is spontaneous in the forward direction.

23.10 $\Delta G^\circ = \Delta H^\circ - T\,\Delta S^\circ = (391.9 \text{ kJ}) - (T)(-0.01029 \text{ kJ/K})$

ΔH° and ΔS° are both unfavorable; therefore ΔG° will remain positive and the reaction will remain nonspontaneous with changes in temperature.

23.11 (a)

$$\Delta S^\circ = \frac{\Delta H^\circ - \Delta G^\circ}{T} = \frac{[(235.8 \text{ kJ}) - (195.8 \text{ kJ})](1000 \text{ J}/1 \text{ kJ})}{298.15 \text{ K}} = 134 \text{ J/K}$$

$\Delta G^\circ_{1000} = \Delta H^\circ - T\,\Delta S^\circ = (235.8 \text{ kJ}) - (1000 \text{ K})(1 \text{ kJ}/1000 \text{ J})(134 \text{ J/K})$

$= 102 \text{ kJ}$

(b)
$$\log\left[\frac{K_2}{K_1}\right] = \frac{-\Delta H^\circ}{(2.303)R}\left[\frac{1}{T_2} - \frac{1}{T_1}\right]$$

$$\log\left[\frac{K_2}{5.0 \times 10^{-35}}\right] = \frac{-(235.8 \text{ kJ})(1000 \text{ J/1 kJ})}{(2.303)(8.314 \text{ J/K mol})}\left[\frac{1}{500.\text{ K}} - \frac{1}{298 \text{ K}}\right]$$
$$= 16.7$$

$$\frac{K_2}{5.0 \times 10^{-35}} = 5 \times 10^{16}$$
$$K_2 = 3 \times 10^{-18}$$

(c) $\Delta G^\circ = \Delta H^\circ - T\,\Delta S^\circ = 0$

$$T = \frac{\Delta H^\circ}{\Delta S^\circ} = \frac{(235.8 \text{ kJ})(1000 \text{ J/1 kJ})}{(134 \text{ J/K})} = 1760 \text{ K}$$

Solutions to Odd-Numbered Questions and Problems

23.1 The quantitative measure of the randomness or disorder in a system is called entropy. The order of increasing randomness of the following systems is

(b) 1 mol of solid A < (c) 1 mol of liquid A < (a) 1 mol of gas A

23.3 Heating a gas increases its entropy because the molecules have more kinetic energy, which is reflected in more movement, hence, in an increase in randomness.

23.5 The decrease in entropy, indicated by the negative sign, is the result of the ions in the crystal structure having less kinetic energy at the lower temperature. As a result of the smaller amount of energy, the average vibrational motion of the ions in the lattice decreases and hence there is less chaotic motion in the crystal. Less chaotic motion means less entropy.

23.7 $$\Delta S^\circ = \frac{\Delta H^\circ}{T} = \frac{2100 \text{ J/mol}}{275.6 \text{ K}} = 7.6 \text{ J/K mol}$$

23.9 $$\Delta S^\circ(\text{fusion}) = \frac{\Delta H^\circ(\text{fusion})}{T} = \frac{6010 \text{ J/mol}}{273 \text{ K}} = 22.0 \text{ J/K mol}$$

$$\Delta S^\circ(\text{vaporization}) = \frac{\Delta H^\circ(\text{vaporization})}{T} = \frac{40{,}660 \text{ J/mol}}{373 \text{ K}} = 109 \text{ J/K mol}$$

The second value is so much larger because the change in randomness is larger in going from a liquid to a gas than in going from a solid to a liquid.

23.11 The value of $S°$ for $Cl_2(g)$ is greater than that of $Br_2(l)$ because a gas has more randomness than a liquid.

23.13 (a) $O_3(g)$, (b) $FeCl_3(s)$, (c) $H_2O(g)$

23.15 The contributions to the absolute entropy of $O_2(g)$ at 298 K ($S°_{298}$) include (1) $S°_0$, (2) heating γ-solid from 0 K to 24 K, (3) changing γ-solid to β-solid at 24 K, (4) heating β-solid from 24 K to 44 K, (5) changing β-solid to α-solid at 44 K, (6) heating α-solid from 44 K to 54 K, (7) fusion of α-solid at 54 K, (8) heating the liquid from 54 K to 90.2 K, (9) vaporization of the liquid at 90.2 K, and (10) heating the gas from 90.2 K to 298 K.

23.17 $$S°_{298} = S°_0 + \Delta S°_{0-35.61} + \Delta S°_{s-s} + \Delta S°_{35.61-63.14} + \Delta S°_{fus} + \Delta S°_{63.14-77.32} + \Delta S°_{vap} + \Delta S°_{77.32-298}$$

$$= 0.0 + 27.2 + 6.4 + 23.4 + 11.4 + 11.4 + 72.2 + 39.2$$

$$= 191.2 \text{ J/K}$$

23.19 (a) Two moles of solid and three moles of gas produce two moles of solid, so the entropy change is (i) large and negative.

(b) Two moles of solid and three moles of liquid produce three moles of solid, so the entropy change is (iii) small.

(c) One mole of ions and one mole of gas produce one mole of solid and two moles of ions, so the entropy change is (i) large and negative.

(d) One mole of solid and three moles of gas produce one mole of gas, so the entropy change is (i) large and negative.

(e) Four moles of liquid or solute (4 mol ions and 2 mol liquid) produce two moles of gas and two moles of solute (4 mol ions), so the entropy change is (ii) large and positive.

(f) Three moles of solid or solute (4 mol ions and 1 mol solid) produce one mole of solute (3 mol ions) and one mole of gas, so the entropy change is (ii) large and positive.

(g) Four moles of gas produce two moles of gas and one mole of solute (2 mol ions, H^+ and HSO_4^-), so the entropy change is (i) large and negative.

23.21 (a) $\Delta S° = [(1 \text{ mol})S°(g)] - [(1 \text{ mol})S°(s)]$

$= [(1 \text{ mol})(260.69 \text{ J/K mol})] - [(1 \text{ mol})(116.135 \text{ J/K mol})]$

$= 144.56 \text{ J/K}$

(b) $\Delta S° = [(2\ \text{mol})S°(\text{I})] - [(1\ \text{mol})S°(\text{I}_2)]$
$= [(2\ \text{mol})(180.791\ \text{J/K mol})] - [(1\ \text{mol})(260.69\ \text{J/K mol})]$
$= 100.89\ \text{J/K}$

These entropy changes are similar in sign and magnitude because both reactions essentially produce a net increase of one mole of gas.

23.23 (a) $\Delta S° = [(1\ \text{mol})S°(\text{Fe}_2\text{O}_3)] - [(2\ \text{mol})S°(\text{Fe}) + (\frac{3}{2}\ \text{mol})S°(\text{O}_2)]$
$= [(1\ \text{mol})(87.40\ \text{J/K mol})] - [(2\ \text{mol})(27.28\ \text{J/K mol}) + (\frac{3}{2}\ \text{mol})(205.138\ \text{J/K mol})]$
$= -274.87\ \text{J/K}$

(b) $\Delta S° = [(1\ \text{mol})S°(\text{Fe}_3\text{O}_4)] - [(3\ \text{mol})S°(\text{Fe}) + (2\ \text{mol})S°(\text{O}_2)]$
$= [(1\ \text{mol})(146.4\ \text{J/K mol})] - [(3\ \text{mol})(27.28\ \text{J/K mol}) + (2\ \text{mol})(205.138\ \text{J/K mol})]$
$= -345.7\ \text{J/K}$

Neither of these reactions has a favorable entropy change.

23.25

$\text{I}_2(s) \rightarrow \text{I}_2(g)$	$\Delta S° = 144.56\ \text{J/K}$
$\text{I}_2(g) \rightarrow 2\text{I}(g)$	$\Delta S° = 100.89\ \text{J/K}$
$\text{I}_2(s) \rightarrow 2\text{I}(g)$	$\Delta S° = 245.45\ \text{J/K}$

This is a favorable entropy change.

23.27

$3[\text{Fe}_2\text{O}_3(s) \rightarrow 2\text{Fe}(\alpha\text{-solid}) + \frac{3}{2}\text{O}_2(g)$	$\Delta S° = 274.87\ \text{J/K}]$
$2[3\text{Fe}(\alpha\text{-solid}) + 2\text{O}_2(g) \rightarrow \text{Fe}_3\text{O}_4(s)$	$\Delta S° = -345.7\ \text{J/K}]$
$3\text{Fe}_2\text{O}_3(s) \rightarrow 2\text{Fe}_3\text{O}_4(s)$	$\Delta S° = 133.2\ \text{J/K}$

This is a favorable entropy change.

23.29 In terms of enthalpy and entropy, free energy is defined as $G = H - TS$. Values of ΔG are (a) negative for a spontaneous process, (b) positive for a nonspontaneous process, and (c) zero for a process at equilibrium.

23.31 The conditions that would predict (a) process(es) that is (are)

(a) always spontaneous is (iii) $\Delta H < 0$, $\Delta S > 0$, because $\Delta G = (-) - (+)(+) = -$

(b) always nonspontaneous is (ii) $\Delta H > 0$, $\Delta S < 0$ because $\Delta G = (+) - (+)(-) = +$

(c) spontaneous or nonspontaneous, depending on the temperature and magnitudes of ΔH and ΔS, are (i) $\Delta H > 0$, $\Delta S > 0$ because $\Delta G = (+) - (+)(+) = ?$ and (iv) $\Delta H < 0$, $\Delta S < 0$ because $\Delta G = (-) - (+)(-) = ?$

23.33 $\Delta G° = \Delta H° - T\,\Delta S° = (-285.4 \text{ kJ}) - (298.15 \text{ K})(1 \text{ kJ}/1000 \text{ J})(137.55 \text{ J/K})$
$= -326.4 \text{ kJ}$

The negative $\Delta G°$ implies the reaction is spontaneous. Because the value of $\Delta H°$ is negative and the value of $\Delta S°$ is positive, both driving forces are favorable.

23.35 $\Delta G° = \Delta H° - \text{T}\,\Delta S°$

$$= (-36.40 \text{ kJ/mol}) - (298.15 \text{ K})(57.238 \text{ J/K mol})\left[\frac{1 \text{ kJ}}{1000 \text{ J}}\right]$$
$= -53.47 \text{ kJ/mol}$

23.37
$$\Delta S° = \frac{\Delta H° - \Delta G°}{T} = \frac{[(-241.818 \text{ kJ/mol}) - (-228.572 \text{ kJ/mol})]\left[\frac{1000 \text{ J}}{1 \text{ kJ}}\right]}{298.15 \text{ K}}$$
$= -44.43 \text{ J/K mol}$

23.39 The larger negative value of ΔG°_f for liquid water would predict it to be more stable at 25°C.

23.41 $\Delta G° = \Delta H° - T\,\Delta S° = (-57.20 \text{ kJ}) - (298.15 \text{ K})(1 \text{ kJ}/1000 \text{ J})(-175.83 \text{ J/K})$
$= -4.78 \text{ kJ}$

The negative value of $\Delta G°$ implies the reaction is spontaneous. The $\Delta H°$ is the driving force for the reaction.

23.43

Reaction	
$HF(aq) \rightarrow H^+ + F^-$	$\Delta G° = 18.03$ kJ
$H^+ + OH^- \rightarrow H_2O(l)$	$\Delta G° = -79.885$ kJ
$HF(aq) + OH^- \rightarrow H_2O(l) + F^-$	$\Delta G° = -61.86$ kJ

23.45

Reaction	
$H_2(g) + S(s) \rightarrow H_2S(g)$	$\Delta G° = -33.56$ kJ
$H_2S(g) \rightarrow H_2S(aq)$	$\Delta G° = 5.73$ kJ
$H_2S(aq) \rightarrow HS^- + H^+$	$\Delta G° = 39.91$ kJ
$HS^- \rightarrow H^+ + S^{2-}$	$\Delta G° = 73.7$ kJ
$H_2(g) + S(s) \rightarrow 2H^+ + S^{2-}$	$\Delta G° = 85.8$ kJ

23.47 $\Delta G° = [(1 \text{ mol})\Delta G^\circ_f(SiO_2) + (2 \text{ mol})\Delta G^\circ_f(H_2O)]$
$- [(1 \text{ mol})\Delta G^\circ_f(SiH_4) + (2 \text{ mol})\Delta G^\circ_f(O_2)]$
$= [(1 \text{ mol})(-856.64 \text{ kJ/mol}) + (2 \text{ mol})(-237.129 \text{ kJ/mol})]$
$- [(1 \text{ mol})(56.9 \text{ kJ/mol}) + (2 \text{ mol})(0)]$
$= -1387.8 \text{ kJ}$

23.49 $\Delta G° = [(1\ \text{mol})\Delta G_f°(Na^+) + (1\ \text{mol})\Delta G_f°(I^-)] - [(1\ \text{mol})\Delta G_f°(NaI)]$

$= [(1\ \text{mol})(-261.905\ \text{kJ/mol}) + (1\ \text{mol})(-51.57\ \text{kJ/mol})] - [(1\ \text{mol})(-286.06\ \text{kJ/mol})]$

$= -27.42\ \text{kJ}$

23.51 $\Delta G° = [(4\ \text{mol})\Delta G_f°(Fe) + (3\ \text{mol})\Delta G_f°(CO_2)] - [(2\ \text{mol})\Delta G_f°(Fe_2O_3) + (3\ \text{mol})\Delta G_f°(C)]$

$= [(4\ \text{mol})(0) + (3\ \text{mol})(-395\ \text{kJ/mol})] - [(2\ \text{mol})(-637\ \text{kJ/mol}) + (3\ \text{mol})(0)]$

$= 90\ \text{kJ}$

$\Delta G° = [(2\ \text{mol})\Delta G_f°(Cu) + (1\ \text{mol})\Delta G_f°(CO_2)] - [(2\ \text{mol})\Delta G_f°(CuO) + (1\ \text{mol})\Delta G_f°(C)]$

$= [(2\ \text{mol})(0) + (1\ \text{mol})(-395\ \text{kJ/mol})] - [(2\ \text{mol})(-92\ \text{kJ/mol}) + (1\ \text{mol})(0)]$

$= -210\ \text{kJ}$

CuO can be reduced by using carbon in a wood fire.

23.53 The relationship between the enthalpy and the entropy changes for a process at equilibrium is $\Delta H = T\,\Delta S$.

23.55 $$\log K = \frac{-\Delta G°}{(2.303)RT} = \left[\frac{-(-4.73\ \text{kJ})}{(2.303)(8.314\ \text{J/K mol})(298.15\ \text{K})}\right]\left[\frac{1000\ \text{J}}{1\ \text{kJ}}\right] = 0.829$$

$K = 6.74$

A value larger than unity is expected because $\Delta G° < 0$.

23.57 $$Q = \frac{P_{HCl}}{P_{H_2}^{1/2}\ P_{Cl_2}^{1/2}} = \frac{(0.31)}{(3.5)^{1/2}(1.5)^{1/2}} = 0.14$$

$\Delta G = \Delta G° + (2.303)RT \log Q$

$= (-95.299\ \text{kJ}) + (2.303)(8.314\ \text{J/K mol})(298.15\ \text{K})\left[\frac{1\ \text{kJ}}{1000\ \text{J}}\right] \log(0.14)$

$= -100.2\ \text{kJ}$

The process is more favorable under these conditions than under standard state conditions.

23.59 (a) $\Delta G° = [(1\ \text{mol})\Delta G_f°(NH_3) + (1\ \text{mol})\Delta G_f°(HCl)] - [(1\ \text{mol})\Delta G_f°(NH_4Cl)]$

$= [(1\ \text{mol})(-16.45\ \text{kJ/mol}) + (1\ \text{mol})(-95.299\ \text{kJ/mol})] - [(1\ \text{mol})(-202.87\ \text{kJ/mol})]$

$= 91.12\ \text{kJ}$

(b) $\log K_p = \frac{-\Delta G°}{(2.303)RT}$

$$= \left[\frac{-(91.12 \text{ kJ})}{(2.303)(8.314 \text{ J/K mol})(298.15 \text{ K})}\right]\left[\frac{1000 \text{ J}}{1 \text{ kJ}}\right] = -15.96$$

$$K_p = 1.1 \times 10^{-16}$$

(c) $K_p = P_{NH_3} P_{HCl} = P^2_{NH_3} = 1.1 \times 10^{-16}$

$$P_{NH_3} = 1.1 \times 10^{-8} \text{ bar}$$

23.61 Because $\Delta S°$ is positive for both reactions, $\Delta G°$ becomes more negative as the temperature increases. Because $\Delta S°$ is much larger for the formation of CO, this reaction has the larger temperature dependence.

23.63 $\Delta S° = \frac{\Delta H° - \Delta G°}{T} = \frac{[(142.7 \text{ kJ}) - (163.2 \text{ kJ})](1000 \text{ J/1 kJ})}{298.15 \text{ K}}$

$$= -68.8 \text{ J/K}$$

$$\Delta G°_{1000} = \Delta H° - T\,\Delta S° = (142.7 \text{ kJ}) - (1000. \text{ K})(1 \text{ kJ/1000 J})(-68.8 \text{ J/K})$$

$$= 211.5 \text{ kJ}$$

23.65 $\log\left[\frac{K_2}{K_1}\right] = \frac{-\Delta H°}{(2.303)R}\left[\frac{1}{T_2} - \frac{1}{T_1}\right]$

$$\log\left[\frac{K_2}{2.2 \times 10^{-9}}\right] = \frac{-(18.9 \text{ kJ})(1000 \text{ J/1 kJ})}{(2.303)(8.314 \text{ J/K mol})}\left[\frac{1}{(75 + 273.15 \text{ K})} - \frac{1}{298.15 \text{ K}}\right]$$

$$= 0.475$$

$$\frac{K_2}{2.2 \times 10^{-9}} = 2.99$$

$$K_2 = (2.2 \times 10^{-9})(2.99) = 6.6 \times 10^{-9}$$

23.67 $\Delta H° = \frac{-(2.303)R \log(K_2/K_1)}{(1/T_2 - 1/T_1)}$

$$= \frac{-(2.303)(8.314 \text{ J/K mol})(1 \text{ kJ/1000 J}) \log(1.35 \times 10^{11}/2.10 \times 10^{18})}{(1/500. \text{ K}) - (1/298 \text{ K})}$$

$$= -102 \text{ kJ/mol}$$

23.69 (a) $\Delta S° = [(2 \text{ mol})S°(O_3)] - [(3 \text{ mol})S°(O_2)]$

$$= [(2 \text{ mol})(238.93 \text{ J/K mol})] - [(3 \text{ mol})(205.138 \text{ J/K mol})]$$

$$= -137.55 \text{ J/K}$$

The production of ozone from oxygen is a decrease in randomness.

(b) $\Delta H° = [(2\ \text{mol})\Delta H_f°(O_3)] - [(3\ \text{mol})\Delta H_f°(O_2)]$

$= [(2\ \text{mol})(142.7\ \text{kJ/mol})] - [(3\ \text{mol})(0)] = 285.4\ \text{kJ}$

This is an unfavorable enthalpy change.

(c) $\Delta G° = \Delta H° - T\,\Delta S°$

$$= (285.4\ \text{kJ}) - (298.15\ \text{K})(-137.55\ \text{J/K})\left[\frac{1\ \text{kJ}}{1000\ \text{J}}\right] = 326.4\ \text{kJ}$$

$$\Delta G_f° = \frac{326.4\ \text{kJ}}{2\ \text{mol}} = 163.2\ \text{kJ/mol}$$

This reaction is not spontaneous under standard conditions.

23.71 (a) $\Delta H° = [(1\ \text{mol})\Delta H_f°(H^+) + (1\ \text{mol})\Delta H_f°(OH^-)] - [(1\ \text{mol})\Delta H_f°(H_2O)]$

$= [(1\ \text{mol})(0) + (1\ \text{mol})(-229.994\ \text{kJ/mol})]$

$- [(1\ \text{mol})(-285.830\ \text{kJ/mol})]$

$= 55.836\ \text{kJ}$

(b) $\Delta G° = [(1\ \text{mol})\Delta G_f°(H^+) + (1\ \text{mol})\Delta G_f°(OH^-)] - [(1\ \text{mol})\Delta G_f°(H_2O)]$

$= [(1\ \text{mol})(0) + (1\ \text{mol})(-157.244\ \text{kJ/mol})]$

$- [(1\ \text{mol})(-237.129\ \text{kJ/mol})]$

$= 79.885\ \text{kJ}$

(c)
$$\log K = \frac{-\Delta G°}{(2.303)RT} = \left[\frac{-(79.885\ \text{kJ})}{(2.303)(8.314\ \text{J/K mol})(298.15\ \text{K})}\right]\left[\frac{1000\ \text{J}}{1\ \text{kJ}}\right]$$

$= -13.99$

$K = 1.0 \times 10^{-14}$

(d)
$$\log\left[\frac{K_2}{K_1}\right] = \frac{-\Delta H°}{(2.303)R}\left[\frac{1}{T_2} - \frac{1}{T_1}\right]$$

$$\log\left[\frac{K_2}{1.0 \times 10^{-14}}\right] = \left[\frac{-(55.836\ \text{kJ})}{(2.303)(8.314\ \text{J/K mol})}\right]\left[\frac{1000\ \text{J}}{1\ \text{kJ}}\right]\left[\frac{1}{308\ \text{K}} - \frac{1}{298.15\ \text{K}}\right]$$

$= 0.31$

$$\frac{K_2}{1.0 \times 10^{-14}} = 2.0$$

$K_2 = 2.0 \times 10^{-14}$

(e) $K = [H^+][OH^-] = [H^+]^2$

$[H^+] = \sqrt{K} = \sqrt{2.0 \times 10^{-14}} = 1.4 \times 10^{-7}\ \text{mol/L}$

$\text{pH} = -\log\,[H^+] = -\log(1.4 \times 10^{-7}) = 6.85$

The solution does not become acidic at 35 °C because $[H^+] = [OH^-]$.

CHAPTER 24

OXIDATION-REDUCTION AND ELECTROCHEMISTRY

Solutions to Exercises

24.1 $Cl_2(g) + 2e^- \rightarrow 2Cl^-$

The redox couple for this ion-electron equation is $Cl_2(g)/Cl^-$.

$2Br^- \rightarrow Br_2(l) + 2e^-$

The redox couple for this ion-electron equation is $Br_2(l)/Br^-$.

24.2 (a) The oxidized and reduced substances in acidic media are

$$MnO_4^- \rightarrow Mn^{2+}$$

a. The Mn atoms are balanced.

b. Balance O atoms by adding H_2O on the side deficient in O atoms.

$$MnO_4^- \rightarrow Mn^{2+} + 4H_2O$$

c. Balance H atoms by adding H^+ for acidic solutions.

$$MnO_4^- + 8H^+ \rightarrow Mn^{2+} + 4H_2O$$

d. Balance charge by adding electrons.

$$MnO_4^- + 8H^+ + 5e^- \rightarrow Mn^{2+} + 4H_2O(l)$$

(b) The oxidized and reduced substances in alkaline media are

$$MnO_4^- \rightarrow MnO_2$$

a. The Mn atoms are balanced.

b. Balance O atoms by adding H_2O on the side deficient in O atoms.

$$MnO_4^- \rightarrow MnO_2 + 2H_2O$$

c. Balance H atoms by adding OH^- and H_2O for alkaline solutions.

$$MnO_4^- \rightarrow MnO_2 + 2H_2O + 4OH^-$$

$$MnO_4^- + 4H_2O \rightarrow MnO_2 + 2H_2O + 4OH^-$$

$$MnO_4^- + 2H_2O \rightarrow MnO_2 + 4OH^-$$

d. Balance charge by adding electrons.

$$MnO_4^- + 2H_2O(l) + 3e^- \rightarrow MnO_2(s) + 4OH^-$$

24.3 Step 1. Write the overall unbalanced equation.

$$H_2O_2 + ClO_3^- \rightarrow H_2O + ClO_4^-$$

Step 2. Identify oxidized and reduced substances and write the unbalanced ion-electron equations.

$H_2O_2 \rightarrow H_2O$ $\qquad ClO_3^- \rightarrow ClO_4^-$

Step 3. Balance each ion-electron equation for atoms and charge.

a. Balance for atoms other than O or H.

$H_2O_2 \rightarrow H_2O$ $\qquad ClO_3^- \rightarrow ClO_4^-$

b. Balance O atoms by adding H_2O on the side deficient in O atoms.

$H_2O_2 \rightarrow 2H_2O$ $\qquad H_2O + ClO_3^- \rightarrow ClO_4^-$

c. Balance H atoms by adding H^+ for acidic solutions.

$2H^+ + H_2O_2 \rightarrow 2H_2O$ $\qquad H_2O + ClO_3^- \rightarrow ClO_4^- + 2H^+$

d. Balance charge by adding electrons.

$2H^+ + 2e^- + H_2O_2 \rightarrow 2H_2O(l)$ $\qquad H_2O(l) + ClO_3^- \rightarrow ClO_4^- + 2H^+ + 2e^-$

Step 4. Multiply ion-electron equations by appropriate factors so that electrons gained equals electrons lost.

Step 5. Add the ion-electron equations, cancelling when appropriate.

$$2H^+ + 2e^- + H_2O_2(aq) \rightarrow 2H_2O(l)$$
$$H_2O(l) + ClO_3^- \rightarrow ClO_4^- + 2H^+ + 2e^-$$
$$\overline{H_2O_2(aq) + ClO_3^- \rightarrow ClO_4^- + H_2O(l)}$$

24.4 Step 1. Write the overall unbalanced equation.

$Mg + CrO_4^{2-} \rightarrow Mg(OH)_2 + Cr(OH)_3$

Step 2. Identify oxidized and reduced substances and write the unbalanced ion-electron equations.

$Mg \rightarrow Mg(OH)_2$ $\qquad CrO_4^{2-} \rightarrow Cr(OH)_3$

Step 3. Balance each ion-electron equation for atoms and charge.

a. Balance for atoms other than O or H.

b. Balance O atoms by adding H_2O on side deficient in O atoms.

$2H_2O + Mg \rightarrow Mg(OH)_2$ $\qquad CrO_4^{2-} \rightarrow Cr(OH)_3 + H_2O$

c. Balance H atoms by adding OH^- for alkaline solutions.

$2H_2O + 2OH^- + Mg \rightarrow Mg(OH)_2$ $\qquad CrO_4^{2-} \rightarrow 5OH^- + Cr(OH)_3 + H_2O$

$2H_2O + 2OH^- + Mg \rightarrow Mg(OH)_2 + 2H_2O$ $\qquad 5H_2O + CrO_4^{2-} \rightarrow 5OH^- + Cr(OH)_3 + H_2O$

$2OH^- + Mg \rightarrow Mg(OH)_2$ $\qquad 4H_2O + CrO_4^{2-} \rightarrow 5OH^- + Cr(OH)_3$

d. Balance charge by adding electrons.

$2OH^- + Mg(s) \rightarrow Mg(OH)_2(s) + 2e^-$ $\qquad 4H_2O(l) + CrO_4^{2-} + 3e^- \rightarrow 5OH^- + Cr(OH)_3(s)$

Step 4. Multiply ion-electron equations by appropriate factors so that electrons gained equals electrons lost.

$$(3)[2OH^- + Mg(s) \rightarrow Mg(OH)_2(s) + 2e^-]$$
$$(2)[4H_2O(l) + CrO_4^{2-} + 3e^- \rightarrow 5OH^- + Cr(OH)_3(s)]$$

Step 5. Add the ion-electron equations, cancelling when appropriate.

$$6OH^- + 3Mg(s) \rightarrow 3Mg(OH)_2(s) + 6e^-$$
$$8H_2O(l) + 2CrO_4^{2-} + 6e^- \rightarrow 10\ OH^- + 2Cr(OH)_3(s)$$
$$\overline{3Mg(s) + 2CrO_4^{2-} + 8H_2O(l) \rightarrow 3Mg(OH)_2(s) + 2Cr(OH)_3(s) + 4OH^-}$$

24.5

$$H_2O_2(aq) \rightarrow O_2(g) + 2H^+ + 2e^-$$
$$H_2O_2(aq) + 2H^+ + 2e^- \rightarrow 2H_2O(l)$$
$$\overline{2H_2O_2(aq) \rightarrow 2H_2O(l) + O_2(g)}$$

24.6 anode: $Fe(s) \rightarrow Fe^{2+} + 2e^-$

cathode: $Sn^{4+} + 2e^- \rightarrow Sn^{2+}$

overall: $Fe(s) + Sn^{4+} \rightarrow Fe^{2+} + Sn^{2+}$

24.7

$$(0.015\ A)(15\ min)\left[\frac{60\ s}{1\ min}\right]\left[\frac{1\ C}{1\ A\ s}\right] = 14\ C$$

$$(14\ C)\left[\frac{1\ mol\ e^-}{96{,}500\ C}\right] = 1.5 \times 10^{-4}\ mol\ e^-$$

$$Zn(s) \rightarrow Zn^{2+} + 2e^-$$

$$(1.5 \times 10^{-4}\ mol\ e^-)\left[\frac{1\ mol\ Zn}{2\ mol\ e^-}\right]\left[\frac{65.38\ g\ Zn}{1\ mol\ Zn}\right] = 0.0049\ g\ Zn$$

24.8

$$q = \frac{-w}{E} = \left[\frac{-(-1.0\ kJ)}{1.52\ V}\right]\left[\frac{1000\ J}{1\ kJ}\right]\left[\frac{1\ V\ C}{1\ J}\right] = 660\ C$$

24.9

$$(3)[Zn(s) + 2OH^- \rightarrow Zn(OH)_2(s) + 2e^-] \qquad E° = 1.245\ V$$
$$IO_3^- + 3H_2O(l) + 6e^- \rightarrow I^- + 6OH^- \qquad E° = 0.26\ V$$
$$\overline{3Zn(s) + IO_3^- + 3H_2O(l) \rightarrow I^- + 3Zn(OH)_2(s) \qquad E° = 1.51\ V}$$

24.10 The order of the $E°$ values for the possible reactions of the species under consideration is as follows:

$$NO_3^- + H_2O(l) + 2e^- \rightarrow NO_2^- + 2OH^- \qquad E° = 0.01\ V$$
$$ClO_3^- + H_2O(l) + 2e^- \rightarrow ClO_2^- + 2OH^- \qquad E° = 0.33\ V$$
$$ClO_4^- + H_2O(l) + 2e^- \rightarrow ClO_3^- + 2OH^- \qquad E° = 0.36\ V$$
$$ClO^- + H_2O(l) + 2e^- \rightarrow Cl^- + 2OH^- \qquad E° = 0.89\ V$$

The order of strength as oxidizing agents is $NO_3^- < ClO_3^- < ClO_4^- < ClO^-$.

24.11

$$(2)[CrO_4^{2-} + 4H_2O(l) + 3e^- \rightarrow Cr(OH)_3(s) + 5OH^-] \qquad E° = -0.13\ V$$
$$I^- + 6OH^- \rightarrow IO_3^- + 3H_2O(l) + 6e^- \qquad E° = -0.26\ V$$
$$2CrO_4^{2-} + 5H_2O(l) + I^- \rightarrow 2Cr(OH)_3(s) + 4OH^- + IO_3^- \qquad E° = -0.39\ V$$

Iodide ion cannot be oxidized to iodate ion by reaction with chromate ion in alkaline solution under standard state conditions.

$$Cr_2O_7^{2-} + 14H^+ + 6e^- \rightarrow 2Cr^{3+} + 7H_2O(l) \qquad E° = 1.33\ V$$
$$(3)[2I^- \rightarrow I_2(s) + 2e^-] \qquad E° = -0.5355\ V$$
$$Cr_2O_7^{2-} + 6I^- + 14H^+ \rightarrow 2Cr^{3+} + 3I_2(s) + 7H_2O(l) \qquad E° = 0.79\ V$$

Iodide ion is oxidized to iodine using dichromate ion in acidic solution under standard state conditions.

$$(5)[Cr_2O_7^{2-} + 14H^+ + 6e^- \rightarrow 2Cr^{3+} + 7H_2O(l)] \qquad E° = 1.33\ V$$
$$(3)[I_2(s) + 6H_2O \rightarrow 2IO_3^- + 12H^+ + 10e^-] \qquad E° = -1.195\ V$$
$$5Cr_2O_7^{2-} + 34H^+ + 3I_2(s) \rightarrow 10Cr^{3+} + 6IO_3^- + 17H_2O(l) \qquad E° = 0.14\ V$$

Iodine is then oxidized to iodate ion under standard state conditions.

24.12

$$2[Fe^{3+} + e^- \rightarrow Fe^{2+}]$$
$$2I^- \rightarrow I_2(s) + 2e^-$$
$$2Fe^{3+} + 2I^- \rightarrow 2Fe^{2+} + I_2(s)$$

$$\Delta G° = -nFE° = -(2\ mol\ e^-)(96.5\ kJ/V\ mol\ e^-)(0.236\ V) = -45.5\ kJ$$

24.13 The reaction quotient is

$$Q = \frac{[Mn^{2+}]}{[MnO_4^-][H^+]^8} = \frac{(2.5 \times 10^{-5})}{(0.10)(1.3 \times 10^{-2})^8} = 3.1 \times 10^{11}$$

The Nernst equation gives the reduction potential under these conditions as

$$E = E° - \frac{0.0592}{n} \log Q = (1.51\ V) - \frac{0.0592}{5} \log(3.1 \times 10^{11}) = 1.37\ V$$

Because $E < E°$, MnO_4^- under these conditions is a weaker oxidizing agent.

24.14 $$E° = \frac{(0.0592)\log K}{n} = \frac{(0.0592)\log(9.9 \times 10^5)}{2} = 0.177\ V$$

23.15

$$2IO_3^- + 12H^+ + 10e^- \rightarrow I_2 + 6H_2O(l) \qquad E° = 1.195\ V$$
$$I_2(s) + 2e^- \rightarrow 2I^- \qquad E° = 0.5365\ V$$
$$2IO_3^- + 12H^+ + 12e^- \rightarrow 2I^- + 6H_2O(l)$$

$$E° = \frac{(10)(1.195\ V) + (2)(0.5365\ V)}{12} = 1.085\ V$$

Solutions to Odd-Numbered Questions and Problems

24.1 Oxidation is electron loss and reduction is electron gain. The number of electrons gained is equal to the number of electrons lost in a redox reaction.

24.3 (a) $H_2O_2(aq)/H_2O(l)$ and $I_2(s)/I^-$

$$H_2O_2(aq) + 2H^+ + 2e^- \rightarrow 2H_2O(l)$$
$$2I^- \rightarrow I_2(s) + 2e^-$$
$$\overline{H_2O_2(aq) + 2I^- + 2H^+ \rightarrow 2H_2O(l) + I_2(s)}$$

(b) $I_2(s)/I^-$ and $NO_2^-/NO(g)$

$$(2)[NO_2^- + 2H^+ + e^- \rightarrow NO(g) + H_2O(l)]$$
$$2I^- \rightarrow I_2(s) + 2e^-$$
$$\overline{2NO_2^- + 4H^+ + 2I^- \rightarrow 2NO(g) + I_2(s) + 2H_2O(l)}$$

(c) $Al^{3+}/Al(s)$ and $H_2SO_4(conc)/SO_2(g)$

$$(2)[Al(s) \rightarrow Al^{3+} + 3e^-]$$
$$(3)[H_2SO_4(conc) + 2H^+ + 2e^- \rightarrow SO_2(g) + 2H_2O(l)]$$
$$\overline{2Al(s) + 3H_2SO_4(conc) + 6H^+ \rightarrow 2Al^{3+} + 3SO_2(g) + 6H_2O(l)}$$

(d) $[Zn(OH)_4]^{2-}/Zn(s)$ and $NO_3^-/NH_3(aq)$

$$(4)[Zn(s) + 4OH^- \rightarrow [Zn(OH)_4]^{2-} + 2e^-]$$
$$NO_3^- + 6H_2O(l) + 8e^- \rightarrow NH_3(aq) + 9OH^-$$
$$\overline{4Zn(s) + NO_3^- + 6H_2O(l) + 7OH^- \rightarrow 4[Zn(OH)_4]^{2-} + NH_3(aq)}$$

24.5 (a)

$$I_2(s) + 2e^- \rightarrow 2I^-$$
$$H_2S(aq) \rightarrow S(s) + 2H^+ + 2e^-$$
$$\overline{H_2S(aq) + I_2(s) \rightarrow 2I^- + S(s) + 2H^+}$$

(b)

$$PbO_2(s) + 2Cl^- + 4H^+ + 2e^- \rightarrow PbCl_2(s) + 2H_2O(l)$$
$$2Cl^- \rightarrow Cl_2(g) + 2e^-$$
$$\overline{PbO_2(s) + 4Cl^- + 4H^+ \rightarrow PbCl_2(s) + Cl_2(g) + 2H_2O(l)}$$

(c)

$$2Cu(s) + 2OH^- \rightarrow Cu_2O(s) + H_2O(l) + 2e^-$$
$$Br_2(aq) + 2e^- \rightarrow 2Br^-$$
$$\overline{2Cu(s) + Br_2(aq) + 2OH^- \rightarrow Cu_2O(s) + 2Br^- + H_2O(l)}$$

(d)

$$\begin{array}{rcl} S^{2-} + 8OH^- & \rightarrow & SO_4{}^{2-} + 4H_2O(l) + 8e^- \\ (4)[Cl_2(g) + 2e^- & \rightarrow & 2Cl^-] \\ \hline S^{2-} + 4Cl_2(g) + 8OH^- & \rightarrow & SO_4{}^{2-} + 8Cl^- + 4H_2O(l) \end{array}$$

(e)

$$\begin{array}{rcl} (2)[MnO_4{}^- + 2H_2O(l) + 3e^- & \rightarrow & MnO_2(s) + 4OH^-] \\ (3)[IO_3{}^- + 2OH^- & \rightarrow & IO_4{}^- + H_2O(l) + 2e^-] \\ \hline 2MnO_4{}^- + 3IO_3{}^- + H_2O(l) & \rightarrow & 2MnO_2(s) + 3IO_4{}^- + 2OH^- \end{array}$$

24.7 (a)

$$\begin{array}{rcl} (6)[[Fe(CN)_6]^{4-} & \rightarrow & [Fe(CN)_6]^{3-} + e^-] \\ 2CrO_4{}^{2-} + 5H_2O(l) + 6e^- & \rightarrow & Cr_2O_3(s) + 10OH^- \\ \hline 6[Fe(CN)_6]^{4+} + 2CrO_4{}^{2-} + 5H_2O(l) & \rightarrow & 6[Fe(CN)_6]^{3-} + Cr_2O_3(s) + 10OH^- \end{array}$$

(b)

$$\begin{array}{rcl} I_2(s) + 12OH^- & \rightarrow & 2IO_3{}^- + 6H_2O(l) + 10e^- \\ (5)[ClO^- + H_2O(l) + 2e^- & \rightarrow & Cl^- + 2OH^-] \\ \hline I_2(s) + 2OH^- + 5ClO^- & \rightarrow & 5Cl^- + 2IO_3{}^- + H_2O(l) \end{array}$$

(c)

$$\begin{array}{rcl} CN^- + 2OH^- & \rightarrow & CNO^- + H_2O(l) + 2e^- \\ (2)[[Fe(CN)_6]^{3-} + e^- & \rightarrow & [Fe(CN)_6]^{4-}] \\ \hline 2[Fe(CN)_6]^{3-} + CN^- + 2OH^- & \rightarrow & 2[Fe(CN)_6]^{4-} + CNO^- + H_2O(l) \end{array}$$

(d)

$$\begin{array}{rcl} (2)[Cu(OH)_2(s) + 2e^- & \rightarrow & Cu(s) + 2OH^-] \\ N_2H_4(aq) + 4OH^- & \rightarrow & N_2(g) + 4H_2O(l) + 4e^- \\ \hline 2Cu(OH)_2(s) + N_2H_4(aq) & \rightarrow & N_2(g) + 2Cu(s) + 4H_2O(l) \end{array}$$

24.9 The two classifications of electrochemical cells are voltaic, or galvanic, cells and electrolytic cells. The voltaic cell generates electrical energy from a spontaneous redox reaction and the electrolytic cell uses electrical energy from outside the cell to cause a redox reaction to occur.

24.11 See Fig. 24-1. A salt bridge is not necessary for this reaction.

24.13

$$\begin{array}{rcl} H_2(g) & \rightarrow & 2H^+ + 2e^- \\ (2)[Fe^{3+} + e^- & \rightarrow & Fe^{2+}] \\ \hline 2Fe^{3+} + H_2(g) & \rightarrow & 2H^+ + 2Fe^{2+} \end{array}$$

24.15 $Zn^{2+} + 2e^- \rightarrow Zn(s)$

$$(1 \text{ mol } e^-)\left[\frac{1 \text{ mol } Zn^{2+}}{2 \text{ mol } e^-}\right]\left[\frac{65.38 \text{ g } Zn^{2}}{1 \text{ mol } Zn^{2+}}\right] = 32.69 \text{ g } Zn^{2+}$$

24.17 $Na^+ + e^- \rightarrow Na(s)$

$$\text{electrical charge} = It = (15 \text{ A})(3.0 \text{ h})\left[\frac{3600 \text{ s}}{1 \text{ h}}\right]\left[\frac{1 \text{ C}}{1 \text{ A s}}\right] = 1.6 \times 10^5 \text{ C}$$

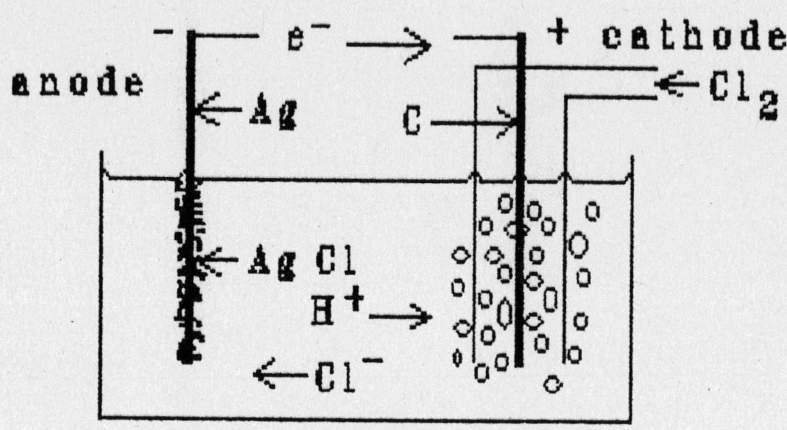

$$(1.6 \times 10^5\ \text{C})\left[\frac{1\ \text{mol}\ e^-}{96{,}500\ \text{C}}\right] = 1.7\ \text{mol}\ e^-$$

$$(1.7\ \text{mol}\ e^-)\left[\frac{1\ \text{mol Na}}{1\ \text{mol}\ e^-}\right]\left[\frac{22.99\ \text{g Na}}{1\ \text{mol Na}}\right] = 39\ \text{g Na}$$

24.19 electrical charge = $It = (1.5\ \text{A})(5.0\ \text{h})\left[\frac{3600\ \text{s}}{1\ \text{h}}\right]\left[\frac{1\ \text{C}}{1\ \text{A s}}\right] = 2.7 \times 10^4\ \text{C}$

$$(2.7 \times 10^4\ \text{C})\left[\frac{1\ \text{mol}\ e^-}{96{,}500\ \text{C}}\right] = 0.28\ \text{mol}\ e^-$$

$$(0.28\ \text{mol}\ e^-)\left[\frac{1\ \text{mol}\ Cl_2}{2\ \text{mol}\ e^-}\right] = 0.14\ \text{mol}\ Cl_2$$

$$V = \frac{nRT}{P} = \frac{(0.14\ \text{mol})(0.0821\ \text{L atm/K mol})(358\ \text{K})}{(745\ \text{Torr})(1\ \text{atm}/760\ \text{Torr})} = 4.2\ \text{L}$$

$$(4.2\ \text{L})(0.75) = 3.1\ \text{L}$$

24.21 $[PtCl_6]^{2-} + 4e^- \rightarrow Pt(s) + 6Cl^-$

$$(0.50\ \text{g Pt})\left[\frac{1\ \text{mol Pt}}{195.08\ \text{g Pt}}\right]\left[\frac{4\ \text{mol}\ e^-}{1\ \text{mol Pt}}\right]\left[\frac{96{,}500\ \text{C}}{1\ \text{mol}\ e^-}\right] = 990\ \text{C}$$

$$I = \frac{\text{charge}}{t} = \frac{(990\ \text{C})(1\ \text{A s}/1\ \text{C})}{(5.0\ \text{h})\left[\frac{3600\ \text{s}}{1\ \text{h}}\right]} = 0.055\ \text{A}$$

24.23 $Ag^+ + e^- \rightarrow Ag(s)$ $\qquad Au^{n+} + ne^- \rightarrow Au(s)$

$$(0.583\ \text{g Ag})\left[\frac{1\ \text{mol Ag}}{107.87\ \text{g Ag}}\right]\left[\frac{1\ \text{mol}\ e^-}{1\ \text{mol Ag}}\right] = 5.40 \times 10^{-3}\ \text{mol}\ e^-$$

$$(0.355\ \text{g Au})\left[\frac{1\ \text{mol Au}}{196.97\ \text{g Au}}\right] = 1.80 \times 10^{-3}\ \text{mol Au}$$

$$\frac{5.40 \times 10^{-3} \text{ mol } e^-}{1.80 \times 10^{-3} \text{ mol Au}} = 3 \text{ mol } e^-/1 \text{ mol Au}$$

Therefore, $n = 3$ and the oxidation state of gold in this salt is +3.

24.25 (a) The order of increasing strength as oxidizing agents is $K^+ < Na^+ < Fe^{2+} < Cu^{2+} < Cu^+ < Ag^+ < F_2$.

(b) The oxidizing agents which will oxidize Cu under standard state conditions include Ag^+ and F_2.

24.27 The strongest reducing agents have the most negative $E°$ values. The activity series in order of decreasing activity as reducing agents is

$$Eu > Ra > Pu > Rh$$

Europium is a more active metal than lithium (more negative $E°$ value). Europium, radium, and plutonium are all more active reducing agents than hydrogen (all negative $E°$ values). Rhodium is a less active reducing agent than hydrogen, but not as "noble" a metal--not as unreactive--as gold.

24.29 The data show that within a narrow range of $E°$ values nitric acid can form several different products and these products can undergo various further redox reactions. One must conclude that it would be very difficult to make qualitative predictions of what the products will be when nitric acid reacts as an oxidizing agent.

24.31 Under standard state conditions,

(a) the strongest reducing agent is Mg(s) because $E°$ for $Mg(OH)_2(s)/Mg(s)$ is the most negative

(b) both Mg(s) and Al(s) could reduce $Zn(OH)_2(s)$ to Zn(s) because both couples have more negative $E°$ values than $Zn(OH)_2(s)/Zn(s)$

(c) Mg(s), Al(s), and Zn(s) could reduce $Fe(OH)_2(s)$ to Fe(s) because all three couples have more negative $E°$ values than $Fe(OH)_2(s)/Fe(s)$

24.33

$$\begin{array}{lll} Zn(s) & \rightarrow Zn^{2+} + 2e^- & E° = 0.76\text{ V} \\ \underline{Cl_2(g) + 2e^-} & \underline{\rightarrow 2Cl^-} & \underline{E° = 1.36\text{ V}} \\ Zn(s) + Cl_2(g) & \rightarrow Zn^{2+} + 2Cl^- & E° = 2.12\text{ V} \end{array}$$

This is a spontaneous reaction under standard state conditions.

24.35 (a)

$$\begin{array}{lll} Cd(s) + 2OH^- & \rightarrow Cd(OH)_2(s) + 2e^- & E° = 0.809\text{ V} \\ \underline{HgO(s) + H_2O(l) + 2e^-} & \underline{\rightarrow Hg(l) + 2OH^-} & \underline{E° = 0.098\text{ V}} \\ Cd(s) + HgO(s) + H_2O(l) & \rightarrow Cd(OH)_2(s) + Hg(l) & E° = 0.907\text{ V} \end{array}$$

The reaction is spontaneous under standard state conditions.

(b)

$$\begin{array}{lll} (2)[Fe(s) \rightarrow Fe^{3+} + 3e^-] & & E° = 0.036\ V \\ (3)[2H^+ + 2e^- \rightarrow H_2(g)] & & E° = 0.000\ V \\ \hline 2Fe(s) + 6H^+ \rightarrow 3H_2(g) + 2Fe^{3+} & & E° = 0.036\ V \end{array}$$

The reaction is spontaneous under standard state conditions.

(c)

$$\begin{array}{ll} (3)[2Cl^- \rightarrow Cl_2(g) + 2e^-] & E° = -1.36\ V \\ Cr_2O_7^{2-} + 14H^+ + 6e^- \rightarrow 2Cr^{3+} + 7H_2O(l) & E° = 1.33\ V \\ \hline Cr_2O_7^{2-} + 14H^+ + 6Cl^- \rightarrow 3Cl_2(g) + 2Cr^{3+} + 7H_2O(l) & E° = -0.03\ V \end{array}$$

The reaction is nonspontaneous under standard state conditions.

(d)

$$\begin{array}{ll} (3)[H_2SO_3(aq) + H_2O(l) \rightarrow SO_4^{2-} + 4H^+ + 2e^-] & E° = -0.17\ V \\ (2)[NO_3^- + 4H^+ + 3e^- \rightarrow NO(g) + 2H_2O(l)] & E° = 0.96\ V \\ \hline 3H_2SO_3(aq) + 2NO_3^- \rightarrow 3SO_4^{2-} + 2NO(g) + 4H^+ + H_2O(l) & E° = 0.79\ V \end{array}$$

The reaction is spontaneous under standard state conditions.

24.37 Several reduction processes could take place at the cathode. These are

$Na^+ + e^- \rightarrow Na(s)$ $E° = -2.714$ V

$2H^+ + 2e^- \rightarrow H_2(g)$ $E° = 0.0000$ V

$2H_2O(l) + 2e^- \rightarrow H_2(g) + 2OH^-$ $E° = -0.8281$ V

$SO_4^{2-} + H_2O(l) + 2e^- \rightarrow SO_3^{2-} + 2OH^-$ $E° = -0.93$ V

The process most likely to occur is the one with the most positive $E°$ value--the reduction of H^+ to H_2. Either the oxidation of water or of sulfate ion can occur at the anode.

$O_2(g) + 4H^+ + 4e^- \rightarrow 2H_2O(l)$ $E° = 1.229$ V

$S_2O_8^{2-} + 2e^- \rightarrow 2SO_4^{2-}$ $E° = 2.01$ V

The process more likely to occur is the one with the lesser positive $E°$ value--the oxidation of water to O_2.

24.39 The chemist had forgotten to consider that H^+, not K^+, is reduced at the cathode to produce $H_2(g)$, which is flammable.

24.41 $\Delta G° = -nFE° = -(1\ \text{mol}\ e^-)(96.485\ \text{kJ/V mol}\ e^-)(1.776\ \text{V}) = -171.4\ \text{kJ}$

24.43

$$\begin{array}{l} H_2(g) \rightarrow 2H^+ + 2e^- \\ \frac{1}{2}O_2(g) + 2H^+ + 2e^- \rightarrow H_2O(l) \\ \hline H_2(g) + \frac{1}{2}O_2(g) \rightarrow H_2O(l) \end{array}$$

$$E° = \frac{-\Delta G°}{nF} = \frac{-(-237.2\ \text{kJ})}{(2\ \text{mol}\ e^-)(96.485\ \text{kJ/V mol}\ e^-)} = 1.229\ \text{V}$$

24.45 $\Delta G° = -nFE° = -(3 \text{ mol } e^-)(96.5 \text{ kJ/V mol } e^-)(-0.036 \text{ V}) = 10. \text{ kJ}$

$\Delta G° = -(1 \text{ mol } e^-)(96.5 \text{ kJ/V mol } e^-)(-0.036 \text{ V}) = 3.5 \text{ kJ}$

24.47 $Q = [Co^{2+}] = 1 \times 10^{-4} \text{ mol/L}$

$$E = E° - \frac{0.0592}{n} \log Q = (0.277) - \frac{0.0592}{2} \log(1 \times 10^{-4}) = 0.40 \text{ V}$$

24.49 $Ag^+ + e^- \rightarrow Ag(s)$

$$Q = \frac{1}{[Ag^+]}$$

$$E = E° - \frac{0.0592}{n} \log Q$$

$$0.35 = 0.80 - \frac{0.0592}{1} \log\left[\frac{1}{[Ag^+]}\right]$$

$$\log\left[\frac{1}{[Ag^+]}\right] = 7.6$$

$$\frac{1}{[Ag^+]} = 4 \times 10^7$$

$$[Ag^+] = 3 \times 10^{-8} \text{ mol/L}$$

24.51

$(2)[H_2(g) \rightarrow 2H^+ + 2e^-]$	$E° = 0.000$ V
$O_2(g) + 4H^+ + 4e^- \rightarrow 2H_2O(l)$	$E° = 1.229$ V
$2H_2(g) + O_2(g) \rightarrow 2H_2O(l)$	$E° = 1.229$ V

$$Q = \frac{1}{P_{H_2}^2 \; P_{O_2}} = \frac{1}{(5.0)^2(9.0)} = 4.4 \times 10^{-3}$$

$$E = E° - \frac{0.0592}{n} \log Q = 1.229 - \frac{0.0592}{4} \log(4.4 \times 10^{-3}) = 1.264 \text{ V}$$

24.53

$$Q = \frac{P_{H_2} \; [M^{2+}]}{[H^+]^2} = \frac{(1.0)(0.10)}{(1.0)^2} = 0.10$$

$$E° = E + \frac{0.0592}{n} \log Q = (0.500) + \frac{0.0592}{2} \log(0.10) = 0.470 \text{ V}$$

$M \rightarrow M^{2+} + 2e^-$	$E° = E°_M$
$2H^+ + 2e^- \rightarrow H_2(g)$	$E° = 0.000$ V
$M + 2H^+ \rightarrow M^{2+} + H_2(g)$	$E° = E°_M + 0.000 = E° = 0.470$ V

Because $E°_M = 0.470$ V for oxidation, the standard reduction potential is −0.470 V.

24.55 (a)

$$\begin{array}{lcll} 2Hg(l) & \rightarrow & Hg_2^{2+} + 2e^- & E° = -0.788\ V \\ (2)[Fe^{3+} + e^- & \rightarrow & Fe^{2+}] & E° = 0.771\ V \\ \hline 2Hg(l) + 2Fe^{3+} & \rightarrow & 2Fe^{2+} + Hg_2^{2+} & E° = -0.017\ V \end{array}$$

(b) (inert)|Hg(l)|Hg_2^{2+}‖Fe^{2+}, Fe^{3+}|(inert)

(c) -0.017 V; the reaction is not spontaneous under standard conditions.

(d) $$Q = \frac{[Fe^{2+}]^2[Hg_2^{2+}]}{[Fe^{3+}]^2} = \frac{(0.10)^2(0.0010)}{(1.00)^2} = 1.0 \times 10^{-5}$$

$$E = E° - \frac{0.0592}{n} \log Q = (-0.017) - \frac{0.0592}{2} \log(1.0 \times 10^{-5}) = 0.131\ V$$

The reaction is more favorable under these conditions compared to standard state conditions.

24.57

$$\begin{array}{lcll} (2)[K(s) & \rightarrow & K^+ + e^-] & E° = 2.925\ V \\ 2H_2O(l) + 2e^- & \rightarrow & H_2(g) + 2OH^- & E° = -0.8281\ V \\ \hline 2K(s) + 2H_2O(l) & \rightleftarrows & 2K^+ + 2OH^- + H_2(g) & E° = 2.097\ V \end{array}$$

$$\log K = \frac{n}{0.0592} E° = \frac{(2)(2.097)}{(0.0592)} = 70.8$$

$K = 6 \times 10^{70}$

24.59

$$\begin{array}{lcll} PbSO_4(s) + 2e^- & \rightarrow & Pb(s) + SO_4^{2-} & E° = -0.359\ V \\ Pb(s) + 2I^- & \rightarrow & PbI_2(s) + 2e^- & E° = 0.365\ V \\ \hline PbSO_4(s) + 2I^- & \rightleftarrows & PbI_2(s) + SO_4^{2-} & E° = 0.006\ V \end{array}$$

$$\log K = \frac{n}{0.0592} E° = \frac{(2)(0.006)}{(0.0592)} = 0.2$$

$K = 2$

24.61

$$\begin{array}{lcll} AgBr(s) + e^- & \rightarrow & Ag(s) + Br^- & E° = 0.071\ V \\ Ag(s) & \rightarrow & Ag^+ + e^- & E° = -0.799\ V \\ \hline AgBr(s) & \rightleftarrows & Ag^+ + Br^- & E° = -0.728\ V \end{array}$$

$$\log K = \frac{n}{0.0592} E° = \frac{(1)(-0.728)}{0.0592} = -12.3$$

$K = 5 \times 10^{-13}$

24.63

$$\begin{array}{lcll} Yb^{3+} + 3e^- & \rightarrow & Yb(s) & E° = -2.267\ V \\ Yb(s) & \rightarrow & Yb^{2+} + 2e^- & E° = 2.797\ V \\ \hline Yb^{3+} + e^- & \rightarrow & Yb^{2+} & \end{array}$$

$$E° = \frac{(3)(-2.267\ V) + (2)(2.797\ V)}{(1)} = -1.207\ V$$

24.65 $\Delta G° = -nFE° = -(1\ mol\ e^-)(96.485\ kJ/V\ mol\ e^-)(0.7991\ V) = -77.10\ kJ$

$\Delta G° = -(2.303)RT \log K = -(2.303)(8.314\ J/K\ mol)(1\ kJ/1000\ J)(298\ K) \times \log(1.8 \times 10^{-10})$

$= 55.6\ kJ$

$Ag^+ + e^- \rightarrow Ag(s)$	$\Delta G° = -77.10\ kJ$
$AgCl(s) \rightarrow Ag^+ + Cl^-$	$\Delta G° = 55.6\ kJ$
$AgCl(s) + e^- \rightarrow Ag(s) + Cl^-$	$\Delta G° = -21.5\ kJ$

$$E° = -\frac{\Delta G°}{nF} = \frac{-(-21.5\ kJ)}{(1\ mol\ e^-)(96.5\ kJ/mol\ e^-\ V)} = 0.223\ V$$

24.67 Figure 24.8 of the text shows that $Fe(OH)_3$ is formed by Fe^{3+} until the pH becomes less than about 1. The correct equation would be

$$Fe(s) + 3OH^- \rightarrow Fe(OH)_3(s) + 3e^-$$

24.69 A storage cell works by a favorable redox reaction occurring during discharge and electrolysis occurring during recharge.

24.71 The physical size of a commercial cell does not govern the potential that it will deliver because the potential is dependent only on the half-reactions. The size determines the amount of electrical work that will be produced.

24.73 (a)

$(6)[Ag(s) + Cl^- \rightarrow AgCl(s) + e^-]$	(b) $E° = -0.22\ V$
$Cr_2O_7^{2-} + 14H^+ + 6e^- \rightarrow 2Cr^{3+} + 7H_2O(l)$	$E° = 1.33\ V$
$6Ag(s) + 6Cl^- + Cr_2O_7^{2-} + 14H^+ \rightarrow 2Cr^{3+} + 7H_2O(l) + 6AgCl(s)$	$E° = 1.11\ V$

(c) $$Q = \frac{[Cr^{3+}]^2}{[Cl^-]^6[Cr_2O_7^{2-}][H^+]^{14}} = \frac{(0.1)^2}{(1)^6(0.001)(1)^{14}} = 10$$

$$E = E° - \frac{0.0592}{n}\log Q = (1.11) - \frac{(0.0592)}{(6)}\log(10) = 1.10\ V$$

(d) $$\log K = \frac{n}{0.0592}E° = \frac{(6)(1.11)}{(0.0592)} = 113$$

$K = 10^{113}$

24.75 (a)

$Sn(s) \rightarrow Sn^{2+} + 2e^-$	$E° = 0.136\ V$
$(2)[AgCl(s) + e^- \rightarrow Ag(s) + Cl^-]$	$E° = 0.2222\ V$
$Sn(s) + 2AgCl(s) \rightarrow Sn^{2+} + 2Ag(s) + 2Cl^-$	$E° = 0.358\ V$

(b) $Q = [Sn^{2+}][Cl^-]^2 = (0.10)(0.20)^2 = 4.0 \times 10^{-3}$

$$E = E° - \frac{0.0592}{n} \log Q = (0.358) - \frac{(0.0592)}{(2)} \log(4.0 \times 10^{-3}) = 0.429 \text{ V}$$

(c) $Q = (0.010)(0.020)^2 = 4.0 \times 10^{-6}$

$$E = (0.358) - \frac{0.0592}{(2)} \log(4.0 \times 10^{-6}) = 0.518 \text{ V}$$

(d) At equilibrium the cell potential is 0 V.

(e) $$\log K = \frac{n}{0.0592} E° = \frac{(2)(0.358)}{(0.0592)} = 12.1$$

$$K = 1 \times 10^{12}$$

(f) $K = [Sn^{2+}][Cl^-]^2 = (x)(2x)^2 = 4x^3 = 1 \times 10^{12}$

$$x = [Sn^{2+}] = 6 \times 10^3 \text{ mol/L}$$

24.77 The oxidation number of U in

(a) UO_2 is +4

(b) UF_4 is +4

(c) U is 0

(d) The reducing agent is Mg(s).

(e) The substance reduced is UF_4(s).

(f) $$(1.00 \text{ g } UF_4)\left[\frac{1 \text{ mol } UF_4}{314.03 \text{ g } UF_4}\right]\left[\frac{4 \text{ mol } e^-}{1 \text{ mol } UF_4}\right]\left[\frac{96{,}500 \text{ C}}{1 \text{ mol } e^-}\right] = 1230 \text{ C}$$

$$I = \frac{\text{charge}}{t} = \frac{(1230 \text{ C})(1 \text{ A s/1 C})}{(1.00 \text{ min})(60 \text{ s/1 min})} = 20.5 \text{ A}$$

(g) $$(1.00 \text{ g U})\left[\frac{1 \text{ mol U}}{238.03 \text{ g U}}\right]\left[\frac{1 \text{ mol } UF_4}{1 \text{ mol U}}\right]\left[\frac{4 \text{ mol HF}}{1 \text{ mol } UF_4}\right] = 0.0168 \text{ mol HF}$$

$$V = \frac{nRT}{P} = \frac{(0.0168 \text{ mol})(0.0821 \text{ L atm/K mol})(298 \text{ K})}{(10.0 \text{ atm})} = 0.0411 \text{ L HF}$$

(h) $$(1.00 \text{ g U})\left[\frac{1 \text{ mol U}}{238.03 \text{ g U}}\right]\left[\frac{2 \text{ mol Mg}}{1 \text{ mol U}}\right]\left[\frac{24.31 \text{ g Mg}}{1 \text{ mol Mg}}\right] = 0.204 \text{ g Mg}$$

Yes, 1.00 g Mg would be enough to produce 1.00 g U.

CHAPTER 25

MOLECULAR ORBITAL THEORY

Solutions to Exercises

25.1 For H_2^-, the molecular orbital configuration is $\sigma_{1s}^2\ \sigma_{1s}^*$. The bond order $= \frac{1}{2}(2 - 1) = \frac{1}{2}$. Because H_2^- has a bond order greater than zero, its formation would be possible.

25.2 The electron configurations are

N_2^+ $1s^2\ 1s^2\ \sigma_{2s}^2\ \sigma_{2s}^{*2}\ \pi_{2p_y}^2\ \pi_{2p_z}^2\ \sigma_{2p_x}^1$

N_2 $1s^2\ 1s^2\ \sigma_{2s}^2\ \sigma_{2s}^{*2}\ \pi_{2p_y}^2\ \pi_{2p_z}^2\ \sigma_{2p_x}^2$

N_2^- $1s^2\ 1s^2\ \sigma_{2s}^2\ \sigma_{2s}^{*2}\ \pi_{2p_y}^2\ \pi_{2p_z}^2\ \sigma_{2p_x}^2\ \pi_{2p_y}^{*1}$

	(a) bond orders	(b) magnetic properties	(c) bond lengths	(d) bond dissociation energies
N_2^+	$\frac{1}{2}(7 - 2) = 2\frac{1}{2}$	paramagnetic	same as N_2^-	about same as N_2^-
N_2	$\frac{1}{2}(8 - 2) = 3$	diamagnetic	least	greatest
N_2^-	$\frac{1}{2}(8 - 3) = 2\frac{1}{2}$	paramagnetic	same as N_2^+	about same as N_2^+

25.3 Because of the unpaired electron in the σ_{2p_x} orbital, BO will be paramagnetic.

Solutions to Odd-Numbered Questions and Problems

25.1 A molecular orbital is the space in which an electron with a specific energy is most likely to be found in the vicinity of two or more nuclei that are bonded together. Molecular orbital calculations can be used to develop electron density maps and energy level diagrams for molecules. Electron density maps are used for molecular structures and energy level diagrams are used for energies of bond formation and for spectral studies.

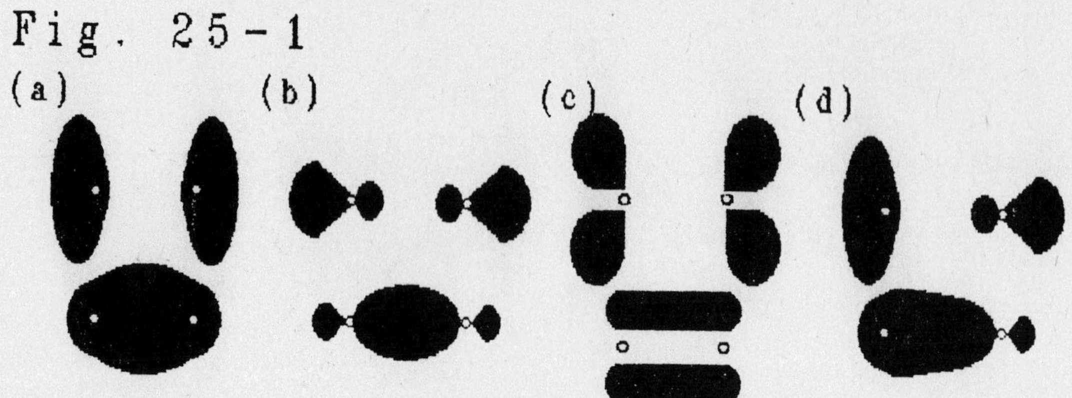

25.3 The energy of a bonding molecular orbital is lower than the energies of the original atomic orbitals. The energy of a antibonding molecular orbital is higher than the energies of the original atomic orbitals.

25.5 See Fig. 25-1.

25.7 Electrons are pictured as entering the orbitals one by one according to the following rules:

1. Electrons first occupy the molecular orbitals of lowest energy.
2. Each molecular orbital can accommodate a maximum of two electrons.
3. Molecular orbitals of equal energy are occupied by single electrons with parallel spins before electron pairing begins.

25.9 Bond order = $\frac{1}{2}$[(no. of bonding electrons) - (no. of antibonding electrons)]

(a) H_2^+ σ_{1s}^1 exists bond order = $\frac{1}{2}(1 - 0) = \frac{1}{2}$

(b) H_2 σ_{1s}^2 exists bond order = $\frac{1}{2}(2 - 0) = 1$

(c) H_2^- $\sigma_{1s}^2\ \sigma_{1s}^{*1}$ exists bond order = $\frac{1}{2}(2 - 1) = \frac{1}{2}$

(d) H_2^{2-} $\sigma_{1s}^2\ \sigma_{1s}^{*2}$ does not exist bond order = $\frac{1}{2}(2 - 2) = 0$

25.11 Bond order = $\frac{1}{2}$[(no. of bonding electrons) - (no. of antibonding electrons)]

(a) Be_2 $1s^2\ 1s^2\ \sigma_{2s}^2\ \sigma_{2s}^{*2}$ does not exist

bond order = $\frac{1}{2}(2 - 2) = 0$

(b) C_2 $1s^2\ 1s^2\ \sigma_{2s}^2\ \sigma_{2s}^{*2}\ \pi_{2p_y}^2\ \pi_{2p_z}^2$ exists

bond order = $\frac{1}{2}(6 - 2) = 2$

(c) Ne_2 $1s^2\ 1s^2\ \sigma_{2s}^2\ \sigma_{2s}^{*2}\ \sigma_{2p_x}^2\ \pi_{2p_y}^2\ \pi_{2p_z}^2\ \pi_{2p_y}^{*2}\ \pi_{2p_z}^{*2}\ \sigma_{2p_x}^{*2}$ does not exist

bond order = $\frac{1}{2}(8 - 8) = 0$

25.13 $\Delta H_0^\circ = [(1\ \text{mol})\Delta H_{f,0}^\circ(O^+) + (1\ \text{mol})\Delta H_{f,0}^\circ(O)] - [(1\ \text{mol})\Delta H_{f,0}^\circ(O_2^+)]$

$= [(1\ \text{mol})(1560.716\ \text{kJ/mol}) + (1\ \text{mol})(246.785\ \text{kJ/mol})]$

$- [(1\ \text{mol})(1164.70\ \text{kJ/mol})]$

$= 642.80\ \text{kJ}$

25.15 $N_2^+(g) \rightarrow N^+(g) + N(g)$

$\Delta H_0^\circ = [(1\ \text{mol})\Delta H_{f,0}^\circ(N^+) + (1\ \text{mol})\Delta H_{f,0}^\circ(N)] - [(1\ \text{mol})\Delta H_{f,0}^\circ(N_2^+)]$

$= [(1\ \text{mol})(1873.156\ \text{kJ/mol}) + (1\ \text{mol})(470.842\ \text{kJ/mol})]$

$- (1\ \text{mol})(1503.378\ \text{kJ/mol})]$

$= 840.620\ \text{kJ}$

25.17 Bond order = $\frac{1}{2}$[(no. of bonding electrons) - (no. of antibonding electrons)]

O_2 $1s^2\ 1s^2\ \sigma_{2s}^2\ \sigma_{2s}^{*2}\ \sigma_{2p_x}^2\ \pi_{2p_y}^2\ \pi_{2p_z}^2\ \pi_{2p_y}^{*1}\ \pi_{2p_z}^{*1}$

bond order = $\frac{1}{2}(8 - 4) = 2$

O_2^+ $1s^2\ 1s^2\ \sigma_{2s}^2\ \sigma_{2s}^{*2}\ \sigma_{2p_x}^2\ \pi_{2p_y}^2\ \pi_{2p_z}^2\ \pi_{2p_y}^{*1}$

bond order = $\frac{1}{2}(8 - 3) = \frac{5}{2} = 2\frac{1}{2}$

O_2^- $1s^2\ 1s^2\ \sigma_{2s}^2\ \sigma_{2s}^{*2}\ \sigma_{2p_x}^2\ \pi_{2p_y}^2\ \pi_{2p_z}^2\ \pi_{2p_y}^{*2}\ \pi_{2p_z}^{*1}$

bond order = $\frac{1}{2}(8 - 5) = \frac{3}{2} = 1\frac{1}{2}$

O_2^{2+} $1s^2\ 1s^2\ \sigma_{2s}^2\ \sigma_{2s}^{*2}\ \sigma_{2p_x}^2\ \pi_{2p_y}^2\ \pi_{2p_z}^2$

bond order = $\frac{1}{2}(8 - 2) = 3$

O_2^{2-} $1s^2\ 1s^2\ \sigma_{2s}^2\ \sigma_{2s}^{*2}\ \sigma_{2p_x}^2\ \pi_{2p_y}^2\ \pi_{2p_z}^2\ \pi_{2p_y}^{*2}\ \pi_{2p_z}^{*2}$

bond order = $\frac{1}{2}(8 - 6) = 1$

The order of decreasing bond strength is $O_2^{2+} > O_2^{+} > O_2 > O_2^{-} > O_2^{2-}$. The species O_2^{2+} has the shortest bond length.

25.19 NO $1s^2\ 1s^2\ \sigma_{2s}^2\ \sigma_{2s}^{*2}\ \pi_{2p_y}^2\ \pi_{2p_z}^2\ \sigma_{2p_x}^2\ \pi_{2p_y}^{*1}$

The molecule NO would be paramagnetic. The ion NO^+ would have one less antibonding electron than NO, so it would be more stable.

25.21 $AB(g) \rightarrow A(g) + B(g)$

bond energy = $[(1\ mol)\Delta H_f^\circ(A) + (1\ mol)\Delta H_f^\circ(B)] - [(1\ mol)\Delta H_f^\circ(AB)]$

$\Delta H^\circ_{CN^-} = [(1\ mol)(711\ kJ/mol) + (1\ mol)(478\ kJ/mol)] - [(1\ mol)(67\ kJ/mol)]$
$= 1122\ kJ$

$\Delta H^\circ_{CN} = [(1\ mol)(711\ kJ/mol) + (1\ mol)(471\ kJ/mol)] - [(1\ mol)(434\ kJ/mol)]$
$= 748\ kJ$

$\Delta H^\circ_{CN^+} = [(1\ mol)(1798\ kJ/mol) + (1\ mol)(471\ kJ/mol)]$
$- [(1\ mol)(1793\ kJ/mol)]$
$= 476\ kJ$

CHAPTER 26

METALS AND METALLURGY: THE *s*- AND *p*-BLOCK METALS

Solutions to Odd-Numbered Questions and Problems

26.1 When we think of metals, we are inclined to consider substances that are dense, high melting, malleable, ductile, and good conductors of heat and electricity. They have a tendency to form positive ions and to form salts.

26.3 A mineral is a naturally occurring inorganic substance with a characteristic crystal structure and composition. Not every mineral has a definite, fixed composition because a variety of metal cations can occupy similar sites in the anionic framework.

26.5 $$\text{mass \% Cr} = \frac{104\text{ g Cr}}{224\text{ g FeCr}_2\text{O}_4} \times 100 = 46.4\ \%$$

$$\text{mass \% Cr} = \frac{104\text{ g Cr}}{192\text{ g MgCr}_2\text{O}_4} \times 100 = 54.2\ \%$$

ave mass % Cr= 50.3 %

26.7 Metallurgy is the science and technology of metals--their production from the compounds in which they occur in nature, their purification, and the study of their properties.

26.9 Gangue may be separated from the desired mineral by crushing the ore, adding the crushed ore to a large tank of water containing a wetting agent (a selected surface active agent) and a frothing agent, and collecting the mineral in the froth that is skimmed off.

26.11 Pyrometallurgy differs from the other processes in its high temperatures. Common reducing agents in smelting are carbon or metals. The remaining gangue is removed during smelting by adding something that will react with it (a flux) to give a material (slag) that is liquid at the temperature of the smelting furnace.

26.13 (a) $Fe_3O_4(s) + 4CO(g) \rightarrow 3Fe(l) + 4CO_2(g)$ (v) reduction

(b) $MgCO_3(s) + SiO_2(s) \rightarrow MgSiO_3(l) + CO_2(g)$ (iv) adding a flux

(c) $4Au(s) + 8CN^- + 2H_2O(l) + O_2(g) \rightarrow 4[Au(CN)_2]^- + 4OH^-$ (iii) leaching

26.15 $2NaCl(aq) + 2H_2O(l) \rightarrow H_2(g) + Cl_2(g) + 2NaOH(aq)$

$2Cl^- \rightarrow Cl_2(g) + 2e^-$

$2e^- + 2H_2O(l) \rightarrow H_2(g) + 2OH^-$

$$(1\ \text{mol}\ e^-)\left[\frac{1\ \text{mol}\ Cl_2}{2\ \text{mol}\ e^-}\right]\left[\frac{70.90\ \text{g}\ Cl_2}{1\ \text{mol}\ Cl_2}\right] = 35.45\ \text{g}\ Cl_2$$

$$(1\ \text{mol}\ e^-)\left[\frac{1\ \text{mol}\ H_2}{1\ \text{mol}\ e^-}\right]\left[\frac{2.02\ \text{g}\ H_2}{1\ \text{mol}\ H_2}\right] = 1.01\ \text{g}\ H_2$$

$$(1\ \text{mol}\ e^-)\left[\frac{2\ \text{mol}\ OH^-}{2\ \text{mol}\ e^-}\right]\left[\frac{1\ \text{mol NaOH}}{1\ \text{mol}\ OH^-}\right]\left[\frac{40.00\ \text{g NaOH}}{1\ \text{mol NaOH}}\right] = 40.00\ \text{g NaOH}$$

26.17 Purified Al_2O_3 is dissolved in a molten electrolyte. Electrolysis produces Al at the cathode and the carbon anodes are consumed by the O_2 formed.

26.19 Bauxite is the ore that contains a high concentration of hydrated aluminum oxide. Its usual impurities are silica (SiO_2) and iron(III) oxide. The bauxite is digested under pressure with hot sodium hydroxide solution, which dissolves the aluminum oxide as $Na[Al(OH)_4]$. Insoluble materials are removed by filtration, and the filtrate is then diluted with water and cooled, precipitating the aluminum hydroxide. The precipitate is separated by filtration and heated to form Al_2O_3.

26.21 Sulfur dioxide is the undesirable gaseous product formed during the smelting of copper and other sulfide minerals. It has an unpleasant odor, is harmful to one's health, kills plants, and is converted in moist air to H_2SO_4. The removal of sulfur dioxide from stack gases is done either by catalytically oxidizing the sulfur dioxide in the stack to sulfuric acid (via sulfur trioxide) or by reducing it to elemental sulfur by passing it over red hot carbon.

26.23 During the electrolytic purification of copper if the voltage across the cell were

(a) too high, Ag and Au might also be oxidized, or water might be decomposed to H_2 and O_2

(b) too low, Cu would not be oxidized

26.25 Assume 100.0 g of each ore.

(a) $$(3.80\text{ g azurite})\left[\frac{1\text{ mol azurite}}{344.67\text{ g azurite}}\right]\left[\frac{3\text{ mol Cu}}{1\text{ mol azurite}}\right]\left[\frac{63.55\text{ g Cu}}{1\text{ mol Cu}}\right] = 2.10\text{ g Cu}$$

(b) $$(4.85\text{ g CuFeS}_2)\left[\frac{1\text{ mol CuFeS}_2}{183.54\text{ g CuFeS}_2}\right]\left[\frac{1\text{ mol Cu}}{1\text{ mol CuFeS}_2}\right]\left[\frac{63.55\text{ g Cu}}{1\text{ mol Cu}}\right] = 1.68\text{ g Cu}$$

The ore containing (a) 3.80 mass % azurite would yield the larger quantity of Cu on a mass basis.

26.27 $2ZnS(s) + 3O_2(g) \rightarrow 2ZnO(s) + 2SO_2(g)$

$ZnO(s) + C(s) \xrightarrow{\Delta} Zn(g) + CO(g)$

26.29 The names and formulas of some common iron minerals are pyrite, FeS_2; hematite, Fe_2O_3; magnetite, Fe_3O_4 ($FeO \cdot Fe_2O_3$); and siderite, $FeCO_3$. The oxidation state of iron is +2 in pyrite and siderite, +3 in hematite, and both +2 and +3 in magnetite.

26.31 The overall equation for the process is

$$\begin{aligned}
(3)[2C(s) + O_2(g) &\rightarrow 2CO(g)] \\
(2)[Fe_2O_3(s) + 3CO(g) &\rightarrow 2Fe(s) + 3CO_2(g)] \\
\hline
2Fe_2O_3(s) + 6C(s) + 3O_2(g) &\rightarrow 4Fe(s) + 6CO_2(g)
\end{aligned}$$

$$(1.00\text{ T Fe})\left[\frac{9.07 \times 10^5\text{ g}}{1\text{ T}}\right]\left[\frac{1\text{ mol Fe}}{55.85\text{ g Fe}}\right]\left[\frac{3\text{ mol O}_2}{4\text{ mol Fe}}\right]\left[\frac{32.00\text{ g O}_2}{1\text{ mol O}_2}\right]$$

$$\times \left[\frac{100\text{ g air}}{21\text{ g O}_2}\right]\left[\frac{1\text{ T}}{9.07 \times 10^5\text{ g}}\right] = 2.0\text{ T air}$$

26.33 The pig iron withdrawn from the blast furnace contains small amounts of carbon, sulfur, phosphorus, silicon, and manganese. The last three are converted by air or oxygen to oxides, which react with appropriate fluxes to give slags. Sulfur enters the slag as a sulfide, and carbon is burned to carbon monoxide or carbon dioxide.

$MnO(s) + SiO_2(s) \rightarrow MnSiO_3(l)$

$SiO_2(s) + MgO(s) \xrightarrow{\Delta} MgSiO_3(l)$

$P_4O_{10}(s) + 6CaO(s) \xrightarrow{\Delta} 2Ca_3(PO_4)_2(l)$

26.35 An alloy is a mixture of two or more metals, or metals plus nonmetals, yielding a substance with metallic properties. The three classes of alloys are heterogeneous mixtures, solid solutions (homogeneous mixtures), and intermetallic compounds. In general, the properties of alloys are unique--they are not the average of the properties of the metals of which they are composed.

26.37 Assume exactly 100 g of each compound.

	Ag	Cd
mass	24.2 g	75.8 g
molar mass	107.87 g/mol	112.4 g/mol
no. of moles	$(24.2\text{ g})\left[\frac{1\text{ mol}}{107.87\text{ g}}\right] = 0.224\text{ mol}$	$(75.8\text{ g})\left[\frac{1\text{ mol}}{112.4\text{ g}}\right] = 0.674\text{ mol}$
mole ratio n/n_{Ag}	$\frac{0.224}{0.224} = 1.00$	$\frac{0.674}{0.224} = 3.01$
Rel. moles of atoms	1	3

Empirical formula is $AgCd_3$.

$$\frac{\text{no. of valence } e^-}{\text{no. of atoms}} = \frac{(1) + (3)(2)}{(1) + (3)} = \frac{7}{4}, \text{ therefore, an } \epsilon\text{-alloy}$$

	Ag	Cd
mass	49.0 g	51.0 g
molar mass	107.87 g/mol	112.4 g/mol
no. of moles	$(49.0\text{ g})\left[\frac{1\text{ mol}}{107.87\text{ g}}\right] = 0.454\text{ mol}$	$(51.0\text{ g})\left[\frac{1\text{ mol}}{112.4\text{ g}}\right] = 0.454\text{ mol}$
mole ratio n/n_{Ag}	$\frac{0.454}{0.454} = 1.00$	$\frac{0.454}{0.454} = 1.00$
Rel. moles of atoms	1	1

Empirical formula is AgCd.

$$\frac{\text{no. of valence } e^-}{\text{no. of atoms}} = \frac{(1) + (2)}{(1) + (1)} = \frac{3}{2} = \frac{21}{14}, \text{ therefore, a } \beta\text{-alloy}$$

	Ag	Cd
mass	37.5 g	62.5 g
molar mass	107.87 g/mol	112.4 g/mol
no. of moles	$(37.5\text{ g})\left[\frac{1\text{ mol}}{107.87\text{ g}}\right] = 0.348\text{ mol}$	$(62.5\text{ g})\left[\frac{1\text{ mol}}{112.4\text{ g}}\right] = 0.556\text{ mol}$
mole ratio n/n_{Ag}	$\frac{0.348}{0.348} = 1.00$	$\frac{0.556}{0.348} = 1.60$

Rel. moles of atoms	5	8

Empirical formula is Ag_5Cd_8.

$$\frac{\text{no. of valence } e^-}{\text{no. of atoms}} = \frac{(5)(1) + (8)(2)}{(5) + (8)} = \frac{21}{13}, \text{ therefore a } \gamma\text{-alloy}$$

26.39 The elements in Groups I and II are too reactive ever to be found in the free state in nature. Their primary sources are seawater and brines of soluble salts. The metals are obtained by electrolysis of molten chlorides.

26.41 The *s* block metals have the general outer electron configurations of ns^x. The predicted oxidation states could be 0 and $+x$. The type of bonding expected would be ionic bonding in most of their compounds because of the low electronegativity.

26.43 $4Li(s) + O_2(g) \rightarrow 2Li_2O(s)$ $\qquad$ $2M(s) + O_2(g) \rightarrow 2MO(s)$

$2Na(s) + O_2(g) \rightarrow Na_2O_2(s)$ $\qquad$ M = Be, Mg, Ca, Sr

$M(s) + O_2(g) \rightarrow MO_2(s)$ $\qquad$ $Ba(s) + O_2(g) \rightarrow BaO_2(s)$

M = K, Rb, Cs

The difference can be expressed as the basis of electronegativity, size, and charge.

26.45 (a) $2NaCl(aq) + 2H_2O(l) \xrightarrow{\text{electricity}} Cl_2(aq) + H_2(g) + 2NaOH(aq)$

Solid NaOH is obtained by evaporation of the solvent.

(b) $CaCO_3(s) + 2NaCl(aq) \xrightarrow[\text{process}]{\text{Solvay}} Na_2CO_3(s) + CaCl_2(aq)$

(c) $Na_2CO_3(aq) + H_2O(l) + CO_2(g) \rightarrow 2NaHCO_3(s)$

(d) $CaCO_3(s) \xrightarrow{\Delta} CaO(s) + CO_2(g)$

(e) $CaO(s) + H_2O(l) \rightarrow Ca(OH)_2(s)$

26.47 Upon heating, LiOH is the only alkali metal hydroxide that decomposes, as do the alkaline earth hydroxides, to form water and the oxide. Likewise, Li_2CO_3 is the only alkali metal carbonate that decomposes, as do the alkaline earth carbonates, to form carbon dioxide and the oxide.

26.49 (a) $NaOH(s) \xrightarrow{H_2O} NaOH(6\ M)$

$\Delta H^\circ = [(1\ mol)\Delta H^\circ_f(6\ M)] - [(1\ mol)\Delta H^\circ_f(s)]$

$= [(1\ mol)(-469.23\ kJ/mol)] - [(1\ mol)(-426.73\ kJ/mol)] = -42.50\ kJ$

Dissolving 1 mol NaOH (40 g) in 10 mol H_2O (180 g) gives about 225 g of a 6 M NaOH solution. The heat released by the dissolution process is

$q_{lost} = -42.50\ kJ = -42{,}500\ J$

and the heat gained by the solution is

$$q_{gained} = (\text{mass})(\text{specific heat})\Delta T = (225\ g)(4\ J/K\ g)\Delta T$$

where the specific heat is roughly that of water.

$$q_{lost} + q_{gained} = 0$$

$$(-42{,}500\ J) + (225\ g)(4\ J/K\ g)\Delta T = 0$$

$$\Delta T = \frac{42{,}500\ J}{(225\ g)(4\ J/K\ g)} = 50\ K$$

The solution will become quite warm.

(b) $NaOH(6\ M) \xrightarrow{H_2O} NaOH(0.1\ M)$

$$\Delta H° = [(1\ mol)\Delta H_f°(0.1\ M)] - [(1\ mol)\Delta H_f°(6\ M)]$$

$$= [(1\ mol)(-469.10\ kJ/mol)] - [(1\ mol)(-469.23\ kJ/mol)] = 0.13\ kJ$$

$$(0.13\ kJ)\left[\frac{1000\ J}{1\ kJ}\right] = 130\ J$$

26.51 The *p*-block metals have the general outer electron configuration of $ns^2\ np^x$; oxidation states of 0, $+x$, $+(x + 2)$, and $-(6 - x)$; and ionic bonding in compounds with metals and covalent bonding in compounds with nonmetals. These elements have moderate values of electronegativity. They have more than one positive oxidation state because loss of the *p* electrons gives the lower oxidation state and loss of *p* and *s* electrons gives the higher oxidation state.

26.53 The solutions are acidic because the aluminum ion undergoes hydrolysis.

26.55 (a) $2Al(s) + 6H^+ \rightarrow 2Al^{3+} + 3H_2(g)$

$2Al(s) + 2OH^- + 6H_2O(l) \rightarrow 2[Al(OH)_4]^- + 3H_2(g)$

(b) $Al_2O_3(s) + 6H^+ \rightarrow 2Al^{3+} + 3H_2O(l)$

$Al_2O_3(s) + 2OH^- + 3H_2O(l) \rightarrow 2[Al(OH)_4]^-$

26.57 The hydration energy is great enough to form the $Al^{3+}(aq)$ ion.

26.59 $3PbO_2(s) + 4Al(s) \rightarrow 3Pb(s) + 2Al_2O_3(s)$

$$\Delta H° = [(3\ mol)\Delta H_f°(Pb) + (2\ mol)\Delta H_f°(Al_2O_3)] - [(3\ mol)\Delta H_f°(PbO_2) + (4\ mol)\Delta H_f°(Al)]$$

$$= [(3\ mol)(0) + (2\ mol)(-1675.7\ kJ/mol)] - [(3\ mol)(-277.4\ kJ/mol) + (4\ mol)(0)]$$

$$= -2519.2\ kJ$$

The energy change is favorable.

26.61 (a) Some uses of Al are as structural materials, wire, utensils, and foil.

(b) Some uses of $AlCl_3$ are as catalysts and antiperspirants.

(c) Some uses of Al_2O_3 include as abrasives, jewels, and catalyst support.

(d) Some uses of $Al_2(SO_4)_3$ are in water purification and pulp and paper sizing.

26.63

$$\begin{array}{lll} Sn(s) \rightarrow Sn^{2+} + 2e^- & & E° = 0.136\ V \\ Pb^{2+} + 2e^- \rightarrow Pb(s) & & E° = -0.126\ V \\ \hline Sn(s) + Pb^{2+} \rightarrow Pb(s) + Sn^{2+} & & E° = 0.010\ V \end{array}$$

The tin electrode is the anode (site of oxidation).

$$Q = \frac{[Sn^{2+}]}{[Pb^{2+}]}$$

$$E = E° - \frac{0.0592}{n} \log Q = 0.010\ V - \frac{0.0592}{2} \log \frac{[Sn^{2+}]}{[Pb^{2+}]}$$

The cell voltage can be increased ($E > E°$) by making $[Sn^{2+}] < [Pb^{2+}]$.

26.65 The redox reactions are (a) and (c). The oxidizing agents are (a) O_2 and (c) H_2O. The reducing agents are (a) Sn and (c) K.

26.67 The types of each of these reactions are

(a) nonredox--decomposition to give compounds

(b) nonredox--combination of compounds

(c) nonredox--combination of compounds

26.69 Using Table 17.10, the first "yes" answer is

(a) 3; questions 4 and 5 are "no," but a combination can occur

$NaOH(aq) + CO_2(g,xs) \rightarrow NaHCO_3(aq)$

(b) 2, a redox reaction of an element (disproportionation)

$Cl_2(g) + 2NaOH(aq) \rightarrow NaOCl(aq) + NaCl(aq) + H_2O(l)$

(c) 1, a nonredox decomposition

$MgCO_3 \cdot CaCO_3(s) \xrightarrow{\Delta} MgO(s) + CaO(s) + 2CO_2(g)$

(d) 6, a nonredox combination

$Na_2O(s) + H_2O(l) \rightarrow 2NaOH(aq)$

26.71 Only reactions (a) and (b) will occur.

(a) $Sn(s) + 2HCl(aq) \rightarrow SnCl_2(aq) + H_2(g)$

(b) $Pb^{2+} + Mg(s) \rightarrow Pb(s) + Mg^{2+}$

CHAPTER 27

NONMETALS: HALOGENS AND NOBLE GASES

Solutions to Exercises

27.5 The chlorite ion is a stronger oxidizing agent than the chlorate ion.

Solutions to Odd-Numbered Questions and Problems

27.1 Electron configuration of the atomic halogens

F [He]$2s^2 2p^5$

Cl [Ne]$3s^2 3p^5$

Br [Ar]$3d^{10} 4s^2 4p^5$

I [Kr]$4d^{10} 5s^2 5p^5$

At [Xe]$4f^{14} 5d^{10} 6s^2 6p^5$

The Lewis symbol of a halogen atom is $:\ddot{\underset{..}{X}}\cdot$. The usual oxidation state of each of the halogens in direct combination with metals, semiconducting metals, and most nonmetals is -1. The halogens (except F) can have positive oxidation states (+1, +3, +5, +7) in some compounds, such as the interhalogens and oxoacids.

27.3 London forces are the only significant intermolecular forces in molecular halogens. Their strength increases going down the family. The molecular halogen and its physical state: F_2, pale yellow gas; Cl_2, yellow-green gas; Br_2, red-brown liquid; I_2, violet-black solid.

27.5 (a) The predicted physical state is solid, (b) the predicted melting point is 300 °C, (c) the predicted ionic radius is 0.25 nm, (d) the predicted bond energy is 100 kJ/mol.

27.7 To determine whether or not an aqueous solution contains NaI in addition to NaBr, add Br_2 and CCl_4. If any NaI were present, I_2 would form, as evidenced by a purple color in the CCl_4 layer.

27.9 Assume exactly 100 g of the white powder.

	Sb	Cl
mass	53 g	47 g
molar mass	121.75 g/mol	35.45 g/mol
no. of moles	$(53\text{ g})\left[\frac{1\text{ mol}}{121.75\text{ g}}\right] = 0.44\text{ mol}$	$(47\text{ g})\left[\frac{1\text{ mol}}{35.45\text{ g}}\right] = 1.33\text{ mol}$
mol ratio n/n_{Sb}	$\frac{0.44}{0.44} = 1.0$	$\frac{1.33}{0.44} = 3.00$
Rel. moles of atoms	1	3

Empirical formula is $SbCl_3$.

$2Sb(s) + 3Cl_2(g) \rightarrow 2SbCl_3(s)$

27.11 To make the reaction of H_2 and Br_2 to form HBr, proceed as fast as possible, keep the concentrations of H_2 and Br_2 as high as possible, and reduce the concentration of HBr.

27.13 In the reaction of the halogens with water, F_2 forms HF and O_2,

$$2F_2(g) + 2H_2O(l) \rightarrow 4HF(aq) + O_2(g)$$

and Cl_2, Br_2, and I_2 form HX(aq) and HOX(aq),

$$X_2 + H_2O(l) \rightarrow H^+ + X^- + HXO(aq) \qquad X = Cl, Br, I$$

Thus fluorine is quite different from the others in its reactions.

27.15 $$n = \frac{PV}{RT} = \frac{(1.00\text{ atm})(50.0\text{ L})}{(0.0821\text{ L atm/K mol})(298\text{ K})} = 2.04\text{ mol }Cl_2$$

$$3Cl_2(g) + 6KOH(aq, conc) \rightarrow KClO_3(aq) + 5KCl(aq) + 3H_2O(l)$$

$$(2.04\text{ mol }Cl_2)\left[\frac{1\text{ mol }KClO_3}{3\text{ mol }Cl_2}\right]\left[\frac{122.55\text{ g }KClO_3}{1\text{ mol }KClO_3}\right] = 83.3\text{ g }KClO_3$$

$$Cl_2(g) + 2KOH(aq, dil) \rightarrow KClO(aq) + KCl(aq) + H_2O(l)$$

$$(2.04\text{ mol }Cl_2)\left[\frac{1\text{ mol }KClO}{1\text{ mol }Cl_2}\right]\left[\frac{90.55\text{ g }KClO}{1\text{ mol }KClO}\right] = 185\text{ g }KClO$$

27.17 Only iodine occurs as the oxoanion because iodine, with its larger size, shares electrons more readily than the other halogens and forms the stable IO_3^- ion.

27.19 Although fluorine is a highly reactive substance, it can be safely stored in certain containers because protective fluoride coatings are formed.

27.21 $$(0.100\text{ mol }MnO_2)\left[\frac{1\text{ mol }Cl_2}{1\text{ mol }MnO_2}\right] = 0.100\text{ mol }Cl_2$$

$$(0.100 \text{ mol Cl}_2)\left[\frac{70.90 \text{ g Cl}_2}{1 \text{ mol Cl}_2}\right] = 7.09 \text{ g Cl}_2$$

$$V = \frac{nRT}{P} = \frac{(0.100 \text{ mol})(0.0821 \text{ L atm/K mol})(300.\text{ K})}{(765 \text{ Torr})(1 \text{ atm/760 Torr})} = 2.45 \text{ L}$$

27.23 (a) The compound Ca(OCl)Cl is calcium hypochlorite chloride, $CaCl_2$ is calcium chloride, and $Ca(OCl)_2$ is calcium hypochlorite.

(b) The oxidation state of Cl is 0 in Cl_2, +1 in OCl^- of Ca(OCl)Cl and $Ca(OCl)_2$, -1 in Cl^- of Ca(OCl)Cl and $CaCl_2$.

(c) The oxidizing agent is Cl_2 in both reactions.

(d) The reducing agent is Cl_2 in both reactions.

(e) The element being oxidized is Cl.

(f) The element being reduced is Cl.

(g) As the chlorine is being generated, the pH increases.

27.25 mass Br_2 = 57.46 g - 20.65 g = 36.81 g

$$(36.81 \text{ g Br}_2)\left[\frac{1 \text{ mol Br}_2}{159.80 \text{ g Br}_2}\right] = 0.2304 \text{ mol Br}_2$$

$$(20.65 \text{ g Ar})\left[\frac{1 \text{ mol Ar}}{39.95 \text{ g Ar}}\right] = 0.5169 \text{ mol Ar}$$

Using Dalton's law

$$P_{Br_2} = P_{total}X_{Br_2} = (734 \text{ Torr})\left[\frac{0.2304}{0.2304 + 0.5169}\right] = 226 \text{ Torr}$$

27.27 The overall equation for the process is

$$(2)[NaBr + H_2SO_4 \rightarrow HBr + NaHSO_4]$$
$$2HBr + MnO_2 + H_2SO_4 \rightarrow Br_2 + MnSO_4 + 2H_2O$$
$$\overline{2NaBr + 3H_2SO_4 + MnO_2 \rightarrow Br_2 + MnSO_4 + 2H_2O + 2NaHSO_4}$$

$$(100.0 \text{ g NaBr})\left[\frac{1 \text{ mol NaBr}}{102.89 \text{ g NaBr}}\right] = 0.9719 \text{ mol NaBr}$$

$$(0.9719 \text{ mol NaBr})\left[\frac{1 \text{ mol Br}_2}{2 \text{ mol NaBr}}\right] = 0.4860 \text{ mol Br}_2$$

$$(0.4860 \text{ mol Br}_2)\left[\frac{159.80 \text{ g Br}_2}{1 \text{ mol Br}_2}\right] = 77.66 \text{ g Br}_2$$

27.29 Iodized salt is table salt to which sodium or potassium iodide is added. Iodide ion is necessary for the production of thyroxine in the thyroid gland.

27.31 (a) $^{209}_{83}Bi + ^{4}_{2}He \rightarrow 2^{1}_{0}n + ^{211}_{85}At$

(b) $^{211}_{85}At \rightarrow ^{4}_{2}He + ^{207}_{83}Bi$ $\qquad$ $^{211}_{85}At + ^{0}_{-1}e \rightarrow ^{211}_{84}Po$

(c) $k = \frac{0.693}{t_{1/2}} = \frac{0.693}{7.21\ h} = 0.0961\ h^{-1}$

$$\log \frac{m}{m_0} = -\frac{kt}{2.303}$$

$$\log\left[\frac{m}{0.05\ \mu g}\right] = \frac{-(0.0961\ h^{-1})(1\ wk)}{(2.303)}\left[\frac{7\ day}{1\ wk}\right]\left[\frac{24\ h}{1\ day}\right] = -7.01$$

$$\frac{m}{0.05\ \mu g} = 9.8 \times 10^{-8}$$

$$m = (5 \times 10^{-9}\ \mu g)\left[\frac{1\ g}{10^{6}\ \mu g}\right]\left[\frac{10^{15}\ fg}{1\ g}\right] = 5\ fg$$

27.33 The compounds HCl, HBr, and HI are gases at room temperature and pressure and fume in moist air as they react to form droplets of the strong nonoxidizing hydrohalic acids. From HCl to HBr to HI, their melting points, boiling points, heats of fusion and vaporization, strengths as reducing agents, and bond lengths increase, whereas bond energies decrease. Because of extensive hydrogen bonding, HF has significantly higher values of melting point, boiling point, and heats of fusion and vaporization than HCl, and acts as a weak acid in aqueous solution.

27.35 $\Delta H° = [(1\ mol)\Delta H°_f(F^-) + (1\ mol)\Delta H°_f(H_2O)] - [(1\ mol)\Delta H°_f(HF) + (1\ mol)\Delta H°_f(OH^-)]$

$= [(1\ mol)(-332.63\ kJ/mol) + (1\ mol)(-285.830\ kJ/mol)] - [(1\ mol)(-320.08\ kJ/mol) + (1\ mol)(-229.994\ kJ/mol)]$

$= -68.39\ kJ$

$HF(aq) + OH^- \rightarrow F^- + H_2O(l)$	$\Delta H° = -68.39\ kJ$
$H_2O(l) \rightarrow H^+ + OH^-$	$\Delta H° = 55.835\ kJ$
$HF(aq) \rightarrow H^+ + F^-$	$\Delta H° = -12.56\ kJ$

27.37 The type of bonding between atoms of (a) Na and F is ionic, (b) Ca and F is ionic, (c) I and Cl is polar covalent, and (d) Br and Br is covalent.

27.39 An element that can exhibit multiple oxidation states is in its highest oxidation state when it combines with fluorine but in lower oxidation states with bromine or iodine.

27.41 The oxidation number of Cl is -1 in (b) BrCl, (c) ICl, (f) ICl_3, (i) ICl_2^-, (j) ICl_4^-, (k) $BrCl_2^-$, (l) $IClBr^-$, and (m) $IFCl_3^-$; is +1 in (a) ClF and (g) ClF_2^-; is +3 in (d) ClF_3 and (h) ClF_4^-; and is +5 in (e) ClF_5.

27.43 The Lewis structures are

```
(a)            (b)                (c)              (d)            (e)
 ..  ..         ..  .. ..  ..      ..         ..    ..    ..       ..    ..
:Br - F:       :F - Br - F:       :F         F:   :Br - Cl:      :I - Br:
 ..  ..         ..    |    ..      ..  \ .. / ..    ..    ..       ..    ..
                     :F:                 Br
                      ..                / | \
                                      :F:   :F:
                                      ..  :F:  ..
                                           ..
```

The respective shapes and polarities are (a) linear, polar; (b) T-shape, polar; (c) square pyramidal, polar; (d) linear, polar; (e) linear, polar. The three dimensional sketches are shown in Fig. 27-1.

27.45 Unlike the other halogens, fluorine forms no oxoacids nor oxoanions because fluorine forms only a -1 oxidation state, which cannot combine with oxygen to form oxoanions.

27.47 The oxochloro acids and anions generally act as oxidizing agents. The reduction potential decreases with increasing oxidation state.

27.49 The thermal decomposition of alkali metal halogen oxoanions yields O_2 and the alkali metal halide salt. The halogen oxoanions of the less reactive metals yield O_2, X_2, and the metal oxide.

27.51 $10.00 \text{ g} - 7.00 \text{ g} = 3.00 \text{ g } O_2$

$$(3.00 \text{ g } O_2)\left[\frac{1 \text{ mol } O_2}{32.0 \text{ g } O_2}\right] = 9.38 \times 10^{-2} \text{ mol } O_2$$

$$(9.38 \times 10^{-1} \text{ mol } O_2)\left[\frac{2 \text{ mol } KClO_3}{3 \text{ mol } O_2}\right]$$

$$= 6.25 \times 10^{-2} \text{ mol } KClO_3 \text{ reacted to form } O_2$$

$$(10.00 \text{ g } KClO_3)\left[\frac{1 \text{ mol } KClO_3}{122.55 \text{ g } KClO_3}\right] = 8.160 \times 10^{-2} \text{ mol } KClO_3 \text{ reacting}$$

$$\frac{6.25 \times 10^{-2}}{8.160 \times 10^{-2}} \times 100 = 76.6\ \% \text{ undergoes reaction to form } O_2$$

27.53 $\text{molality} = \dfrac{\text{no. of moles ions}}{\text{solv mass in kg}}$

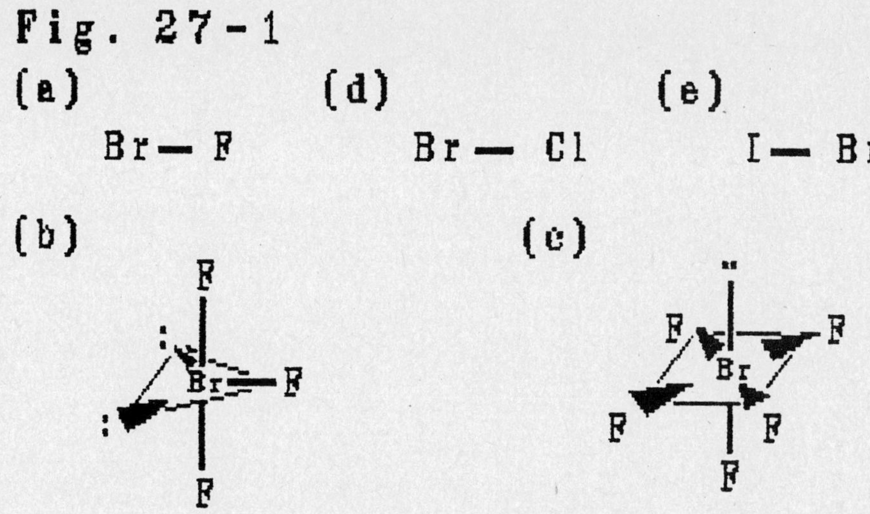

for $CaCl_2$:

$$\text{no. of moles} = (59.5\text{ g }CaCl_2)\left[\frac{1\text{ mol }CaCl_2}{110.98\text{ g }CaCl_2}\right]\left[\frac{3\text{ mol ions}}{1\text{ mol }CaCl_2}\right]$$

$$= 1.61\text{ mol ions}$$

$$\text{molality} = \left[\frac{1.61\text{ mol ions}}{100\text{ g }H_2O}\right]\left[\frac{1000\text{ g}}{1\text{ kg}}\right] = 16.1\text{ mol/kg}$$

$$\Delta T_f = K_f\, m = (1.86\text{ °C kg/mol})(16.1\text{ mol/kg}) = 29.9\text{ °C}$$

$$T_{f,soln} = T_{f,solvent} - \Delta T_f = 0.0\text{ °C} - 29.9\text{ °C} = -29.9\text{ °C}$$

for NaCl:

$$\text{no. of moles} = (35.7\text{ g NaCl})\left[\frac{1\text{ mol NaCl}}{58.44\text{ g NaCl}}\right]\left[\frac{2\text{ mol ions}}{1\text{ mol NaCl}}\right]$$

$$= 1.22\text{ mol ions}$$

$$\text{molality} = \left[\frac{1.22\text{ mol ions}}{100\text{ g }H_2O}\right]\left[\frac{1000\text{ g}}{1\text{ kg}}\right] = 12.2\text{ mol/kg}$$

$$\Delta T_f = (1.86\text{ °C kg/mol})(12.2\text{ mol/kg}) = 22.7\text{ °C}$$

$$T_{f,soln} = 0.0\text{ °C} - 22.7\text{ °C} = -22.7\text{ °C}$$

$CaCl_2$ is more effective in lowering the freezing point of water.

27.55 (a) $3Br_2(l) + 6OH^- \rightarrow BrO_3^- + 5Br^- + 3H_2O(l)$

(b) $3BrO^- \xrightarrow{\Delta} BrO_3^- + 2Br^-$

(c) $Br_2(l) + 5F_2(g) \rightarrow 2BrF_5(g)$

(d) $2Br^- + Cl_2(aq) \rightarrow Br_2(aq) + 2Cl^-$

27.57 The redox reactions are (a), (b), and (d). The oxidizing agent in each is (a) BrO_3^-, (b) Br_2, and (d) $KClO_3$. The reducing agent in each is (a) Br^-, (b) I^-, and (d) $KClO_3$.

27.59 The reaction types according to the classifications in Tables 17.2 and 17.7 are

(a) redox--displacement of one element from a compound by another element

(b) nonredox--partner-exchange reaction

(c) nonredox--partner-exchange reaction between ions in aqueous solution

(d) redox--combination of two elements to give a compound

27.61 In using Table 17.10 to predict the major products of the following reactions, the first "yes" answer is for question

(a) 1, so the reaction is a redox decomposition reaction

$$2KHF_2(l) \xrightarrow{\text{electricity}} H_2(g) + F_2(g) + 2KF(s)$$

(b) 2, so the reaction is a redox combination reaction

$$S_8(s) + 24Br_2(l) + 32H_2O(l) \rightarrow 8H_2SO_4(aq) + 48HBr(g)$$

(c) 3; the answer to question 4 is "no" and to 5 is "yes", so the reaction is a redox reaction of ions in aqueous solution

$$IO_3^- + 5I^- + 6H^+ \rightarrow 3I_2(s) + 3H_2O(l)$$

(d) 2, so the reaction is a redox combination reaction

$$2Fe(s) + 3Cl_2(g) \rightarrow 2FeCl_3(s)$$

27.63 The general trends down the noble gas family for the following properties are (a) increase in atomic radii, (b) decrease in ionization energies, (c) increase in gas densities, and (d) increase in boiling points.

27.65 The electron configuration of xenon is $[Kr]4d^{10}5s^25p^6$. It might be predicted to be inert because it has a complete octet. The Lewis structure of XeF_4 is

```
         ..
        :F:
     ..  |..  ..
    :F - Xe - F:
     ..  |..  ..
        :F:
         ..
```

This exception to the octet rule can be explained in terms of the empty 5*d* orbitals, which can accommodate the extra electrons.

27.67 XeF_2 is linear and XeF_4 is square planar. Both molecules have only London forces.

27.69 $XeF_4(s) + F_2(g) \rightarrow XeF_6(s)$

$$(3.62 \text{ g } XeF_4)\left[\frac{1 \text{ mol } XeF_4}{207.29 \text{ g } XeF_4}\right] = 0.0175 \text{ mol } XeF_4$$

$$(0.0175 \text{ mol } XeF_4)\left[\frac{1 \text{ mol } XeF_6}{1 \text{ mol } XeF_4}\right] = 0.0175 \text{ mol } XeF_6$$

$$(0.0175 \text{ mol } XeF_6)\left[\frac{245.29 \text{ g } XeF_6}{1 \text{ mol } XeF_6}\right] = 4.29 \text{ g } XeF_6$$

27.71 $2XeF_6(s) + SiO_2(s) \rightarrow 2XeOF_4(l) + SiF_4(g)$

$$(1.00 \text{ g } XeF_6)\left[\frac{1 \text{ mol } XeF_6}{245.29 \text{ g } XeF_6}\right] = 0.00408 \text{ mol } XeF_6$$

$$(0.00408 \text{ mol } XeF_6)\left[\frac{1 \text{ mol } SiF_4}{2 \text{ mol } XeF_6}\right] = 0.00204 \text{ mol } SiF_4$$

$$P = \frac{nRT}{V} = \frac{(0.00204 \text{ mol})(0.0821 \text{ L atm/K mol})(298 \text{ K})}{(1.00 \text{ L})} = 0.0499 \text{ atm}$$

27.73 $\Delta H° = -141 \text{ kJ} = -[(6 \text{ mol})(XeF)] + [(4 \text{ mol})(Xe\text{-}F) + (1 \text{ mol})(F\text{-}F)]$

$-141 \text{ kJ} = -(2)(Xe\text{-}F) + 159 \text{ kJ}$

$$Xe\text{-}F = \frac{-300.\text{ kJ}}{-2 \text{ mol}} = 150.\text{ kJ/mol}$$

CHAPTER 28

NONMETALS: NITROGEN, PHOSPHORUS, AND SULFUR

Solutions to Odd-Numbered Questions and Problems

28.1 [He]$2s^2 2p^3$, ±3 and +5; [Ne]$3s^2 3p^3$, ± 3 and +5; [Ne]$3s^2 3p^4$, -2, +4, and +6.

28.3 $4N \rightarrow 2N_2$ $\Delta H° = (2 \text{ mol})(-946 \text{ kJ/mol}) = -1890 \text{ kJ}$

$4N \rightarrow N_4$ $\Delta H° = (6 \text{ mol})(-159 \text{ kJ/mol}) = -954 \text{ kJ}$

Four gaseous nitrogen atoms would form two nitrogen molecules.

$4P \rightarrow 2P_2$ $\Delta H° = (2 \text{ mol})(-485 \text{ kJ/mol}) = -970. \text{ kJ}$

$4P \rightarrow P_4$ $\Delta H° = (6 \text{ mol})(-243 \text{ kJ/mol}) = -1460 \text{ kJ}$

Four gaseous phosphorus atoms would form one tetrahedral molecule.

28.5 (a) $1s^2\ 1s^2\ \sigma_{2s}^2\ \sigma_{2s}^{*2}\ \pi_{2p_y}^2\ \pi_{2p_z}^2\ \sigma_{2p_x}^2$

(b) The nitrogen molecule has one σ and 2 π bonds.

(c) See Fig. 28-1.

(d) The molecule will be diamagnetic.

28.7 $T = 750 + 273 = 1020 \text{ K}$

$$M = \frac{mRT}{PV} = \frac{(0.5350 \text{ g})(0.0821 \text{ L atm/K mol})(1020 \text{ K})}{(502 \text{ Torr})(1 \text{ atm}/760 \text{ Torr})(1.00 \text{ L})} = 67.8 \text{ g/mol}$$

$(32.06 \text{ g/mol})(x) = 67.8 \text{ g/mol}$

$x = 2$

The major component of the vapor is S_2.

28.9 What happens to a sample of sulfur

(a) that is heated very slowly at 1 atm from 25 °C to 500 °C is that rhombic solid warms up, rhombic solid changes to monoclinic solid at 95.39 °C, monoclinic solid warms up, solid melts at 115.21 °C, liquid warms up, liquid vaporizes at 444.6 °C, and gas warms up.

(b) that is heated at 1420 atm from 25 °C to 153 °C is that rhombic solid warms up, solid melts, some liquid crystallizes in monoclinic form,

Fig. 28 -1

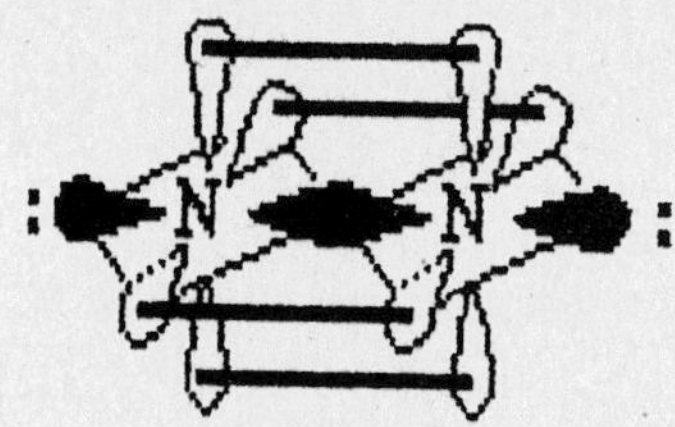

resulting in an equilibrium mixture of monoclinic solid, rhombic solid, and liquid.

(c) that has been melted and is then poured slowly into boiling water is that the liquid cools, solidifies at 115.21 °C to monoclinic solid, cools to 100 °C.

(d) that is in the gaseous phase between 95.31 °C and 115.18 °C and that has the pressure upon it increased to 2000 atm is that the gas deposits as monoclinic solid and monoclinic solid changes to rhombic solid.

(e) in the monoclinic form between 5.1×10^{-6} atm and 3.2×10^{-5} atm when it is heated is that monoclinic solid warms up and monoclinic solid sublimes.

28.11 See Tables 28.3, 28.4, and 28.5 for some of the major uses of elemental nitrogen, phosphorus, and sulfur, respectively.

28.13 In combination reactions with other elements, S is more reactive than P or N. Sulfur and phosphorus combine with most elements. Sulfur combines with certain metals at room temperature to form sulfides, burns in air to produce mainly SO_2, and combines with halogens to form products that vary with the conditions. Reaction conditions determine whether phosphorus forms products in the +3 or +5 oxidation state. Nitrogen combines only with Li at room temperature. At elevated temperatures it combines with reactive metals and certain *p*-block elements. White phosphorus is much more reactive than other allotropic forms.

28.15 The nitride and phosphide ions are much stronger bases than the sulfide ion, reacting vigorously and completely with water to give ammonia and phosphine, respectively.

$N^{3-} + 3H_2O(l) \rightarrow NH_3(aq) + 3OH^-$

$P^{3-} + 3H_2O(l) \rightarrow PH_3(g) + 3OH^-$

$S^{2-} + H_2O(l) \rightleftharpoons HS^- + OH^-$

$HS^- + H_2O(l) \rightleftharpoons H_2S(aq) + OH^-$

28.17 sulfuric acid: $2SO_2(g) + O_2(g) \xrightarrow{V_2O_5} 2SO_3(g)$

$SO_3(g) + H_2SO_4(l) \rightleftharpoons H_2S_2O_7(l)$

$H_2S_2O_7(l) + H_2O(l) \rightarrow 2H_2SO_4(aq)$

nitric acid: $4NH_3(g) + 5O_2(g) \xrightarrow[700\ °C]{Pt-Rh} 4NO(g) + 6H_2O(g)$

$2NO(g) + O_2(g) \rightarrow 2NO_2(g)$

$3NO_2(g) + H_2O(l) \rightarrow 2HNO_3(aq) + NO(g)$

phosphoric acid: $P_4(s) + 5O_2(g) \rightarrow P_4O_{10}(s)$

$P_4O_{10}(s) + 6H_2O(l) \rightarrow 4H_3PO_4(aq)$

or

$Ca_3(PO_4)_2(s) + 3H_2SO_4(aq) \rightarrow 2H_3PO_4(aq) + 3CaSO_4(s)$

28.19 Let $x = [H^+]$ at equilibrium.

	HSO_4^-	$\rightleftharpoons$	H^+	+	SO_4^{2-}
initial	0.010		0.010		0
change	$-x$		$+x$		x
equilibrium	$0.010 - x$		$0.010 + x$		x

$$K_{a_2} = \frac{[H^+][SO_4^{2-}]}{[HSO_4^-]} = \frac{(0.010 + x)(x)}{(0.010 - x)} = 1.0 \times 10^{-2}$$

$1.0 \times 10^{-2}\ x + x^2 = 1.0 \times 10^{-4} - 1.0 \times 10^{-2}\ x$

$x^2 + 2.0 \times 10^{-2}\ x - 1.0 \times 10^{-4} = 0$

$x = 4 \times 10^{-3}$ mol/L

$[SO_4^{2-}] = 4 \times 10^{-3}$ mol/L, $[HSO_4^-] = 6 \times 10^{-3}$ mol/L, $[H^+] = 0.014$ mol/L

pH = $-\log 0.014 = 1.85$

28.21 In any concentration greater than about 2 mol/L, HNO_3 can function as an oxidizing agent. The product to which it is reduced depends on the concentration of the acid, the temperature of the reaction, and the nature of the reducing agent (the stronger the reducing agent and the more dilute the acid, the lower the oxidation state of the reaction product will be).

28.23 The three classes of nitrides and their bonding are salt-like nitrides with ionic bonding, covalent nitrides with covalent bonding either in molecules or in network compounds, and interstitial nitrides with metallic bonding.

28.25 (a) dinitrogen monoxide, nitrogen(I) oxide, or nitrous oxide

```
                              ..
:N = N = O: ↔ :N ≡ N - O:
 ..       ..              ..
sp²  sp         sp  sp
```

(b) nitrogen monoxide, nitrogen(II) oxide, or nitric oxide

```
:N = O:
 •    ..
 sp²
```

(c) nitrous acid

```
    ..   ..
H - O - N = O:
    ..       ..
       sp²
```

(d) nitric acid

```
                                 ..
        :O:                     :O:
    ..   ‖   ..             ..   |   ..
H - O - N - O: ↔  H - O - N = O:
    ..       ..             ..
        sp²                     sp²
```

(e) nitrite ion

```
   ..  ..  ..          ..  ..  ..
[ :O - N = O: ]⁻ ↔ [ :O = N - O: ]⁻
   ..                          ..
      sp²                  sp²
```

(f) nitrate ion

```
      ..                 ..
     :O:   ]⁻          :O:   ]⁻          :O:   ]⁻
  ..  |  ..      ↔  ..  |  ..      ↔  ..  ‖  ..
[ :O - N = O: ]    [ :O = N - O: ]    [ :O - N - O: ]
  ..                          ..         ..       ..
      sp²                 sp²                 sp²
```

28.27 Nitrogen dioxide, a brown gas, is in equilibrium with its dimer, dinitrogen tetroxide, a colorless gas. Because the formation of N_2O_4 is favored at low temperatures, the brown color is lost when cooled to 0 °C.

28.29 Let x = $[H^+]$ at equilibrium.

	$HNO_2(aq)$ + $H_2O(l)$	→	H_3O^+	+	NO_2^-
initial	0.10		0		0.10
change	$-x$		$+x$		$+x$
equilibrium	$0.10 - x$		x		$0.10 + x$

$$K_a = \frac{[H^+][NO_2^-]}{[HNO_2]} = \frac{(x)(0.10 + x)}{(0.10 - x)} = 7.2 \times 10^{-4}$$

Assume that $0.10 \pm x \approx 0.10$.

$x = 7.2 \times 10^{-4}$ mol/L

The approximation is valid because $0.10 \pm 0.00072 \approx 0.10$.

$[H^+] = 7.2 \times 10^{-4}$ mol/L

$pH = -\log [H^+] = -\log(7.2 \times 10^{-4}) = 3.14$

28.31 $NH_4NO_3(s) \rightarrow N_2(g) + 2H_2O(g) + \frac{1}{2}O_2(g)$

$$(1.00 \text{ kg } NH_4NO_3)\left[\frac{1000 \text{ g}}{1 \text{ kg}}\right]\left[\frac{1 \text{ mol } NH_4NO_3}{80.06 \text{ g } NH_4NO_3}\right]\left[\frac{3.5 \text{ mol gases}}{1 \text{ mol } NH_4NO_3}\right] = 43.7 \text{ mol gases}$$

$T = 827 + 273 = 1100.$ K

$$V = \frac{nRT}{P} = \frac{(43.7 \text{ mol})(0.0821 \text{ L atm/K mol})(1100. \text{ K})}{(1.00 \text{ atm})} = 3950 \text{ L}$$

28.33 (a) The oxidation state of N is -2 in N_2H_4, +3 in HNO_2, and $-\frac{1}{3}$ (average) in HN_3.

(b) The oxidizing agent is HNO_2.

(c) Let $x = [H^+]$ at equilibrium

	$HN_3(aq) + H_2O(l)$	$\rightleftharpoons$ H_3O^+	$+ N_3^-$
initial	0.01	0	0
change	$-x$	$+x$	$+x$
equilibrium	$0.01 - x$	x	x

$$K_a = \frac{[H^+][N_3^-]}{[HN_3]} = \frac{(x)(x)}{(0.01 - x)} = 2.4 \times 10^{-5}$$

Assume $0.01 - x \approx 0.01$.

$x^2 = (2.4 \times 10^{-5})(0.01) = 2 \times 10^{-7}$

$x = 4 \times 10^{-4}$ mol/L

The approximation is valid: $\% \text{ error} = \frac{4 \times 10^{-4}}{0.01} \times 100 = 4\ \%$

$[H^+] = 4 \times 10^{-4}$ mol/L

$pH = -\log [H^+] = -\log (4 \times 10^{-4}) = 3.4$

28.35 Molecules of P_4, P_4O_6, and P_4O_{10} have a structural feature in which the phosphorus atoms are in a tetrahedral arrangement (see Figures 28.1 and 28.6). The common structural feature for all of the acids containing phosphorus(V) is a tetrahedron of oxygen atoms about the phosphorus atom.

28.37 All three hydrogen atoms of H_3PO_4 are ionizable because they are bonded to oxygen atoms. One hydrogen atom of H_3PO_3 is bonded directly to the phosphorus atom. Because the hydrogen-phosphorus bond is essentially nonpolar, this hydrogen atom is not acidic.

28.39 $$(1.00 \text{ lb additive})\left[\frac{454 \text{ g}}{1 \text{ lb}}\right]\left[\frac{9.5 \text{ g P}}{100.0 \text{ g additive}}\right]\left[\frac{1 \text{ mol P}}{30.97 \text{ g P}}\right]\left[\frac{1 \text{ mol } Na_5P_3O_{10}}{3 \text{ mol P}}\right]$$

$$\times \left[\frac{367.86 \text{ g } Na_5P_3O_{10}}{1 \text{ mol } Na_5O_3P_{10}}\right] = 170 \text{ g } Na_5P_3O_{10}$$

28.41 For PO_4^{3-}: Let x = $[OH^-]$ at equilibrium.

	$PO_4^{3-} + H_2O(l)$	$\rightleftarrows$	HPO_4^{2-}	$+ OH^-$
initial	0.1		0	0
change	$-x$		$+x$	$+x$
equilibrium	$0.1 - x$		x	x

$$K_b = \frac{K_w}{K_{a_3}} = \frac{1.00 \times 10^{-14}}{1 \times 10^{-12}} = 1 \times 10^{-2} = \frac{[HPO_4^{2-}][OH^-]}{[PO_4^{3-}]} = \frac{(x)(x)}{(0.1 - x)}$$

$$x^2 = 0.001 - (0.01)x$$

$$x^2 + (0.01)x - 0.001 = 0$$

$$x = \frac{-(0.01) \pm \sqrt{(0.01)^2 - (4)(1)(-0.001)}}{(2)(1)} = \frac{-(0.01) \pm (0.06)}{2}$$

$$= 0.03 \text{ mol/L}$$

$[OH^-]$ = 0.03 mol/L

pOH = $-\log [OH^-] = -\log(0.03) = 1.5$

pH = 14.0 - pOH = 14.0 - 1.5 = 12.5

For HPO_4^{2-}: As an acid, $K_{a_3} = 1 \times 10^{-12}$, and for HPO_4^{2-} as a base,

$$K_b = \frac{K_w}{K_{a_2}} = \frac{1.00 \times 10^{-14}}{6.6 \times 10^{-8}} = 1.5 \times 10^{-7}$$

Therefore, HPO_4^{2-} acts as a weak base.

Let x = $[OH^-]$ at equilibrium.

	$HPO_4^{2-} + H_2O(l)$	$\rightleftarrows$	$H_2PO_4^-$	$+ OH^-$
initial	0.1		0	0
change	$-x$		$+x$	$+x$
equilibrium	$0.1 - x$		x	x

$$K_b = \frac{[H_2PO_4^{2-}][OH^-]}{[HPO_4^{2-}]} = \frac{(x)(x)}{(0.1 - x)} = 1.5 \times 10^{-7}$$

Assume that $0.1 - x \approx 0.1$.

$x^2 = (1.5 \times 10^{-7})(0.1) = 2 \times 10^{-8}$

$x = 1 \times 10^{-4}$ mol/L

The approximation is valid because $0.1 - 0.0001 \approx 0.1$.

$[OH^-] = 1 \times 10^{-4}$ mol/L

$pOH = -\log(1 \times 10^{-4}) = 4.0$

$pH = 14.0 - 4.0 = 10.0$

For $H_2PO_4^-$: Let $x = [H^+]$ at equilibrium.

	$H_2PO_4^-$ + $H_2O(l)$ →	H_3O^+ +	HPO_4^{2-}
initial	0.1	0	0
change	$-x$	$+x$	$+x$
equilibrium	$0.1 - x$	x	x

$$K_{a_2} = \frac{[H^+][HPO_4^{2-}]}{[HPO_4^-]} = \frac{(x)(x)}{(0.1 - x)} = 6.6 \times 10^{-8}$$

Assume that $0.1 - x \approx 0.1$.

$x^2 = (6.6 \times 10^{-8})(0.1) = 7 \times 10^{-9}$

$x = 8 \times 10^{-5}$ mol/L

The approximation is valid because $0.1 - 8 \times 10^{-5} \approx 0.1$.

$[H^+] = 8 \times 10^{-5}$ mol/L

$pH = -\log(8 \times 10^{-5}) = 4.1$

28.43 Those of the alkali metals, ammonium ion, calcium, strontium, and barium are considered soluble. They give solutions that are distinctly alkaline as a result of the hydrolysis of the sulfide ion.

28.45

	$2SO_2(g)$ +	O_2 ⇌	$2SO_3(g)$
initial	1.00	5.00	0
change	$-(0.81)(1.00)$	$-(0.81/2)(1.00)$	$+(0.81)(1.00)$
equilibrium	0.19	4.59	0.81

$$K = \frac{[SO_3]^2}{[SO_2]^2[O_2]} = \frac{(0.81)^2}{(0.19)^2(4.59)} = 4.0$$

28.47 The partial equation is

$$CuFeS_2 \xrightarrow{\text{roasting}} 2SO_2$$

giving 2 mol SO_2/1 mol Cu.

$$(1.0 \text{ kg ore})\left[\frac{1000 \text{ g}}{1 \text{ kg}}\right]\left[\frac{0.2 \text{ g Cu}}{100.0 \text{ g ore}}\right]\left[\frac{1 \text{ mol Cu}}{63.55 \text{ g Cu}}\right]\left[\frac{2 \text{ mol } SO_2}{1 \text{ mol Cu}}\right]\left[\frac{64.07 \text{ g } SO_2}{1 \text{ mol } SO_2}\right] = 4.0 \text{ g } SO_2$$

28.49 All four reactions are redox. The oxidizing agents are (a) P_4, (b) NO_2, (c) I_3^-, and (d) S_8. The reducing agents are (a) P_4, (b) NO_2, (c) $S_2O_3^{2-}$, and (d) S_8.

28.51 The reactions are classified according to types listed in Tables 17.2 and 17.7 as

(a) redox--combination of element with compound to give other compounds

(b) nonredox--partner-exchange reaction between ions in aqueous solution (formation of a precipitate)

(c) redox--combination of element with compound to give another compound and nonredox--combination of compounds

(d) redox--decomposition

28.53 Using Table 17.10, the first "yes" answer is for question

(a) 6, a nonredox partner-exchange reaction

$$Li_3N(s) + 3H_2O(l) \rightarrow 3LiOH(s) + NH_3(g)$$

(b) 1, a redox decomposition reaction

$$NH_4NO_3(s) \xrightarrow{\Delta} N_2O(g) + 2H_2O(g)$$

or

$$2NH_4NO_3(s) \xrightarrow{\Delta} 2N_2(g) + 4H_2O(g) + O_2(g)$$

(c) 3; question 4 is "no"; question 5 is "yes", a redox reaction of ions in aqueous solution

$$2HNO_2(aq) + 2I^- + 2H^+ \rightarrow I_2(s) + 2NO(g) + 2H_2O(l)$$

(d) 7, a redox reaction

$$H_2SO_4(18\ M) + H_2S(g) \rightarrow S(s) + SO_2(g) + 2H_2O(l)$$

28.55 (a) $S(s) + 3F_2(g) \rightarrow SF_6(g)$ or $S_8(s) + 24F_2(g) \rightarrow 8SF_6(g)$

(b) $2H_2S(g) + 3O_2(g, \text{excess}) \xrightarrow{\Delta} 2SO_2(g) + 2H_2O(g)$

(c) $3Cu(s) + 8H^+ + 2NO_3^- \rightarrow 3Cu^{2+} + 2NO(g) + 4H_2O(l)$

(d) $2Na_2HPO_4(s) \xrightarrow{\Delta} Na_4P_2O_7(s) + H_2O(g)$

28.57 The equation for the reaction is

$$P_4S_3(s) + 8O_2(g) \rightarrow P_4O_{10}(s) + 3SO_2(g)$$

$$\begin{aligned}\Delta H^\circ &= [(1\ \text{mol})\Delta H_f^\circ(P_4O_{10}) + (3\ \text{mol})\Delta H_f^\circ(SO_2)] \\ &\qquad - [(1\ \text{mol})\Delta H_f^\circ(P_4S_3) + (8\ \text{mol})\Delta H_f^\circ(O_2)] \\ &= [(1\ \text{mol})(-2984\ \text{kJ/mol}) + (3\ \text{mol})(-297\ \text{kJ/mol})] \\ &\qquad - [(1\ \text{mol})(-154\ \text{kJ/mol}) + (8\ \text{mol})(0)] \\ &= -3721\ \text{kJ}\end{aligned}$$

CHAPTER 29

CARBON AND THE SEMICONDUCTING ELEMENTS

Solutions to Odd-Numbered Questions and Problems

29.1 The electron configuration of the C atom is $1s^2 2s^2 2p^4$ and of the Si atom is $1s^2 2s^2 2p^6 3s^2 3p^4$. The formulas of the compounds formed between the elements and fluorine are CF_4 and SiF_4, respectively. The $[SiF_6]^{2-}$ ion can form because two 3*d* orbitals of Si can be used in bonding; no *d* orbitals are available for C, so $[CF_6]^{2-}$ is unknown.

29.3 $$(3106 \text{ carats})\left[\frac{200 \text{ mg}}{1 \text{ carat}}\right]\left[\frac{1 \text{ g}}{1000 \text{ mg}}\right]\left[\frac{1 \text{ cm}^3}{3.51 \text{ g}}\right] = 177 \text{ cm}^3$$

29.5 C(graphite) → C(diamond)

C(graphite) + O_2(g) → CO_2(g)	$\Delta H°$ =	-393.5 kJ
CO_2(g) → C(diamond) + O_2(g)	$\Delta H°$ =	395.4 kJ
C(graphite) → C(diamond)	$\Delta H°$ =	1.9 kJ

$$\Delta S° = [(1 \text{ mol})\Delta S°(\text{diamond})] - [(1 \text{ mol})\Delta S°(\text{graphite})]$$
$$= [(1 \text{ mol})(2.38 \text{ J/K mol})] - [(1 \text{ mol})(5.74 \text{ J/K mol})] = -3.36 \text{ J/K}$$
$$\Delta G° = \Delta H° - T\,\Delta S°$$
$$= (1.9 \text{ kJ}) - (298 \text{ K})(-3.36 \text{ J/K})\left[\frac{1 \text{ kJ}}{1000 \text{ J}}\right] = 2.9 \text{ kJ}$$

Because the reaction is not favorable, graphite is more stable than diamond at 1 bar pressure and 25 °C.

29.7 Both C and CO will reduce water to hydrogen in reactions that are important in the synthesis of fuels and are valuable in freeing metals from their oxide ores.

29.9 :C ≡ O: :Ö = C = Ö: $\left[\begin{array}{c} O \\ | \\ O - C - O \end{array}\right]^{2-}$

The substance with the (a) strongest as well as (b) shortest carbon-oxygen bond is CO because of the triple bond.

29.11 The concentration of CO_2 decreases with increasing temperature. As the partial pressure of the CO_2 above the solution increases, the concentration of CO_2 in water increases.

29.13 $2NaOH(aq) + CO_2(s) \rightarrow Na_2CO_3(aq) + H_2O(l)$

$$(2.2 \text{ g } CO_2)\left[\frac{1 \text{ mol } CO_2}{44.01 \text{ g } CO_2}\right] = 0.050 \text{ mol } CO_2$$

$$(100. \text{ mL soln})\left[\frac{1 \text{ L}}{1000 \text{ mL}}\right]\left[\frac{5.0 \text{ mol NaOH}}{1 \text{ L soln}}\right]\left[\frac{1 \text{ mol } CO_2}{2 \text{ mol NaOH}}\right] = 0.25 \text{ mol } CO_2$$

Thus, CO_2 is the limiting reagent and the NaOH is in excess.

$$(0.050 \text{ mol } CO_2)\left[\frac{1 \text{ mol } Na_2CO_3}{1 \text{ mol } CO_2}\right]\left[\frac{105.99 \text{ g } Na_2CO_3}{1 \text{ mol } Na_2CO_3}\right] = 5.3 \text{ g } Na_2CO_3$$

29.15 The names and symbols of the seven semiconducting elements are boron, B; silicon, Si; germanium, Ge; arsenic, As; antimony, Sb; selenium, Se; and tellurium, Te. They resemble metals in appearance, conduct electricity much less effectively than metals, and are more like nonmetals than metals in their chemical reactions.

29.17 None of the semiconducting elements react with water or with nonoxidizing acids. With oxidizing acids, they give oxoacids or oxides, usually in their higher oxidation states.

29.19 Several chemical equations illustrating how the halides of the semiconducting elements resemble the nonmetallic halides in their behavior toward water much more than they do the metallic halides are

$$BX_3 + 3H_2O(l) \rightarrow H_2BO_3(aq) + 3HX(aq)$$
$$SiX_4 + 2H_2O(l) \rightarrow SiO_2(s) + 4HX(aq)$$
$$AsX_3 + 3H_2O(l) \rightarrow H_3AsO_3(aq) + 3HX(aq)$$

29.21 (a) $1s^22s^22p^1$, (b) $1s^2\ 1s^2\ \sigma_{2s}^2\ \sigma_{2s}^{*2}\ \pi_{2p_y}^1\ \pi_{2p_z}^1$, (c) paramagnetic, (d) paramagnetic, (e) π bonding.

29.23 $2H_3BO_3(s) \xrightarrow{\Delta} 3H_2O(g) + B_2O_3(s)$

$B_2O_3(s) + 3Mg(s) \rightarrow 2B(s) + 3MgO(s)$

$2B(s) + 3Cl_2(g) \rightarrow 2BCl_3(g)$

$2BCl_3(g) + 3H_2(g) \xrightarrow{\Delta} 2B(s) + 6HCl(g)$

29.25 $V = \frac{4}{3}\pi r^3$ for sphere; let x = thickness of crust.

$$V(\text{crust}) = \frac{4}{3}\pi[r^3 - (r - x)^3]$$
$$= \frac{4}{3}\pi[r^3 - (r^3 - 3xr^2 + 3x^2r - x^3)] = \frac{4}{3}\pi(3xr^2 - 3x^2r + x^3)$$
$$= \frac{4}{3}\pi[(3)(50\text{ km})(6400\text{ km})^2 - (3)(50\text{ km})^2(6400\text{ km}) + (50\text{ km})^3] \times \left[\frac{1000\text{ m}}{1\text{ km}}\right]^3$$
$$= 3 \times 10^{19}\text{ m}^3$$

$$m = (3 \times 10^{19}\text{ m}^3)\left[\frac{3.5\text{ g}}{1\text{ cm}^3}\right]\left[\frac{100\text{ cm}}{1\text{ m}}\right]^3 = 1 \times 10^{26}\text{ g}$$

$$(1 \times 10^{26}\text{ g crust})\left[\frac{25.7\text{ g Si}}{100\text{ g crust}}\right] = 3 \times 10^{25}\text{ g Si}$$

29.27 $$(1000\text{ kg ore})\left[\frac{1000\text{ g}}{1\text{ kg}}\right]\left[\frac{1\text{ g Ge}}{2.5 \times 10^6\text{ g ore}}\right]\left[\frac{90\text{ g Ge actual}}{100\text{ g Ge theoretical}}\right] \times \left[\frac{1\text{ mol Ge}}{72.59\text{ g Ge}}\right]\left[\frac{1\text{ mol GeO}_2}{1\text{ mol Ge}}\right]\left[\frac{104.59\text{ g GeO}_2}{1\text{ mol GeO}_2}\right] = 0.5\text{ g GeO}_2$$

29.29 The boranes are made up of the elements boron and hydrogen. The formula and structure of the simplest one are B_2H_6 and

```
H   H   H
 \ . . /
  B   B
 / . . \
H   H   H
```

29.31 $$M = \frac{mRT}{VP} = \frac{dRT}{P} = \frac{(0.57\text{ g/L})(0.0821\text{ L atm/K mol})(298\text{ K})}{(0.500\text{ atm})} = 28\text{ g/mol}$$

29.33 The Lewis structures of BCl_3 and $[BCl_4]^-$ and their geometries are

```
 ..       ..              ..        -
:Cl - B - Cl:           :Cl:
 ..   |   ..       ..     |     ..
     :Cl:         :Cl  -  B  -  Cl:
      ..           ..     |     ..
                        :Cl:
                         ..
```

triangular planar (BCl₃) — tetrahedral ([BCl₄]⁻)

29.35 $SiH_4(g) + (2+x)H_2O(l) \rightarrow SiO_2 \cdot xH_2O(aq) + 4H_2(g)$

Monosilane reacts vigorously with water because the hydrated silica oxide is very stable; methane does not form a stable hydrated oxide.

29.37 $O{=}Si{=}O(g) \rightarrow Si(g) + 2O(g)$

$$\Delta H° = [(1\text{ mol})\Delta H_f°(\text{Si}) + (2\text{ mol})\Delta H_f°(\text{O})] - [(1\text{ mol})\Delta H_f°(\text{SiO}_2)]$$
$$= [(1\text{ mol})(455.6\text{ kJ/mol}) + (2\text{ mol})(249.170\text{ kJ/mol})] - [(1\text{ mol})(-322\text{ J/mol})]$$
$$= 1276\text{ kJ}$$

$$BE = \frac{1276\ kJ}{2\ mol} = 638.0\ kJ/mol$$

29.39 (a) Assume exactly 100 g of the oxide.

	As	O
mass	76 g	24 g
molar mass	74.92 g/mol	16.00 g/mol
no. of moles	$(76\ g)\left[\frac{1\ mol}{74.92\ g}\right] = 1.0\ mol$	$(24\ g)\left[\frac{1\ mol}{16.00\ g}\right] = 1.5\ mol$
mole ratio n/n_{As}	$\frac{1.0}{1.0} = 1.0$	$\frac{1.5}{1.0} = 1.5$
Rel. moles of atoms	2	3

Empirical formula is As_2O_3.

(b) Molar mass of empirical formula = (2)(74.92 g) + (3)(16.00 g) = 197.8 g

$$\frac{200\ g/mol}{197.8\ g/mol} = 1$$

The molecular formula of the oxide is As_2O_3.

(c) Assume exactly 100 g of the acid

	As	O	H
mass	59.5 g	38.1 g	2.4 g
molar mass	74.92 g/mol	16.00 g/mol	1.01 g/mol
no. of moles	$(59.5\ g)\left[\frac{1\ mol}{74.92\ g}\right] = 0.794\ mol$	$(38.1\ g)\left[\frac{1\ mol}{16.00\ g}\right] = 2.38\ mol$	$(2.4\ g)\left[\frac{1\ mol}{1.01\ g}\right] = 2.4\ mol$
mole ratio n/n_{As}	$\frac{0.794}{0.794} = 1.00$	$\frac{2.38}{0.794} = 3.00$	$\frac{2.4}{0.794} = 3.0$
Rel. moles of atoms	1	3	3

Empirical formula is H_3AsO_3.

(d) $4As(s) + 3O_2(g) \rightarrow 2As_2O_3(s)$

$As_2O_3(s) + 3H_2O(l) \rightarrow 2H_3AsO_3(aq)$

29.41 $\Delta G° = [(1\ mol)\Delta G_f°(H^+) + (1\ mol)\Delta G_f°(HSe^-)] - [(1\ mol)\Delta G_f°(H_2Se)]$

$= [(1\ mol)(0) + (1\ mol)(44.0\ kJ/mol)] - [(1\ mol)(22.2\ kJ/mol)]$

$= 21.8\ kJ$

$$\log K = \frac{-\Delta G°}{(2.303)RT} = \frac{-(21.8\ \text{kJ})(1000\ \text{J/1 kJ})}{(2.303)(8.314\ \text{J/K mol})(298.15\ \text{K})} = -3.82$$

$$K = 1.5 \times 10^{-4}$$

29.43 The fundamental building unit in all silicon-containing substances in the crust of the earth is the SiO_4 tetrahedron. The SiO_4 tetrahedra are linked to each other to form various polymers, ranging through rings, chains, and sheets to three-dimensional networks like that of silica.

29.45 Silica is a substance of considerable chemical stability, inert to acids except hydrofluoric acid. Hot concentrated sodium hydroxide slowly converts silica to water-soluble silicates.

$SiO_2(s) + 4HF(aq\ or\ g) \rightarrow SiF_4(g) + 2H_2O(l)$

$xSiO_2(s) + 2NaOH(aq) \rightarrow Na_2O\cdot xSiO_2(aq) + H_2O(l)$

29.47 $xNa_2CO_3(l) + xSiO_2(s) \rightarrow Na_{2x}(SiO_3)_x(l) + xCO_2(g)$

$xCaCO_3(s) + xSiO_2(s) \rightarrow Ca_x(SiO_3)_x(l) + xCO_2(g)$

Countless variations in glass composition are possible. The replacement of part of the sodium by potassium makes the glass harder and raises the temperature at which it softens. Replacement of a part of the silicon dioxide with boric oxide gives a glass with a very low coefficient of expansion.

29.49 Clays are formed by the weathering of aluminosilicate minerals. Most clays have iron(III) oxide as an impurity. Ceramic objects are made by shaping, drying, and then heating to a high temperature plastic mixtures of clay, additives such as other silicate minerals or quartz, and water. The glaze material--it may be a metal oxide or salt, or a metal silicate, or a mixture of these--is applied to the surface, and the object is reheated to a temperature where the additive either melts or reacts with the clay to form a glassy coating.

29.51 The hardening process of fresh concrete is exothermic, so when it is covered with straw even though the air temperature drops below freezing, the straw acts as an insulator to retain the heat, and the concrete does not freeze.

29.53 An energy band is the term used for a continuous energy level produced by a large number of closely spaced molecular orbitals. If electrons partially fill this energy band, the element can be expected to conduct electricity.

29.55 An increase in temperature causes an increase in electrical resistance of a metal but a decrease in resistance in semiconductors because in the latter more electrons have sufficient energy to move from the valence band to the conduction band.

29.57 An *n*-type semiconductor operates by negative electrons from the donor impurities serving as the majority of the current carriers, while a *p*-type semiconductor operates by positive holes contributed by the acceptor impurities serving as the majority of the current carriers.

29.59 The redox reactions are (a), (c), (d), (e), and (f). The oxidizing agents are (a) O_2, (c) O_2, (d) Pd^{2+}, (e) HNO_3, and (f) As. The reducing agents are (a) NH_3, (c) Au, (d) CO, (e) Si, and (f) Na.

29.61 The types of each of these reactions are

(a) redox--combination of an element with a compound to give another compound

(b) redox--combination of an element with a compound to give other compounds

(c) nonredox--displacement

(d) redox--displacement of one element from a compound by another element

(e) redox--displacement of one element from a compound by another element

(f) nonredox--partner-exchange reaction (formation of a gas and of water)

(g) nonredox--decomposition to give compounds

(h) redox--does not fit into any of the categories listed in Table 17.7

29.63 Using Table 17.10, the first "yes" answer is for question

(a) 6. The reaction is a nonredox partner-exchange reaction.

$CaC_2(s) + 2H_2O(l) \rightarrow Ca(OH)_2(s) + C_2H_2(g)$

(b) 2. The reaction is a redox combination reaction.

$4Ag(s) + 8CN^- + O_2(g) + H_2O(l) \rightarrow 4[Ag(CN)_2]^- + 4OH^-$

(c) 2. The reaction is a redox reaction.

$CH_4(g) + 2O_2(g) \xrightarrow{\Delta} CO_2(g) + 2H_2O(g)$

or $2CH_4(g) + 3O_2(g) \xrightarrow{\Delta} 2CO(g) + 4H_2O(g)$

(d) 2. The reaction is a redox reaction.

$C(s) + 2H_2SO_4(conc) \xrightarrow{\Delta} CO_2(g) + 2SO_2(g) + 2H_2O(l)$

(e) 6. The reaction is a nonredox partner-exchange reaction.

$BBr_3(g) + 3H_2O(l) \rightarrow H_3BO_3(s) + 3HBr(aq)$

(f) 3. The reaction is a nonredox combination reaction.

$BF_3(g) + NaF(aq) \rightarrow Na[BF_4](aq)$

(g) 2. The reaction is a redox combination reaction.

$Ge(s) + 2Cl_2(g) \rightarrow GeCl_4(l)$

(h) 6. The reaction is a nonredox combination reaction.

$SeO_2(s) + H_2O(l) \rightarrow H_2SeO_3(aq)$

29.65 (a) $CaCO_3(s) + SiO_2(s) \xrightarrow{\Delta} CaSiO_3(l) + CO_2(g)$

(b) $CO_3^{2-} + H_2O(l) \rightleftharpoons HCO_3^- + OH^-$

(c) $Ni(s) + 4CO(g) \rightarrow [Ni(CO)_4](g)$

(d) $CaCO_3(s) + 2H^+ \rightarrow Ca^{2+} + H_2O(l) + CO_2(g)$

(e) $CO_2(g) + C(s) \xrightarrow{\Delta} 2CO(g)$ or $2CO_2(g) \xrightarrow{\Delta} 2CO(g) + O_2(g)$

29.67 $2As^{3+} + 3H_2S(aq) \rightarrow As_2S_3(s) + 6H^+$

$2Sb^{3+} + 3H_2S(aq) \rightarrow Sb_2S_3(s) + 6H^+$

$Sb_2S_3(s) + 8HCl(aq) \rightarrow 2[SbCl_4]^- + 3H_2S(aq) + 2H^+$

$As_2S_3(s) + 2H_2O_2(aq) + 12OH^- \rightarrow 2AsO_4^{3-} + 8H_2O(l) + 3S^{2-}$

$3Ag^+ + AsO_4^{3-} \rightarrow Ag_3AsO_4(s)$

$2[SbCl_4]^- + 8NH_3(aq) + 2H^+ + 3H_2S(aq) \rightarrow Sb_2S_3(s) + 8NH_4Cl(aq)$

CHAPTER 30

d- AND *f*-BLOCK ELEMENTS

Solutions to Odd-Numbered Questions and Problems

30.1 The energy levels that are being filled with electrons for each of the three *d*-transition series are 3*d* for Sc-Cu, 4*d* for Y-Ag, and 5*d* for La-Au. Many chemists consider zinc, cadmium, and mercury to be representative metals because of their $d^{10}s^2$ configurations, which give them many properties of the representative elements.

30.3 Some of the physical properties of the *d*-transition elements include being hard, strong, and dense; having metallic luster and high melting and boiling points; and being good conductors of heat and electricity. Some of the characteristic chemical properties of *d*-transition elements and their ions include high catalytic activity and formation of complex ions.

30.5 The crystal structure and electronic structure of the alloy are different than those of the individual metals.

30.7 The maximum oxidation of chromium is less stable.

30.9 Elements that are ferromagnetic can exhibit magnetism in the absence of an external magnetic field. Ferromagnetism differs from paramagnetism in that the latter can exhibit magnetism only in the presence of an external magnetic field.

30.11 The chief ore of chromium is chromite, $FeCr_2O_4$. Chromium metal is obtained by reduction of the oxide with carbon, silicon, or aluminum.

$$2Al(s) + Cr_2O_3(s) \rightarrow 2Cr(l) + Al_2O_3(l)$$

30.13 $Cr(VI) + 6e^- \rightarrow Cr(s)$

$$\left[\frac{3 \times 10^{-4}\text{ in}}{1\text{ h}}\right](1750\text{ in}^2)\left[\frac{2.54\text{ cm}}{1\text{ in}}\right]^3\left[\frac{7.20\text{ g Cr}}{1\text{ cm}^3}\right]\left[\frac{1\text{ mol Cr}}{52.00\text{ g Cr}}\right]\left[\frac{6\text{ mol }e^-}{1\text{ mol Cr}}\right]$$

$$= 7\text{ mol }e^-/\text{h}$$

$$\left[\frac{7\text{ mol }e^-}{1\text{ h}}\right]\left[\frac{96{,}500\text{ C}}{1\text{ mol }e^-}\right]\left[\frac{100\text{ C}}{20\text{ C}}\right]\left[\frac{1\text{ A s}}{1\text{ C}}\right]\left[\frac{1\text{ h}}{3600\text{ s}}\right] = 900\text{ A}$$

30.15 In acidic solution, the dichromate ion, which does act as a strong oxidizing agent, is the major species present, not the chromate ion.

30.17 The amphoteric behavior of chromium(III) hydroxide is illustrated by the following chemical equations:

$$Cr(OH)_3(s) + 3H^+ \rightarrow Cr^{3+} + 3H_2O(l)$$
$$Cr(OH)_3(s) + 3OH^- \rightarrow [Cr(OH)_6]^{3-}$$

30.19 Because manganese is quite reactive, it is expected to occur in nature as the oxide. Its primary ore is the mineral known as pyrolusite.

30.21 Pure manganese(IV) oxide has several percent of the oxygen atoms missing, leaving holes in the crystal structure. Oxide ions from adjacent sites are able to move into these holes, thus leaving new holes; this process makes the oxide conductive.

30.23 Potassium manganate(VI) is prepared by fusing manganese dioxide with a basic material and an oxidizing agent. Potassium permanganate is prepared from potassium manganate(VI) by acidification in the presence of a strong oxidizing agent.

$$MnO_2(s) + 2KOH(l) + KNO_3(l) \xrightarrow{\Delta} K_2MnO_4(s) + KNO_2(s) + H_2O(g)$$
$$2MnO_4^{2-} + O_3(g) + 2H^+ \rightarrow 2MnO_4^- + O_2(g) + H_2O(l)$$

30.25 (a) The manganate ion disproportionates to MnO_4^- ion and MnO_2.
(b) The manganate ion is oxidized to MnO_4^- ion by O_3.

30.27 Iron is most stable and most common in the +2 (ferrous) and +3 (ferric) states.

30.29 $$n = \frac{PV}{RT} = \frac{(762\text{ Torr})(1\text{ atm}/760\text{ Torr})(215\text{ mL})(1\text{ L}/1000\text{ mL})}{(0.0821\text{ L atm/K mol})(293.2\text{ K})} = 0.00896\text{ mol }O_2$$

$$(0.00896\text{ mol }O_2)\left[\frac{2\text{ mol O}}{1\text{ mol }O_2}\right] = 0.0179\text{ mol O}$$

$$(1.00\text{ g Fe})\left[\frac{1\text{ mol Fe}}{55.85\text{ g Fe}}\right] = 0.0179\text{ mol Fe}$$

The formula of the oxide formed is FeO.

30.31 In the compound $KFe[Fe(CN)_6]$ the oxidation state of one iron atom is +2 and that of the other is +3. The oxidation state of each changes in light because electrons can easily migrate from one iron atom to another.

30.33 These are all thermal decomposition reactions.

(a) *Gentle* heating of a hydrate is likely to drive off water, leaving behind iron(II) nitrate, $Fe(NO_3)_2$.

(b) *Strongly* heating the nitrate of a metal such as iron (Table 17.7) causes thermal decomposition to give nitrogen(IV) oxide (NO_2), oxygen, and the metal oxide. We might expect that strong heating in the presence of air would oxidize the iron present, resulting in the formation of Fe_2O_3, with iron oxidized to the +3 state.

(c) Heating a metal carbonate causes decomposition to give carbon dioxide and the metal oxide. We cannot be *sure* which iron oxide is the product; FeO would form first and at this moderate temperature might not be further oxidized. (The product is FeO.)

30.35

$$(5)[Fe^{2+} \rightarrow Fe^{3+} + e^-]$$
$$MnO_4^- + 8H^+ + 5e^- \rightarrow Mn^{2+} + 4H_2O(l)$$
$$\overline{5Fe^{2+} + MnO_4^- + 8H^+ \rightarrow Mn^{2+} + 5Fe^{3+} + 4H_2O(l)}$$

$$(42.79 \text{ mL soln})\left[\frac{1 \text{ L}}{1000 \text{ mL}}\right]\left[\frac{0.0205 \text{ mol } MnO_4^-}{1 \text{ L soln}}\right]\left[\frac{5 \text{ mol } Fe^{2+}}{1 \text{ mol } MnO_4}\right]\left[\frac{55.85 \text{ g } Fe^{2+}}{1 \text{ mol } Fe^{2+}}\right] = 0.245 \text{ g } Fe^{2+}$$

$$\frac{0.245 \text{ g } Fe^{2+}}{0.302 \text{ g sample}} \times 100 = 81.1\ \%$$

30.37 Let x = [H^+] at equilibrium.

	$Fe^{2+} + H_2O(l)$	$\rightarrow$ $Fe(OH)^+(s)$	$+ H^+$
initial	0.100 M	0	0
change	$-x$	x	x
equilibrium	$(0.100 - x)$	x	x

$$K_a = \frac{[Fe(OH)^+][H^+]}{[Fe^{2+}]} = \frac{(x)(x)}{(0.100 - x)} = 1.2 \times 10^{-6}$$

Assume $(0.100 - x) \approx 0.100$.

$$\frac{x^2}{0.100} = 1.2 \times 10^{-6}$$

$$x = 3.5 \times 10^{-4} \text{ mol/L} = [H^+]$$

$$pH = -\log [H^+] = -\log(3.5 \times 10^{-4}) = 3.46$$

30.39 Both cobalt and nickel find extensive commercial uses in alloys and as catalysts. Nickel is widely used as a protective plate on iron.

30.41 (a) Upon dissolving, blue $CoCl_2$ forms the pink hydrated cobalt(II) ion, $[Co(H_2O)_6]^{2+}$. (b) Addition of the ammonia solution precipitated the pink cobalt(II) hydroxide, $Co(OH)_2$. (c) The tan hexamminecobalt(II) complex ion, $[Co(NH_3)_6]^{2+}$, was formed.

30.43 $(n\text{-}1)d^{10}ns^1$; +1, +2, +3; +2 for Cu, +1 for Ag, and +1 and +3 for Au.

30.45 On heating with air, copper (which is the most reactive of the three) forms copper(II) oxide, CuO; silver slowly forms silver(I) oxide, Ag_2O, which, however, decomposes into its elements on strong heating. Gold does not react with oxygen. The metals do not react with nonoxidizing acids. Both copper and silver readily react with oxidizing acids, such as nitric acid, copper going to the +2 state, and silver to the +1 state. Gold is not attacked by nitric acid, but it is dissolved by aqua regia ($HCl\text{-}HNO_3$), with the formation of $[AuCl_4]^-$.

30.47 Silver nitrate is made by dissolving metallic silver in nitric acid. It is the usual source of other silver compounds.

30.49 (a) $2NaOH(aq) + 2AgNO_3(aq) \rightarrow Ag_2O(s) + H_2O(l) + 2NaNO_3(aq)$

(b) $Au(s) + 3HNO_3(aq) + 4HCl(aq) \rightarrow HAuCl_4(aq) + 3NO_2(g) + 3H_2O(l)$

(c) $2Cu^{2+} + 4CN^- \rightarrow 2CuCN(s) + (CN)_2(g)$

$CuCN(s) + CN^- \rightarrow [Cu(CN)_2]^-$

(d) $Cu^{2+} + 4NH_3(aq) \rightarrow [Cu(NH_3)_4]^{2+}$

(e) $AgCl(s) + 2NH_3(aq) \rightarrow [Ag(NH_3)_2]^+ + Cl^-$

(f) $Cu_2SO_4(s) \xrightarrow{H_2O} Cu(s) + CuSO_4(aq)$

(g) $CuSO_4(s) + 5H_2O(l) \rightarrow CuSO_4 \cdot 5H_2O(l)$

30.51 Hg $1s^22s^22p^63s^23p^63d^{10}4s^24p^64d^{10}4f^{14}5s^25p^65d^{10}6s^2$ diamagnetic

Hg^+ $1s^22s^22p^63s^23p^63d^{10}4s^24p^64d^{10}4f^{14}5s^25p^65d^{10}6s^1$ paramagnetic

Hg^{2+} $1s^22s^22p^63s^23p^63d^{10}4s^24p^64d^{10}4f^{14}5s^25p^65d^{10}$ diamagnetic

$[Hg - Hg]^{2+}$

The σ bond formed in Hg_2^{2+} makes this ion diamagnetic.

30.53 There is less difference in the physical and chemical properties between Zn and Cd than between Cd and Hg because Zn and Cd differ by 18 in atomic number, whereas Cd and Hg differ by 32 in atomic number (as a result of a series of *f* electrons.)

30.55 (a) $2ZnS(s) + 3O_2 \xrightarrow{\Delta} 2ZnO(s) + 2SO_2(g)$

$$ZnO(s) + H_2SO_4(dil) \rightarrow ZnSO_4(aq) + H_2O(l)$$
$$Zn^{2+} + 2e^- \rightarrow Zn(s)$$

(b) $$\left[\frac{90.\ \text{kg ZnS}}{100.\ \text{kg ore}}\right](100.\ \text{kg ore})\left[\frac{1000\ \text{g}}{1\ \text{kg}}\right]\left[\frac{1\ \text{mol ZnS}}{97.4\ \text{g ZnS}}\right] = 920\ \text{mol ZnS}$$

$$(920\ \text{mol ZnS})\left[\frac{2\ \text{mol ZnO}}{2\ \text{mol ZnS}}\right]\left[\frac{1\ \text{mol ZnSO}_4}{1\ \text{mol ZnO}}\right]\left[\frac{1\ \text{mol Zn}^{2+}}{1\ \text{mol ZnSO}_4}\right]\left[\frac{2\ \text{mol}\ e^-}{1\ \text{mol Zn}}\right] = 1800\ \text{mol}\ e^-$$

$$\left[\frac{96{,}500\ \text{C}}{1\ \text{mol}\ e^-}\right](1800\ \text{mol}\ e^-) = 1.7 \times 10^8\ \text{C}$$

30.57
$$3(Hg + 2Cl^- \rightarrow HgCl_2 + 2e^-)$$
$$2(3e^- + NO_3^- + 4H^+ \rightarrow NO + 2H_2O)$$
$$3Hg(l) + 2NO_3^- + 8H^+ + 6Cl^- \rightarrow 3HgCl_2(s) + 2NO(g) + 4H_2O(l)$$

NO_3^- is the oxidizing agent.

30.59 The bonding in the compounds of the zinc family is more covalent than that in compounds of the alkaline earth family because of the smaller size of the ions in the zinc family.

30.61 The general outer electron configuration for atoms of the *f*-transition elements is $(n - 2)f^x(n - 1)d^yns^2$. For the lanthanides the 5*d* and 6*s* electrons, and for the actinides the 6*d* and 7*s* electrons, are most available for bond formation.

30.63 The number of complexes formed by the *f*-transition elements is less than the number of complexes formed by the *d*-transition elements because the bonding is mainly ionic and does not involve *f* electrons. The lutetium ion is slightly smaller than the lanthanum ion and so is more prone to form complexes.

30.65 (a) $M(s) + 2HCl(aq) \rightarrow MCl_2(aq) + H_2(g)$ where M = Mn, Fe, Ni

(b) The second student saw the reaction of the oxide coating with the acid.
$CuO(s) + 2HCl(aq) \rightarrow CuCl_2(aq) + H_2O(l)$

(c) $3Cu(s) + 8HNO_3(aq) \rightarrow 3Cu(NO_3)_2(aq) + 2NO(g) + 4H_2O(l)$
Copper does not react with a nonoxidizing acid such as hydrochloric acid because copper lies below hydrogen in the electromotive series.

30.67 The redox reactions are reactions (a), (b), and (c). The oxidizing agents are (a) MnO_4^{2-}, (b) H_2O, and (c) $Cr_2O_7^{2-}$; the reducing agents are (a) MnO_4^{2-}, (b) $[Co(CN)_6]^{4-}$, and (c) NH_4^+.

30.69 The reactions are classified according to the reaction types listed in Tables 17.2 and 17.7 as

(a) redox--electron transfer between cations in aqueous solution

(b) nonredox--displacement

(c) redox--combination of two elements to give a compound

(d) redox--displacement of one element from a compound by another element

30.71 Using Table 17.10, the first "yes" answer is for question

(a) 3. Question 4 is answered "yes." The reaction is a nonredox partner-exchange reaction with the formation of water.

$Cr_2O_7^{2-} + 2OH^- \rightarrow 2CrO_4^{2-} + H_2O(l)$

(b) 3. Question 4 is answered "no" and question 5 is answered "yes," so the reaction is classified as a more complex redox reaction.

$Mn^{2+} + OCl^- + 2OH^- \rightarrow MnO_2(s) + Cl^- + H_2O(l)$

(c) 2. The reaction is a redox displacement reaction.

$4FeS_2(s) + 11O_2(g) \rightarrow 2Fe_2O_3(s) + 8SO_2(g)$

(d) 3. Question 4 is answered "no" and question 5 is answered "yes," so the reaction is classified as a redox reaction of ions in aqueous solution (electron transfer).

$MnO_4^- + 5Fe^{2+} + 8H^+ \rightarrow Mn^{2+} + 5Fe^{3+} + 4H_2O(l)$

30.73 (a) $\left.\begin{matrix} Mn^{2+} \\ Fe^{2+} \end{matrix}\right] \xrightarrow[KClO_3,\ \Delta]{HNO_3} \begin{matrix} MnO_2(s) \\ Fe^{3+} \end{matrix}$

(b) $\left.\begin{matrix} Ag^{+} \\ Cu^{2+} \end{matrix}\right] \xrightarrow{HCl} \begin{matrix} AgCl(s) \\ Cu^{2+} \end{matrix}$

(c) $\left.\begin{matrix} Pr^{3+} \\ Cr^{3+} \end{matrix}\right] \xrightarrow{NaOH} \begin{matrix} Pr(OH)_3(s) \\ Cr(OH)_4^- \end{matrix}$

(d) $\left.\begin{matrix} Cu^{2+} \\ Ni^{2+} \end{matrix}\right] \xrightarrow[H^+]{H_2S} \begin{matrix} CuS(s) \\ Ni^{2+} \end{matrix}$

(e) $\left.\begin{matrix} Fe^{2+} \\ Co^{2+} \end{matrix}\right] \xrightarrow[NH_4^+]{NH_3} \begin{matrix} Fe(OH)_2(s) \\ [Co(NH_3)_6]^{2+} \end{matrix}$

(f) $\left.\begin{matrix} Y^{3+} \\ Ni^{2+} \end{matrix}\right] \xrightarrow[NH_4^+]{NH_3} \begin{matrix} Y(OH)_3(s) \\ [Ni(NH_3)_6]^{2+} \end{matrix}$

30.75 (a) $M^{2+} + H_2S(aq) \rightarrow MS(s) + 2H^+$ where M = Fe, Co, Ni

(b) $FeS(s) + 2HCl(aq) \rightarrow FeCl_2(aq) + H_2S(aq)$

(c) $MS(s) + 2HNO_3(aq) \rightarrow M^{2+} + 2NO_3^- + H_2S(aq)$ where M = Co, Ni

(d) $Fe^{2+} + 2HNO_3(aq) \rightarrow Fe^{3+} + NO(g) + H_2O(l) + NO_3^-$

(e) $Fe^{3+} + NH_4CNS(s) \rightarrow [Fe(CNS)]^{2+} + NH_4^+$

(f) $M^{2+} + 6NH_3(aq) \rightarrow [M(NH_3)_6]^{2+}$ where M = Co, Ni

(g) $[Ni(NH_3)_6]^{2+} + 2DMG^- \rightarrow [Ni(DMG)_2] + 6NH_3(aq)$

(h) $[Co(NH_3)_6]^{2+} + 6HCl(aq) + 4NH_4CNS(s) \rightarrow [Co(CNS)_4]^{2-} + 10NH_4^+ + 6Cl^-$

30.77

$$
\begin{array}{rcll}
Fe(s) + 2OH^- & \rightarrow & Fe(OH)_2(s) + 2e^- & E° = 0.877\ V \\
(2)[NiOOH(s) + H_2O(l) + e^- & \rightarrow & Ni(OH)_2(s) + OH^-] & \underline{E° = x} \\
\hline
Fe(s) + 2NiOOH(s) + 2H_2O(l) & \rightarrow & Fe(OH)_2(s) + 2Ni(OH)_2(s) & E° = 1.40\ V
\end{array}
$$

$0.877\ V + x = 1.40\ V$

$x = 0.52\ V$

CHAPTER 31

COORDINATION CHEMISTRY

Solutions to Exercises

31.1 (a) This is a coordination compound with the complex ion $[Pt(NH_3)_5Cl]^{3+}$ as the cation. The ligands are five ammonia molecules and a chloride ion, so the coordination number of the central platinum ion is 6. To give the complex ion its +3 charge, the platinum must be in its +4 oxidation state.

$$\overset{+4}{[Pt}(NH_3)_5\overset{-1}{Cl}]\overset{(3)(-1)}{Cl_3}$$

(b) This is a complex cation with six water molecules as ligands around the central manganese ion. Its coordination number is 6 and its oxidation number is +3 (because the ligands are neutral).

$$[\overset{+3}{Mn}(H_2O)_6]^{3+}$$

(c) This is a complex anion with six fluoride ions as ligands around the central cobalt ion. Its coordination number is 6 and its oxidation number is +3.

$$[\overset{+3}{Co}\overset{(6)(-1)}{F_6}]^{3-}$$

31.2 (a) In this coordination compound, the complex ion is an anion. Its 6 CN^- ligands are named cyano and the metal ion is named ferrate. The metal has a +2 charge. The complete name is potassium hexacyanoferrate(II).

(b) This complex anion has two ammine ligands and four nitro ligands. The central cobalt ion has a charge of +3 and an "ate" ending. The complete name is diamminetetranitrocobaltate(III) ion. (If the bonding between the central atom or ion and the ONO^- involves an oxygen atom instead of the nitrogen atom, the group is called nitrito and the name of the complex anion is diamminetetranitritocobaltate(III) ion.)

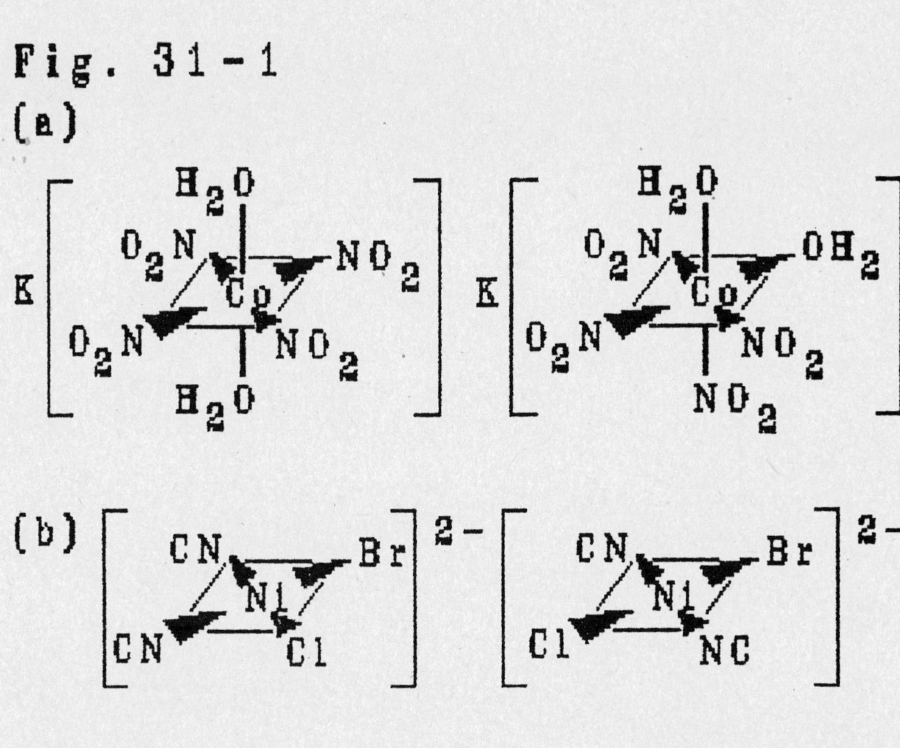

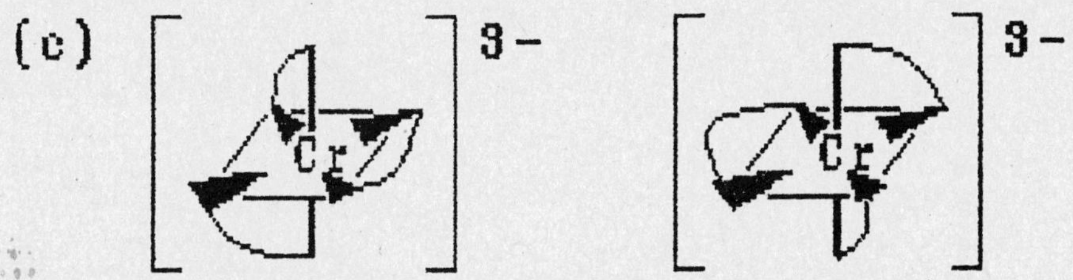

31.3 The formula of

(a) hexaamminechromium(III) chloride is $[Cr(NH_3)_6]Cl_3$

(b) pentaamminechlorochromium(III) chloride is $[Cr(NH_3)_5Cl]Cl_2$

(c) tetraaquanickel(II) ion is $[Ni(H_2O)_4]^{2+}$

(d) diiodocuprate(I) ion is $[CuI_2]^-$

31.4 The sketches of the complexes are shown in Fig. 31-1.

(a) This coordination compound has a complex anion with four nitro ligands and two water ligands; octahedral of the type Ma_4b_2. There is one set of possible *cis-trans* isomers.

(b) This complex anion is square planar of the type Ma_2bc. There is one set of possible *cis-trans* isomers.

(c) This complex anion has three bidentate ligands, so this is an octahedral complex with three chelate rings. It has two optically active isomers.

31.5 In the $[Mn(H_2O)_6]^{3+}$ complex ion, the electron configuration of Mn^{3+} is $[Ar]3d^4$. Because no pairing of electrons occurs, the ligands will use the $4d$ subshell for the sp^3d^2 hybridization. The ion will be paramagnetic.

Solutions to Odd-Numbered Questions and Problems

31.1 Complex ions are formed by the combination of a central metal cation or metal atom and one or more molecules or anions, referred to as ligands. The ligands can be thought of as joined to the central metal ion by coordinate covalent bonds.

31.3

complex	ligands	coordination number	oxidation number
(a) $[Co(NH_3)_2(NO_2)_4]^-$	NH_3, NO_2^-	6	+3
(b) $[Cr(NH_3)_5Cl]Cl_2$	NH_3, Cl^-	6	+3
(c) $K_4[Fe(CN)_6]$	CN^-	6	+2
(d) $[Pd(NH_3)_4]^{2+}$	NH_3	4	+2

31.5 The phenomenon of ring formation by a ligand in a complex is called chelation. Five- and six-membered rings are much less strained than four-membered rings and are very common.

31.7 Assume exactly 100 g compound.

	Fe	C	N	K
mass	17.0 g	21.9 g	25.5 g	35.6 g
molar mass	55.85 g/mol	12.01 g/mol	14.01 g/mol	39.10 g/mol
No. of moles	$(17.0\text{ g})\left[\frac{1\text{ mol}}{55.85\text{ g}}\right]$ = 0.304 mol	$(21.9\text{ g})\left[\frac{1\text{ mol}}{12.01\text{ g}}\right]$ = 1.82 mol	$(25.5\text{ g})\left[\frac{1\text{ mol}}{14.01\text{ g}}\right]$ = 1.82 mol	$(35.6\text{ g})\left[\frac{1\text{ mol}}{39.10\text{ g}}\right]$ = 0.910 mol
mole ratio n/n_{Fe}	$\frac{0.304}{0.304} = 1.00$	$\frac{1.82}{0.304} = 5.99$	$\frac{1.82}{0.304} = 5.99$	$\frac{0.910}{0.304} = 2.99$
Rel. moles of atoms	1	6	6	3

The empirical formula is $FeC_6N_6K_3$.

$$M = \frac{\Pi}{RT} = \frac{(0.095\text{ atm})}{(0.0821\text{ L atm/K mol})(298\text{ K})} = 0.0039\text{ mol/L}$$

$$\frac{0.039\text{ mol/L}}{0.0010\text{ mol/L}} = 3.9 \approx 4\text{ ions}$$

$K_3[Fe(CN)_6]$ is called potassium hexacyanoferrate(III).

31.9 Four moles of ions combine to form one mole of a polyatomic ion, so a decrease in entropy is expected.

$\Delta S° = [(1\ \text{mol})S°([AgI_3]^{2-})] - [(1\ \text{mol})S°(Ag^+) + (3\ \text{mol})S°(I^-)]$

$= [(1\ \text{mol})(253.1\ \text{J/K mol})] - [(1\ \text{mol})(72.68\ \text{J/K mol})$
$+ (3\ \text{mol})(111.3\ \text{J/K mol})]$

$= -153.5\ \text{J/K}$

The calculated value of $\Delta S°$ confirms the predicted entropy decrease.

31.11 The name of

(a) $K_3[Mn(CN)_6]$ is potassium hexacyanomanganate(III)

(b) $[Pd(NH_3)_4](OH)_2$ is tetraamminepalladium(II) hydroxide

(c) $[Ag(CN)_2]^-$ is dicyanoargentate(I) ion

(d) $[Ag(NH_3)_2]^+$ is diamminesilver(I) ion

(e) $K_2[Fe(C_2O_4)_2]\cdot 2H_2O$ is potassium dioxalatoferrate(II) dihydrate

(f) $K_3[Co(C_2O_4)_3]\cdot 6H_2O$ is potassium trioxolatocobaltate(III) hexahydrate

(g) $[Cu(NH_3)_4]^{2+}$ is tetraamminecopper(II) ion

(h) $Na[AgI_2]$ is sodium diiodoargentate(I)

(i) $Na_2[PtI_4]$ is sodium tetraiodoplatinate(II)

31.13 The formula of

(a) diamminedichlorozinc(II) is $[Zn(NH_3)_2Cl_2]$

(b) tin(IV) hexacyanoferrate(II) is $Sn[Fe(CN)_6]$

(c) tetracyanoplatinate(II) ion is $[Pt(CN)_4]^{2-}$

(d) potassium hexacyanochromate(III) is $K_3[Cr(CN)_6]$

(e) tetraammineplatinum(II) ion is $[Pt(NH_3)_4]^{2+}$

(f) hexaamminenickel(II) bromide is $[Ni(NH_3)_6]Br_2$

(g) tetraamminecopper(II) pentacyanohydroxoferrate(III) is $[Cu(NH_3)_4]_3[Fe(CN)_5(OH)]_2$

31.15 See Fig. 31-2. Geometric isomers cannot be formed by (a) tetrahedral Mab_3, (b) tetrahedral Ma_2b_2, nor (c) square planar Mab_3. *Cis* and *trans* geometric isomers can be formed by (d) square planar Ma_2b_2. None of these structures show optical activity.

31.17 The structural formulas for shown in Fig. 31-3.

31.19 The two complexes are (a) identical, (b) geometric isomers, (c) optical isomers, and (d) identical.

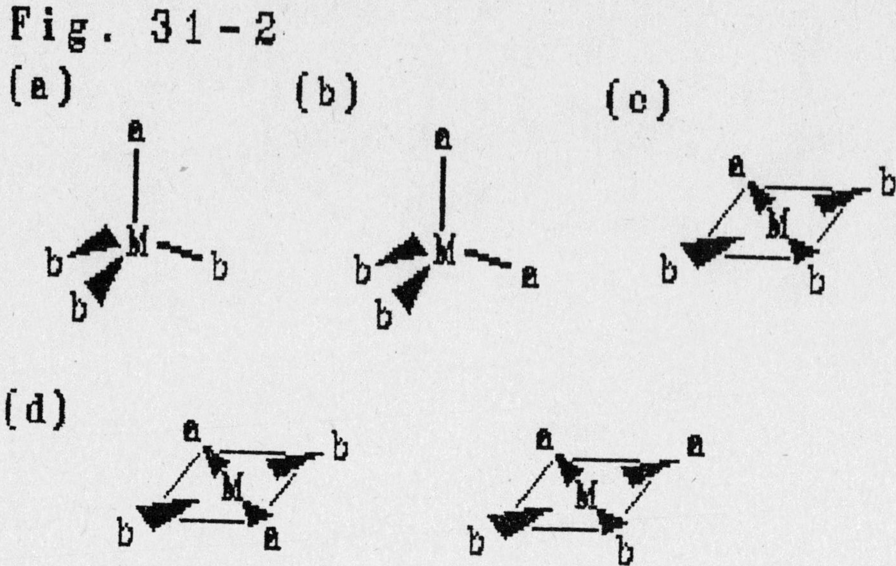

31.21 The thermodynamic stability of a complex is increased with small radius and high charge of the central metal ion. Chelate ring formation greatly increases the stability of complexes.

31.23 The dissociation constant is less for

(a) $[Co(en)_3]^{3+}$ than for $[Co(en)_3]^{2+}$ because the higher oxidation state of Co increases stability

(b) $[Cu(gly)_2]$ than for $[Zn(gly)_2]$ because the smaller ionic size of Cu^{2+} increases stability

(c) $[Cd(en)_2]^{2+}$ than for $[Cd(NH_3)_4]^{2+}$ because ring formation increases stability

31.25 Let x = $[Fe^{3+}]$ at equilibrium.

	$[Fe(C_2O_4)_3]^{3-}$	$\rightleftharpoons$	Fe^{3+}	+ $3C_2O_4{}^{2-}$
initial	0.010		0	1.5
change	$-x$		$+x$	$+3x$
equilibrium	$0.010 - x$		x	$1.5 + 3x$

$$K_d = \frac{[Fe^{3+}][C_2O_4{}^{2-}]^3}{[Fe(C_2O_4)_3{}^{2-}]} = \frac{(x)(1.5 + 3x)^3}{(0.010 - x)} = 3 \times 10^{-21}$$

Assume that $0.010 - x \approx 0.010$ and $1.5 + 3x \approx 1.5$.

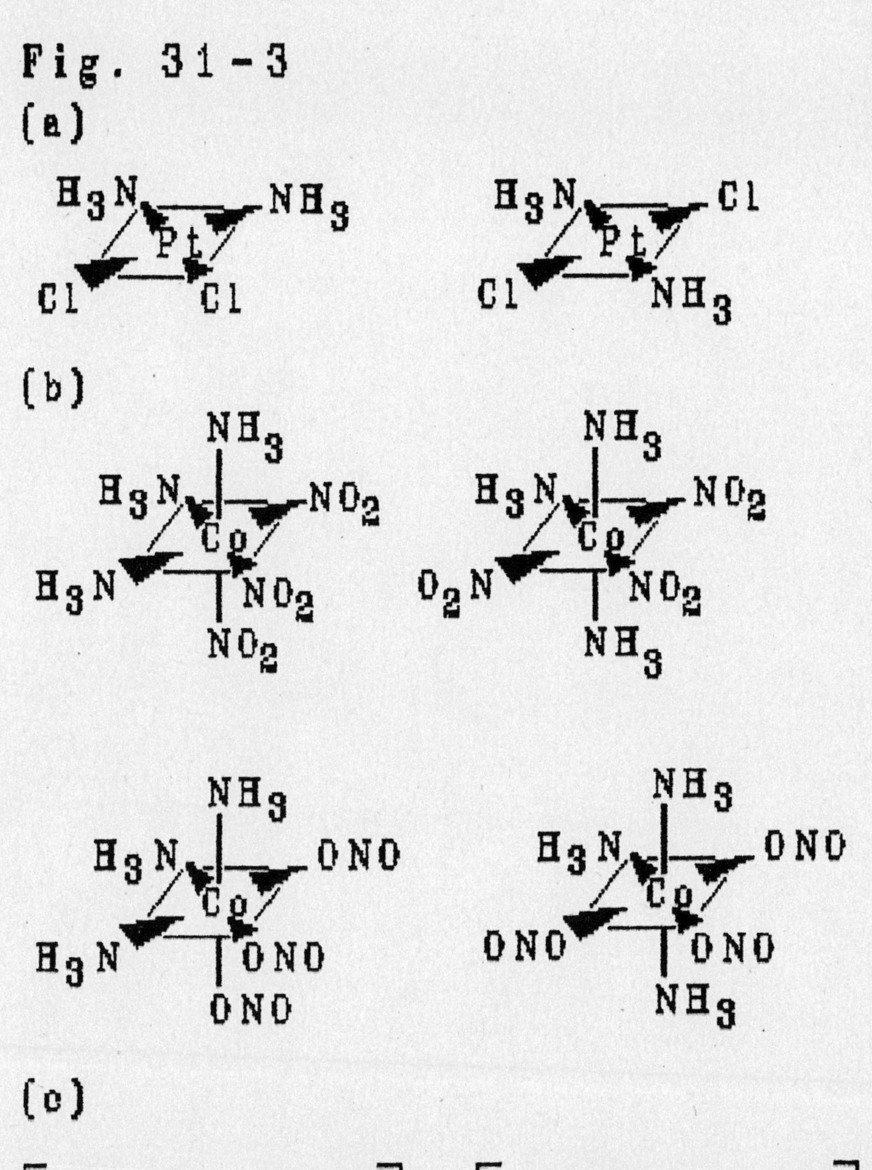

$$\frac{(x)(1.5)^3}{(0.010)} = 3 \times 10^{-21}$$

$$x = 9 \times 10^{-24} \text{ mol/L}$$

The approximations are valid because $0.010 - 9 \times 10^{-24} \approx 0.010$ and $1.5 + (3)(9 \times 10^{-24}) \approx 1.5$.

$[Fe^{3+}] = 9 \times 10^{-24}$ mol/L

31.27 To make the oxidation of water favorable,

$$2H_2O(l) \rightleftharpoons O_2(g) + 4H^+(10^{-7}\ M) + 4e^- \qquad E° = -0.815\ V$$

$E°$ for the reduction half-reaction would have to be greater than +0.815 V (neglecting overvoltage). Only the Co^{3+} species of the first equation will oxidize water.

To make the reduction of water favorable,

$$2H_2O(l) + 2e^- \rightleftharpoons H_2(g) + 2OH^- \qquad E° = -0.828\ V$$

$E°$ for the oxidation half-reaction would have to be greater than 0.828 V (neglecting overvoltage). Only the Co(II) species in $[Co(CN)_5]^{3-}$ of the fourth equation will reduce water.

31.29 The bonding in the metallocene complexes involves sharing of the π electrons by the metal atom as well as by the carbon atoms.

31.31 The three theories used to discuss the bonding in complexes are valence bond approach, in which covalent bonding between ligand and central metal atom uses hybridized atomic orbitals on the metal; crystal field theory, in which ionic bonding is present; and molecular orbital theory, in which molecular orbitals are formed from ligand orbitals and metal atomic orbitals.

31.33 The Co^{3+} ion has the electron configuration $[Ar]3d^6$ (that of the Co atom is $[Ar]3d^7 4s^2$). Using the valence bond theory for a Co^{3+} octahedral complex with ligands that can

(a) occupy only the outer orbitals on the Co^{3+}, the six pairs of electrons furnished by the ligands go into the 4*s*, the three 4*p*, and two of the 4*d* orbitals. The proposed hybridization on Co^{3+} is sp^3d^2.

(b) occupy inner orbitals on the Co^{3+}, the 3*d* electrons on cobalt become paired, allowing the six pairs of electrons from the ligands to enter two 3*d* orbitals, the 4*s* orbital, and the three 4*p* orbitals. The proposed hybridization is d^2sp^3.

31.35 (a) The electron configuration is $[Ar]3d^6 4s^2$ for Fe and $[Ar]3d^5$ for Fe^{3+}. In $[Fe(CN)_6]^{3-}$ four of the 3*d* electrons become paired, allowing the six pairs of electrons from the ligands to enter two 3*d* orbitals, the 4*s* orbital, and the three 4*p* orbitals. Thus there is one unpaired electron in a 3*d* orbital.

(b) In $[Fe(H_2O)_6]^{3+}$ the 3*d* electrons do not pair up, so the six pairs of ligand electrons enter the 4*s* orbital, the three 4*p* orbitals, and two

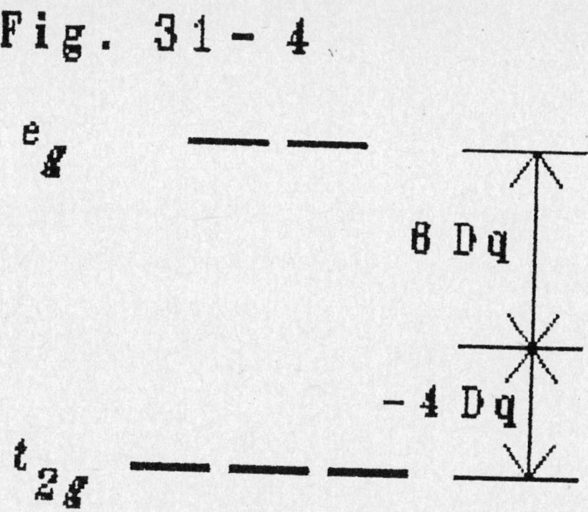

of the $4d$ orbitals. All five electrons in the $3d$ orbitals are unpaired.

(c) The electron configuration is $[Ar]3d^5 4s^2$ for Mn and $[Ar]3d^5$ for Mn^{2+}. In $[Mn(H_2O)_6]^{2+}$ none of the $3d$ electrons pair up, so the six pairs of ligand electrons enter the $4s$ orbital, the three $4p$ orbitals, and two of the $4d$ orbitals. All five electrons in the $3d$ orbitals are unpaired.

(d) The electron configuration is $[Ar]3d^7 4s^2$ for Co and $[Ar]3d^6$ for Co^{3+}. In $[Co(NH_3)_6]^{3+}$ the $3d$ electrons become paired, allowing the six pairs of ligand electrons to enter two $3d$ orbitals, the $4s$ orbital, and the three $4p$ orbitals. There are no unpaired electrons.

31.37 (a) $(t_{2g})^3(e_g)^0$, a low-spin complex ion
CFSE = 6(-4 Dq) + 3 P = -24 Dq + 3 P

(b) $(t_{2g})^4(e_g)^2$, a high-spin complex ion
CFSE = 2(6 Dq) + 4(-4 Dq) + P = -4 Dq + P

31.39 (a) $[Ni(H_2O)_6]^{2+} + 4CN^- \rightarrow [Ni(CN)_4]^{2-} + 6H_2O(l)$

(b) $[Ni(H_2O)_6]^{2+} + 4NH_3(aq) \rightarrow [Ni(NH_3)_4]^{2+} + 6H_2O(l)$

(c) $[Ni(H_2O)_6]^{2+}$ + 4pyridine(aq) $\rightarrow$ $[Ni(pyridine)_4]^{2+} + 6H_2O(l)$

31.41 See Fig. 31-4. (a) $(t_{2g})^3(e_g)^2$, (b) $(t_{2g})^4(e_g)^2$, (c) $(t_{2g})^1(e_g)^0$.

CHAPTER 32

ORGANIC CHEMISTRY

Solutions to Exercises

32.1 The "parent" hydrocarbon in this structure is the ring containing five carbon atoms--the cyclopentane ring. The two groups attached to the ring each have two carbon atoms; these are ethyl groups, indicating that this compound is a diethylcyclopentane. Both ethyl groups are attached to the same carbon atom, which we give the number 1. The complete name is 1,1-diethylcyclopentane.

32.2 The IUPAC name for $CH_3C{\equiv}CCH_2CH_3$ is 2-pentyne and for $ClCH_2C{\equiv}CCH_2CH_3$ is 1-chloro-2-pentyne. No *cis-trans* isomers are possible.

32.3 The structural formula for 1-methylnaphthalene is

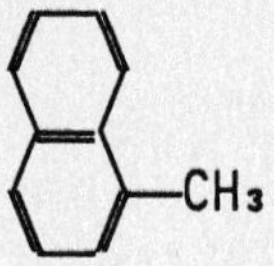

$-CH_3$

32.4 (a) aliphatic, saturated, substituted cycloalkane
(b) aliphatic, saturated, alkane, straight chain
(c) aliphatic, unsaturated, alkene, straight chain, conjugated diene

32.5 3-Chloro-1-butene will exhibit optical isomerism. The Fischer projections for the enantiomeric pairs are

```
                  H                     H
                  |                     |
CH2 = CH - C - CH3     CH3 - C - CH = CH2
                  |                     |
                  Cl                    Cl
```

32.6 There are nine isomers (including *cis-trans* isomers).

```
    Cl                           Cl
    |                            |
H - C = CH - CH2 - CH3     CH2 = C - CH2 - CH3
(cis-trans)
```

```
              Cl                                        Cl
              |                                         |
CH2 = CH - C - CH3                    CH2 = CH - CH2 - C - H
              |                                         |
              H                                         H
       Cl
       |
H - C - CH = CH - CH3                 CH3 - CCl = CH - CH3
       |                              (cis-trans)
       H
(cis-trans)
```

32.7 (a) Cl - CH_2 - Cl is an alkyl chloride where R is CH_2.

(b) C_6H_5—NH - CH_2 - CH_3 is an amine (secondary) where R is C_6H_5 and CH_2CH_3.

Solutions to Odd-Numbered Questions and Problems

32.1 Alkanes are straight and branched saturated hydrocarbons with the general formula C_nH_{2n+2}. A substance having the formula C_6H_{14} could be an alkane because it fits the general formula with $n = 6$.

32.3 An alkene is a hydrocarbon with a covalent double bond. The general formula for an alkene is C_nH_{2n} with $n \geq 2$. The compound C_6H_{12} is possibly an alkene but might also be a cycloalkane because it fits the general formula of both with $n = 6$.

32.5 An alkyl group contains one less hydrogen atom than an alkane. The general formula for the alkyl group derived from an alkane is C_nH_{2n+1} and from a cycloalkane is C_nH_{2n-1}. The name of an alkyl group is obtained by dropping the *ane* of the hydrocarbon name and adding *yl*.

32.7 The structural formulas and names (IUPAC) of the three C_5H_{12} isomers are

```
CH3 - CH2 - CH2 - CH2 - CH3    CH3 - CH2 - CH - CH3              CH3
        n-pentane                           |                      |
                                           CH3              CH3 - C - CH3
                                    2-methylbutane                 |
                                                                  CH3
                                                          2,2-dimethylpropane
```

32.9 The structural formula for 2,2-dimethylpropane and the type of carbon atoms:

```
        1°CH3
 1°       |     1°
 CH3  -   C  -  CH3
        4°|
        1°CH3
```

32.11 The molecular structure of

(a) 1-butyne is

$CH \equiv C - CH_2 - CH_3$

(b) 2-methylpropene is

$CH_2 = C - CH_3$
with CH_3 bonded to C

(c) 2-ethyl-3-methyl-1-butene is

$CH_2 = C - CH - CH_3$
with $CH_2 - CH_3$ bonded to C and CH_3 bonded to CH

(d) 3-methyl-1-butyne is

$CH \equiv C - CH - CH_3$
with CH_3 bonded to CH

32.13 (a) 2,2-dimethylbutane (b) 1,1-dibromoethene

(c) 2,3-dimethylbutane (d) 1,3-cyclohexadiene

32.15 Aromatic hydrocarbons are unsaturated hydrocarbons containing ring systems stabilized by delocalized electrons, as in benzene.

32.17 The molecular structures and names of the three isomeric trimethylbenzenes are

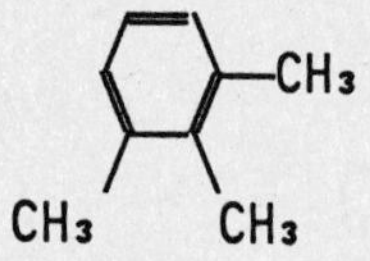

$C_6H_3(CH_3)_3$: 1,2,3-trimethylbenzene 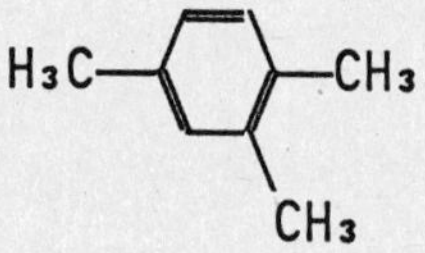 1,2,4-trimethylbenzene 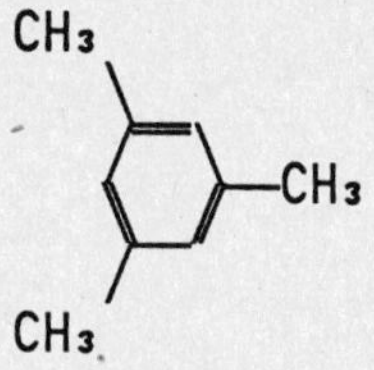 1,3,5-trimethylbenzene

32.19 The molecular structure of

(a) *p*-dinitrobenzene is

$O_2N-C_6H_4-NO_2$

(b) *n*-propylbenzene is

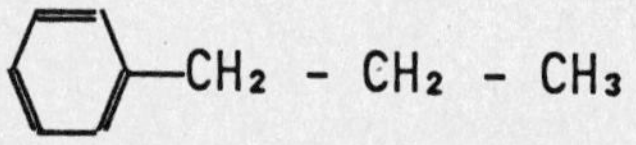

$C_6H_5-CH_2 - CH_2 - CH_3$

(c) 1,3,5-tribromobenzene is

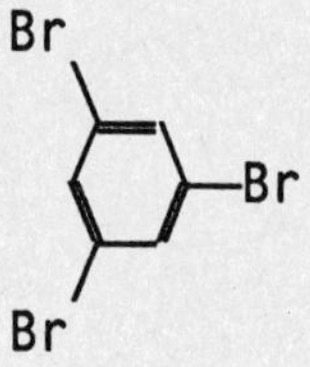

$C_6H_3Br_3$ (Br, Br, Br)

(d) 1,3-diphenylbutane is

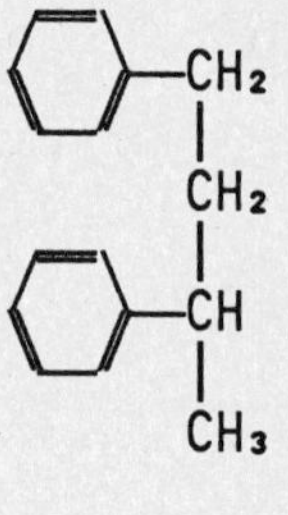

$C_6H_5-CH_2-CH_2-CH(C_6H_5)-CH_3$

32.21 (a) $CH_4(g) \rightarrow C(g) + 4H(g)$

$$\Delta H^\circ = [(1\ mol)\Delta H^\circ_f(C) + (4\ mol)\Delta H^\circ_f(H)] - [(1\ mol)\Delta H^\circ_f(CH_4)]$$

$$= [(1\ mol)(717\ kJ/mol) + (4\ mol)(218\ kJ/mol)] - [(1\ mol)(-75\ kJ/mol)]$$

$$= 1664\ kJ$$

$$BE_{CH} = \frac{1664\ kJ}{4\ mol} = 416\ kJ/mol$$

(b) $C_2H_6(g) \rightarrow 2C(g) + 6H(g)$

$$\Delta H° = [(2\ mol)\Delta H_f°(C) + (6\ mol)\Delta H_f°(H)] - [(1\ mol)\Delta H_f°(C_2H_6)]$$
$$= [(2\ mol)(717\ kJ/mol) + (6\ mol)(218\ kJ/mol)] - [(1\ mol)(-85\ kJ/mol)]$$
$$= 2830\ kJ$$
$$\Delta H° = (1\ mol)BE_{C-C} + (6\ mol)BE_{C-H} = 2830\ kJ$$
$$(1\ mol)BE_{C-C} + (6\ mol)(416\ kJ/mol) = 2830\ kJ$$
$$BE_{C-C} = 330\ kJ/mol$$

(c) $C_2H_4(g) \rightarrow 2C(g) + 4H(g)$

$$\Delta H° = [(2\ mol)\Delta H_f°(C) + (4\ mol)\Delta H_f°(H)] - [(1\ mol)\Delta H_f°(C_2H_4)]$$
$$= [(2\ mol)(717\ kJ/mol) + (4\ mol)(218\ kJ/mol)] - [(1\ mol)(52\ kJ/mol)]$$
$$= 2250\ kJ$$
$$\Delta H° = (1\ mol)BE_{C=C} + (4\ mol)BE_{C-H} = 2250\ kJ$$
$$(1\ mol)BE_{C=C} + (4\ mol)(416\ kJ/mol) = 2250\ kJ$$
$$BE_{C=C} = 590\ kJ/mol$$

(d) $C_2H_2(g) \rightarrow 2C(g) + 2H(g)$

$$\Delta H° = [(2\ mol)\Delta H_f°(C) + (2\ mol)\Delta H_f°(H)] - [(1\ mol)\Delta H_f°(C_2H_2)]$$
$$= [(2\ mol)(717\ kJ/mol) + (2\ mol)(218\ kJ/mol)] - [(1\ mol)(227\ kJ/mol)]$$
$$= 1640\ kJ$$
$$\Delta H° = (1\ mol)BE_{C\equiv C} + (2\ mol)BE_{C-H} = 1640\ kJ$$
$$(1\ mol)BE_{C\equiv C} + (2\ mol)(416\ kJ/mol) = 1640\ kJ$$
$$BE_{C\equiv C} = 810\ kJ/mol$$

32.23 (a) CO_2 (b) CH_3Cl or CH_2Cl_2, etc.

(c) $CH_3 - CH(OSO_3H) - CH_3$ (d) $C_6H_5-NO_2$

(e) $C_6H_5-SO_3H$

32.25 Of the processes utilized in the petroleum industry

(a) isomerization (accomplished by heat and catalysts) converts straight-chain alkanes into branched alkanes, the latter performing better as fuels;

(b) cracking (via heat and catalysts) breaks large molecules above the gasoline range into smaller molecules that are in the gasoline range;

(c) alkylation combines lower molecular mass alkanes and alkenes to form molecules in the gasoline range;

(d) reforming employs catalysts in the presence of hydrogen to convert noncyclic hydrocarbons to aromatic compounds.

32.27 $H_2O(g) + C(s) \rightarrow CO(g) + H_2(g)$

Let $P_{CO} = P_{H_2} = x$.

$$K_p = \frac{P_{CO}\,P_{H_2}}{P_{H_2O}} = \frac{(x)(x)}{15.6} = 3.2$$

$x = 7.1$ bar

$P_{CO} = P_{H_2} = 7.1$ bar

32.29 The three types of isomerism are structural isomerism, *cis-trans* isomerism, and optical isomerism.

32.31 The compounds that can exist as *cis* and *trans* isomers are

(c) 2-bromo-2-butene

```
Br       H            Br        CH3
  \     /               \      /
   C = C       and       C = C
  /     \               /      \
H3C      CH3          H3C        H
     cis                  trans
```

(e) 2,3-dichloro-2-butene

```
Cl       Cl           Cl        CH3
  \     /               \      /
   C = C       and       C = C
  /     \               /      \
H3C      CH3          H3C       Cl
     cis                  trans
```

32.33 A nucleophile is a reagent that has a partially or complete negative charge that enables it to bond with a electron-deficient atom. The nucleophiles are (a), (b), and (d).

32.35 Water is acting as the nucleophile and the $(CH_3)_3C^+$ ion is acting as the electrophile.

32.37 The gain or loss of hydrogen or oxygen atoms is often considered the definition of oxidation-reduction in organic chemistry. The oxidation processes are (a) gain of oxygen and (d) gain of oxygen and the reduction processes are (b) gain of hydrogen and (c) loss of oxygen.

32.39 $Cr_2O_7^{2-} + 14H^+ + 6e^- \rightarrow 2Cr^{3+} + 7H_2O(l)$

$MnO_4^- + 8H^+ + 5e^- \rightarrow Mn^{2+} + 4H_2O(l)$

$$(1.00\ g\ K_2Cr_2O_7)\left[\frac{1\ mol\ K_2Cr_2O_7}{294.20\ g\ K_2Cr_2O_7}\right]\left[\frac{6\ mol\ e^-}{1\ mol\ K_2Cr_2O_7}\right]\left[\frac{1\ mol\ KMnO_4}{5\ mol\ e^-}\right] \times \left[\frac{158.04\ g\ KMnO_4}{1\ mol\ KMnO_4}\right] = 0.645\ g\ KMnO_4$$

32.41 The compounds 2-hexene, styrene, and cyclohexene will undergo additions with Br_2, in which the red-brown color of bromine will disappear.

(a) $CH_3 - CH_2 - CH_2 - CH_2 - CH_2 - CH_3 + Br_2 \rightarrow$ no reaction

$CH_3 - CH = CH - CH_2 - CH_2 - CH_3 + Br_2 \rightarrow CH_3 - CH(Br) - CH(Br) - CH_2 - CH_2 - CH_3$

(b) (benzene) $+ Br_2 \rightarrow$ no reaction

(benzene)$-CH = CH_2 + Br_2 \rightarrow$ (benzene)$-CH(Br) - CH_2(Br)$

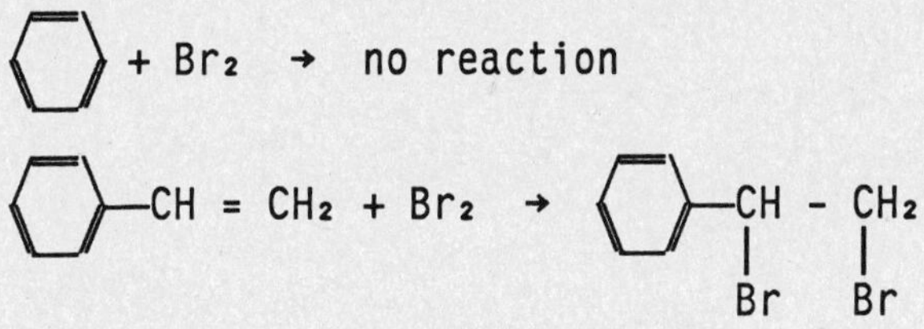

(c) (cyclohexene) $+ Br_2 \rightarrow$ (1,2-dibromocyclohexane: ring with Br, Br)

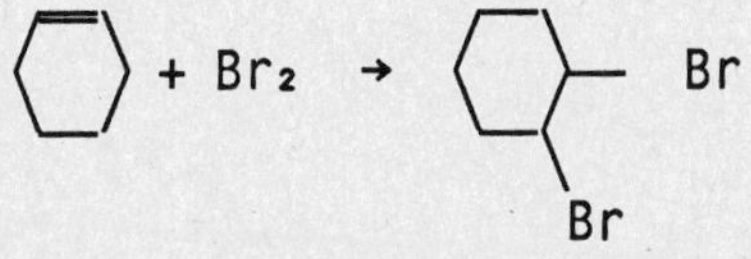

$CH_3 - CH(Br) - CH_3 + Br_2 \rightarrow$ no reaction

32.43 The class of compound formed from $H_a - O - H_b$ when

(a) H_a is replaced by an alkyl group is an alcohol

(b) H_b is replaced by an aromatic group is a phenol

(c) H_a and H_b are replaced by alkyl groups is an ether

(d) H_a is replaced by the $R\text{-}C(=O)\text{-}$ group is a carboxylic acid

(e) H_a is replaced by the $R\text{-}C(=O)\text{-}$ group and H_b is replaced by an alkyl group is an ester

(f) H_a and H_b are replaced by $R-\overset{O}{\overset{\|}{C}}-$ groups is a carboxylic acid anhydride

32.45 The class of organic compound to which the following belong is

(a) $C_6H_5-CH_2-OH$ is an alcohol

(b) $H_2C-C(=O)-O-C(=O)-CH_2$ (cyclic, the two H_2C groups bonded) is an carboxylic acid anhydride

(c) $C_6H_5-O-C_6H_5$ is an ether

(d) $CH_3-\overset{O}{\overset{\|}{C}}-O-C(CH_3)_2-CH_3$ is an ester

(e) 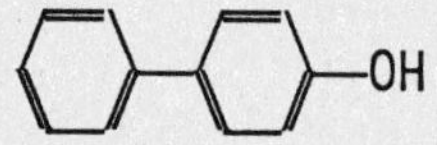 $C_6H_5-C_6H_4-OH$ is a phenol

(f) $CH_3-\overset{O}{\overset{\|}{C}}-CH_2-C_6H_5$ is a ketone

(g) $C_6H_{11}-\overset{O}{\overset{\|}{C}}-OH$ is a carboxylic acid

(h) $C_6H_5-CH_2-\overset{O}{\overset{\|}{C}}-H$ is an aldehyde

(i) $C_6H_{11}-\overset{O}{\overset{\|}{C}}-NH_2$ is an amide

(j) $C_6H_5-CH-CH_2$ (with O bridging CH and CH_2) is a cyclic ether

(k) $C_6H_5-CH_2-CH_2-\overset{O}{\overset{\|}{C}}-Cl$ is an acyl chloride

32.47 The general representation for the formula of an alkyl halide is RX; that of an aryl halide is ArX. The R is a saturated hydrocarbon group and the Ar is an aromatic group.

32.49 The name of

(a) 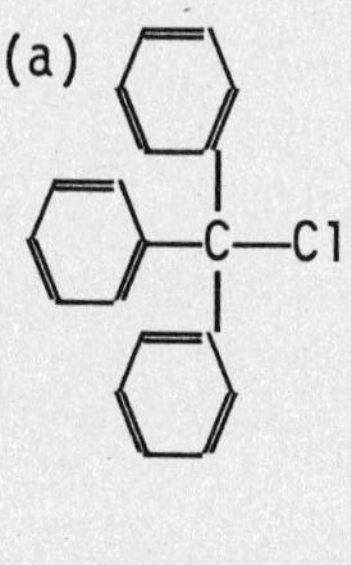 $(C_6H_5)_3C-Cl$ is triphenylchloromethane

(b)
$$\begin{array}{c} CH_3 \\ | \\ CH_3 - CH - CH_2 - Cl \end{array}$$
is 1-chloro-2-methylpropane (or isobutyl chloride)

(c) $CHCl_3$ is trichloromethane (or chloroform)

(d)
$$\begin{array}{ccc} & Cl & H \\ & | & | \\ Cl - & C = & C - Cl \end{array}$$
is trichloroethene (or trichloroethylene)

32.51 The general formula for

(a) a primary alcohol is

$R - CH_2 - OH$

(b) a secondary alcohol is

$$\begin{array}{c} R - CH - OH \\ | \\ R' \end{array}$$

(c) a tertiary alcohol is

$$\begin{array}{c} R'' \\ | \\ R - C - OH \\ | \\ R' \end{array}$$

(d) a phenol is

Ar-OH

32.53 The structural formula of

(a) 1-butanol is $CH_3 - CH_2 - CH_2 - CH_2 - OH$

(b) cyclohexanol is [cyclohexane ring]—OH

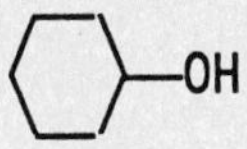

(c) 1,4-pentanediol is
$$\begin{array}{c} CH_3 - CH - CH_2 - CH_2 - CH_2 - OH \\ | \\ OH \end{array}$$

(d) 3-hexyn-1-ol is $CH_3 - CH_2 - C \equiv C - CH_2 - CH_2 - OH$

32.55 The phenols are

(c) catechol

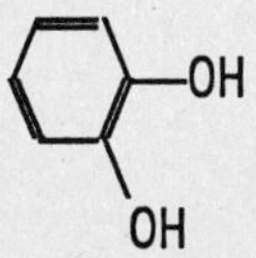

OH
OH

(d) *m*-cyclohexylphenol.

OH

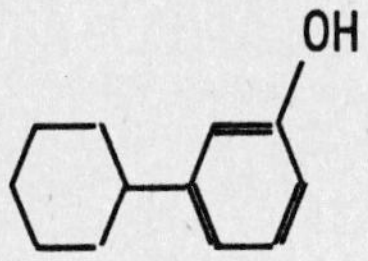

32.57 The general formula of an ether is R-O-R'. If the two hydrocarbon groups are alike, the ether is a simple or symmetrical ether; if they are different, the ether is a mixed or unsymmetrical ether.

32.59 The structural formula of

(a) methoxymethane is $CH_3 - O - CH_3$

(b) 1-ethoxypropane is $CH_3 - CH_2 - O - CH_2 - CH_2 - CH_3$

(c) 1,3-dimethoxybutane is $CH_3 - O - CH_2 - CH_2 - \underset{\displaystyle O - CH_3}{\underset{|}{CH}} - CH_3$

(d) ethoxybenzene is $CH_3 - CH_2 - O-$(benzene ring)

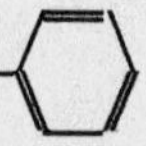

(e) methoxycyclobutane is $CH_3 - O-$(cyclobutane ring)

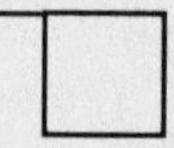

32.61 The general formula for a compound that is

(a) a primary amine is

$R - NH_2$

(b) a secondary amine is

$R - \underset{\displaystyle R'}{\underset{|}{NH}}$

(c) a tertiary amine is

$R - \underset{\displaystyle R'}{\underset{|}{N}} - R''$

The compound $(CH_3)_3CNH_2$ is not a tertiary amine because only one hydrogen atom of the ammonia molecule has been replaced by one hydrocarbon group (a tertiary amine has all three hydrogen atoms replaced).

32.63 Let $x = [C_6H_5NH_3^+]$ at equilibrium.

	$C_6H_5NH_2 + H_2O$	$\rightleftharpoons$ $C_6H_5NH_3^+$	$+ OH^-$
initial	0.100	0	0
change	$-x$	$+x$	$+x$
equilibrium	$0.100 - x$	x	x

$$K_b = \frac{[C_6H_5NH_3^+][OH^-]}{[C_6H_5NH_2]} = \frac{(x)(x)}{(0.100 - x)} = 4.2 \times 10^{-10}$$

Assuming $(0.100 - x) \simeq 0.100$, $x = 6.5 \times 10^{-6}$ mol/L. Recognizing that the approximation is valid, the concentrations at equilibrium are

$$[C_6H_5NH_3^+] = [OH^-] = x = 6.5 \times 10^{-6} \text{ mol/L}$$

$$[C_6H_5NH_2] = 0.100 \text{ mol/L}$$

32.65 The basic arrangement of atoms that is common to both aldehydes and ketones is $-\overset{|}{C} = O$. In an aldehyde a hydrogen atom and a hydrocarbon group are bonded to the carbonyl group; a ketone contains two hydrocarbon groups attached to the carbon atom of the carbonyl group.

32.67 The name of

(a) $CH_3-CH_2-CH_2-CH_2-\overset{\overset{O}{\|}}{C}-H$ is pentanal

(b) $\underset{\underset{Br}{|}}{\overset{\overset{Br}{|}}{HC}}-CH_2-\overset{\overset{O}{\|}}{C}-H$ is 3,3-dibromopropanal

(c) $=O$ is cyclohexanone

(d) $-\overset{\overset{O}{\|}}{C}-CH_2-CH_3$ is phenylethyl ketone

32.69 The chemical equation for the reaction of benzoic acid with sodium hydroxide is

$$-\overset{\overset{O}{\|}}{C}-OH + NaOH \rightarrow -\overset{\overset{O}{\|}}{C}-O^-Na^+ + H_2O$$

sodium benzoate

32.71 (a) $\Delta T = K_f\, m = (1.86\ °C\ kg/mol)(0.10\ mol/kg) = 0.19\ °C$

(b) $\Delta T = (4.90\ °C\ kg/mol)\left[\frac{0.10\ mol/kg}{2}\right] = 0.25\ °C$

32.73 A carboxylic acid and an alcohol form an ester in an esterification reaction. The hydrolysis of an ester is the reverse of the reaction leading to its formation.

32.75 The general formula for an acyl halide is $R-\overset{\overset{O}{\|}}{C}-X$. An acyl chloride reacts rapidly with water, undergoing hydrolysis to give the corresponding carboxylic acid.

32.77 The name of

(a) $CH_3-CH_2-CH_2-CH_2-OH$ is 1-butanol

(b) $-OH$ is cyclopentanol

(c) $CH_3-\underset{\underset{NH_2}{|}}{CH}-CH_3$ is 2-aminopropane (or isopropylamine or 2-propylamine)

(d) $CH_3-\underset{\underset{Cl}{|}}{C}=CH_2$ is 2-chloropropene

(e)  Br–C_6H_4–Br is 1,4-dibromobenzene (or *p*-dibromobenzene)

(f) $(CH_3 - CH_2)_3N$ is triethylamine

(g) 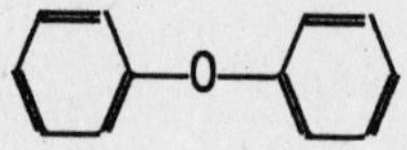 C_6H_5–O–C_6H_5 is diphenyl ether (or phenoxybenzene)

(h) 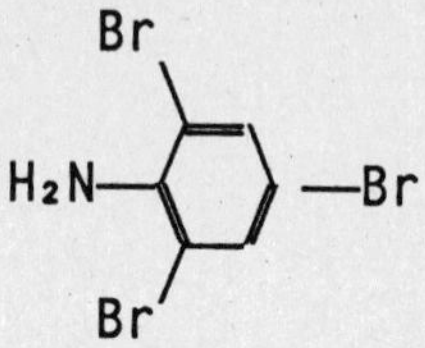 H_2N–$C_6H_2Br_3$ (Br at 2,4,6) is 2,4,6-tribromoaminobenzene (or 2,4,6-tribromoaniline)

32.79 The name of

(a) $CH_3 - CH(CH_3) - CH_2 - OH$ is 2-methyl-1-propanol (or isobutyl alcohol)

(b) $CH_3 - CH_2 - CH_2 - CH_2 - NH_2$ is *n*-butylamine

(c) $CH_3 - CH_2 - CH_2 - CH_2 - CH{=}O$ is pentanal

(d) 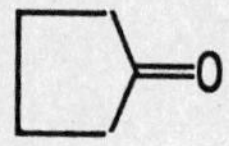 cyclic ketone (ring)=O is cyclopentanone

(e) $CH_3 - CH_2 - CH(O - CH_3) - CH_3$ is 2-methoxybutane

(f) $CH_3 - C(CH_3)_2 - C({=}O) - OH$ is 2,2-dimethylpropanoic acid

32.81 (a) o-HO–C_6H_4–$C({=}O) - OH + 2NaOH \rightarrow$ o-O^-Na^+–C_6H_4–$C({=}O) - O^-Na^+$

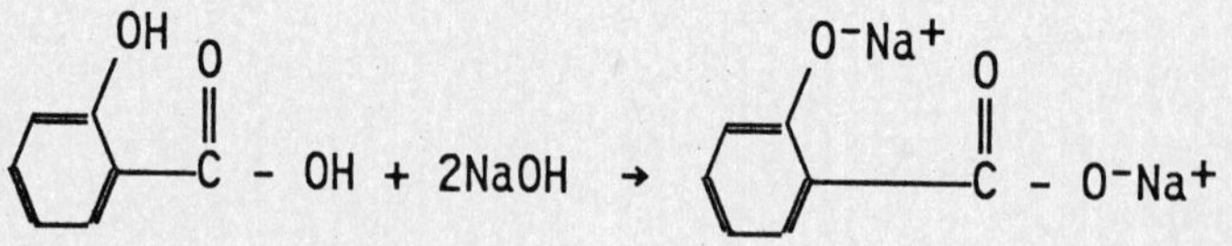

(b) C_6H_5–$C({=}O) - OH + CH_3OH \xrightarrow[\Delta]{H_2SO_4}$ C_6H_5–$C({=}O) - O - CH_3 + H_2O$

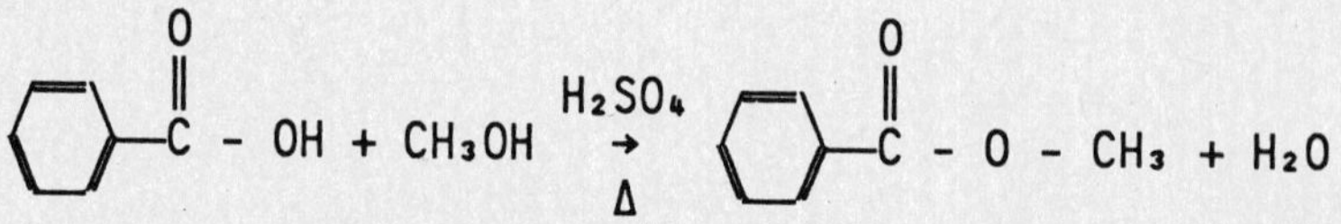

(c) $CH_3 - CH_2 - C({=}O) - H \xrightarrow[H^+]{MnO_4^-} CH_3 - CH_2 - C({=}O) - OH$

32.83 (a) $R - CH_2 - OH \xrightarrow[H_2SO_4]{Na_2Cr_2O_4} R - \overset{\overset{O}{\|}}{C} - H$

$R - \underset{}{\overset{\overset{OH}{|}}{CH}} - R' \xrightarrow[H_2SO_4]{Na_2Cr_2O_4} R - \overset{\overset{O}{\|}}{C} - R'$

(b)

$R - CH_2 - OH \xrightarrow{KMnO_4} R - \overset{\overset{O}{\|}}{C} - OH$

(c)

$R - OH + R' - \overset{\overset{O}{\|}}{C} - OH \rightarrow R' - \overset{\overset{O}{\|}}{C} - O - R + H_2O$

(d) $R - OH + HX \rightarrow R - X + H_2O$

CHAPTER 33

ORGANIC POLYMERS AND BIOCHEMISTRY

Solutions to Odd-Numbered Questions and Problems

33.1 A macromolecule is a polymer made from large numbers of smaller molecules, called monomers, that are chemically bonded to each other. Typical molecular masses are between 1×10^4 u and 1×10^6 u.

33.3 The first commercially important polymers were cellulose and natural rubber. Natural rubber is the elastomer. By converting some of the hydroxyl groups to functional groups that do not form hydrogen bonds, the properties of cellulose can be changed considerably. Sulfur cross-links in natural rubber (vulcanization) changes the properties of rubber considerably.

33.5 The melting process requires that the intermolecular attractions between the polymer chains be overcome by the thermal energy supplied. The individual polymer chains then become free to move independently. If intramolecular bonds in the polymer chain begin to break before the intermolecular attractions are overcome, then decomposition of the polymer chain results.

33.7 Sketches for the three types of polymers are shown in Fig. 33-1. Both linear and branched polymers are generally thermoplastic.

33.9 Two different types of monomers, A and B, might form the following copolymers:

random AAABBAABAAABBAA

alternating ABABABABAB

block AAAABBBBBAAAAABBBB

```
graft AAAAAAAAAAAAAA
          B   B  B
          B   B  B
          B   B  B
          B   B
              B
              B
```

33.11 The segment is

```
....CH2CHCH2CHCH2CHCH2CHCH2CH....
       |     |     |     |     |
      C=O   C=O   C=O   C=O   C=O
       |     |     |     |     |
      CH3   CH3   CH3   CH3   CH3
```

Fig. 33-1

linear polymer

branched polymer

cross-linked polymer

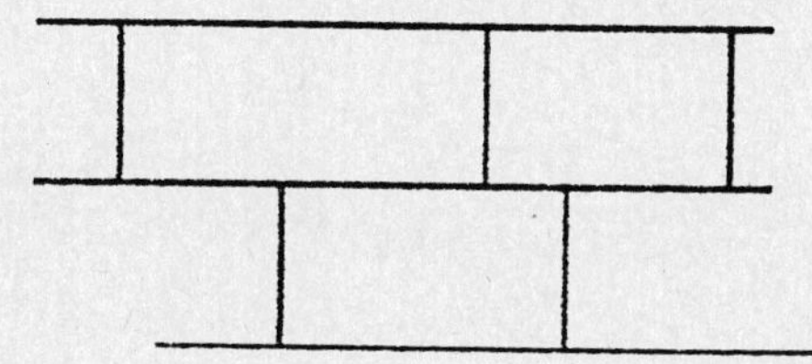

33.13 The repeating unit is

$$-\!\!\left(-CH_2-\!\langle C_6H_{10}\rangle\!-CH_2O\overset{O}{\overset{\|}{C}}-\!\langle C_6H_4\rangle\!-\overset{O}{\overset{\|}{C}}O-\right)_n\!\!-$$

33.15 The polymer chosen for use in

(a) polishing wax is (ii) poly(dimethyl siloxane), T_g = 150 K

(b) molded automobile door handles is (i) polystyrene, T_g = 373 K

33.17 The law of conservation of mass

$$(180.\ \text{g/mol } C_2H_{12}O_6)(x \text{ mol } C_6H_{12}O_6) = 750{,}000 \text{ g/mol} + [(x - 1) \text{ mol } H_2O](18 \text{ g/mol } H_2O)$$

$$(180.)x = 750{,}000 + (18)x - 18$$

$$(162)x = 750{,}000$$

$$x = 4600$$

33.19 The four major classes of organic compounds in living cells are proteins, carbohydrates, lipids, and nucleic acids. Carbohydrates have two principal functions: energy storage and structural support. The chief function of fats in animals is energy storage; many steroids are hormones--

chemical messengers that regulate various physiological processes. Some of the diverse functions of proteins are listed in Table 33.6 of the text. The function of nucleic acids is the preservation of the genetic information, the "reading" of the genetic blueprints, and the translation of the information into actual traits.

33.21 Carbohydrate molecules are either simple sugars (called monosaccharides), disaccharides (composed of two simple sugars), or polymers containing many sugar monomers (called polysaccharides). Monosaccharides are either polyhydroxyl aldehydes (aldoses) or polyhydroxyl ketones (ketoses).

33.23 The chiral carbon atoms are shown by *

```
               O
               ‖
CH2(OH) - C - C(OH)H - C(OH)H - C(OH)H - CH2OH
                *        *        *

 O
 ‖
HC - C(OH)H - C(OH)H - C(OH)H - C(OH)H - CH2OH
       *        *        *        *
```

33.25 (a) Yes, the structures are identical.

(b) Other aldoses include

```
  CH2OH                 CHO
   |                     |
  HCOH                 HOCH
   |                     |
 HOCH                  HOCH
   |          or         |
  HCOH                 HOCH
   |                     |
  HCOH                 HOCH
   |                     |
  CHO                  CH2OH
          L-gulose
```

(c) No, sorbitol is called a polyol. A carbohydrate is a polydroxy ketone or a polyhydroxy aldehyde, or a substance that gives one or more aldoses and/or one or more ketoses on hydrolysis.

33.27 Let x = [α-D-glucose] at equilibrium.

	β-D-glucose ⇄	α-D-glucose
initial	0.10	0
change	$-x$	$+x$
equilibrium	$0.10 - x$	x

$$K = \frac{[\alpha\text{-D-glucose}]}{[\beta\text{-D-glucose}]} = \frac{x}{0.10 - x} = 0.56$$

$$x = 0.056 - (0.56)x$$

$$(1.56)x = 0.056$$

$$x = [\alpha\text{-D-glucose}] = 0.036 \text{ mol/L}$$

$$[\beta\text{-D-glucose}] = 0.064 \text{ mol/L}$$

33.29 The principal types of lipids are the glycerides, phospholipids, waxes, and steroids. They are all nonpolar compounds that contain large hydrocarbon segments and so are hydrophobic. They differ from other large biological molecules in that lipids are not polymeric, although they tend to form aggregates held together by hydrophobic interactions.

33.31 The saturated fatty acids will be those having the R group with the general formula C_nH_{2n+1}: (a), (c), and (d). Those acids containing less hydrogen in the R group are unsaturated: (b) and (e).

33.33 (a) A typical monoglyceride is

```
CH2OH
|
CHOH
|    O
|    ||
CH2OCCH2CH2CH2CH2CH2CH2CH2CH2CH2CH2CH2CH2CH3
```

(b) A typical diglyaride is

```
     O
     ||
CH2OCCH2CH2CH2CH2CH2CH2CH2CH2CH2CH2CH2CH2CH3
|
CHOH
|    O
|    ||
CH2OCCH2CH2CH2CH2CH2CH2CH2CH2CH2CH2CH2CH2CH3
```

(c) A typical triglyceride is

```
     O
     ||
CH2OCCH2CH2CH2CH2CH2CH2CH2CH2CH2CH2CH2CH2CH3
|   O
|   ||
HCOCCH2CH2CH2CH2CH2CH2CH2CH2CH2CH2CH2CH2CH3
|    O
|    ||
CH2OCCH2CH2CH2CH2CH2CH2CH2CH2CH2CH2CH2CH2CH3
```

(d) The triglyceride in (c) is an example of a simple glyceride because all of the fatty acids chains are identical.

(e) A typical mixed glyceride is

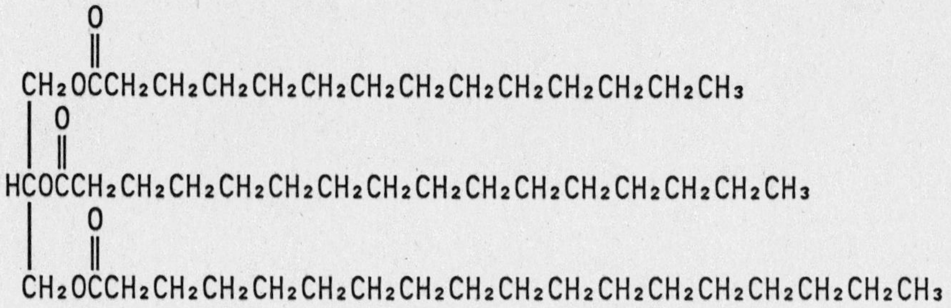

$$\begin{array}{l} \overset{\displaystyle O}{\|} \\ CH_2O\overset{}{C}CH_2CH_2CH_2CH_2CH_2CH_2CH_2CH_2CH_2CH_2CH_2CH_2CH_3 \\ |\quad \overset{\displaystyle O}{\|} \\ HCOCCH_2CH_2CH_2CH_2CH_2CH_2CH_2CH_2CH_2CH_2CH_2CH_2CH_2CH_2CH_3 \\ |\qquad \overset{\displaystyle O}{\|} \\ CH_2OCCH_2CH_2CH_2CH_2CH_2CH_2CH_2CH_2CH_2CH_2CH_2CH_2CH_2CH_2CH_2CH_2CH_3 \end{array}$$

33.35 The formula of one iosmer is

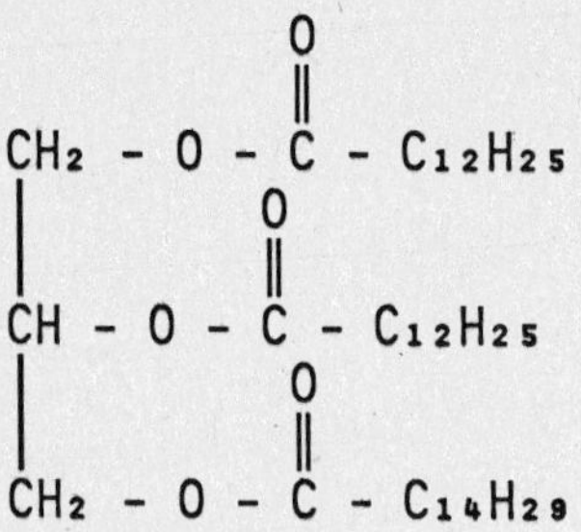

$$\begin{array}{l} CH_2 - O - \overset{\overset{\displaystyle O}{\|}}{C} - C_{12}H_{25} \\ | \\ CH - O - \overset{\overset{\displaystyle O}{\|}}{C} - C_{12}H_{25} \\ | \\ CH_2 - O - \overset{\overset{\displaystyle O}{\|}}{C} - C_{14}H_{29} \end{array}$$

A second isomer could exist with the $C_{14}H_{29}$ group in the middle of the molecule.

33.37 In a phospholipid, two of the hydroxyl groups of glycerol are esterified with fatty acids. The third OH group forms an ester link with phosphoric acid or a derivative--usually one containing an amine group. The two fatty acid chains make up a long nonpolar "tail," while the phosphate carries a negative charge and the amine group is in its cationic form, a dipolar "head." In a membrane, the hydrophilic heads can interact with substances in the aqueous interior of the cell and in the aqueous environment outside. The hydrophobic tails that make up the interior of the membrane do not easily allow the passage of ions or most polar substances.

33.39 Amino acids (except glycine) exhibit optical isomerism because they consist of an α carbon atom to which four different groups are attached: NH_2, COOH, H, and an organic group (except glycine, to which a second H is attached).

33.41 The six dipeptides are

A - B	B - C	C - A
A - C	C - B	B - A

33.43 The two dipeptides possible are formed by the acid group of one amino acid reacting with the amine group on the other amino acid:

```
         H   O        R2                        H   O        R1
         |   ‖        |                         |   ‖        |
   R1 -  C - C - N  - C - COO-            R2 -  C - C - N  - C - COO-
         |       |    |                         |       |    |
        NH3+     H    H                        NH3+     H    H
```

33.45 The link between adjacent monomers in a polypeptide is formed by a condensation reaction between the amine group of one amino acid and the carboxylic acid group of another. The links are called peptide bonds.

33.47 The structure is

```
          O     O     O     O
          ‖     ‖     ‖     ‖
   H2NCH2CNHCHCNHCHCNHCHCNHCHCOOH
             |     |     |     |
            CH2   CH2   CH2   CH2
            O     S     |     |
            H     H     CH    C6H5
                       /  \
                     CH3   CH3
```

33.49 The active site of an enzyme may take the form of a groove along the surface of the molecule, or a deeper crevice or channel into the interior. The bonds between enzyme and substrate are of the same sort that hold the folded protein chain in its three-dimensional shape: hydrogen bonds, hydrophobic interactions, electrostatic attraction between charged groups, even (in some cases) temporary covalent bonds.

33.51 Nucleotides consist of a pentose sugar (ribose or deoxyribose) with a nitrogenous base replacing the OH attached to carbon 1 and a phosphate group joined by an ester link to carbon 5. They are polymerized to form a nucleic acid by ester links between the phosphate of one nucleotide and carbon 3 of the sugar in the following nucleotide.

33.53 If bases were outside, there would be no need for base pairing by hydrogen bonding and two strands of DNA would not have to be complementary. If phosphates were inside, DNA would not be acidic.

33.55 See Fig. 33-2.

33.57 Adenosine triphosphate (or ATP) is an energy-rich phosphate compound consisting of the sugar ribose, the base adenine, and three phosphate groups. Energy release in several of the most steeply downhill steps

Fig. 33-2

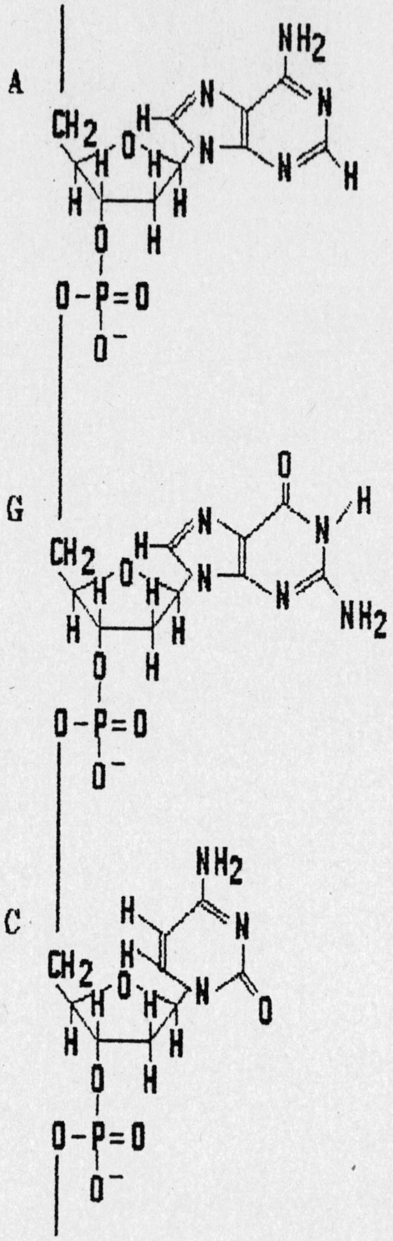

A

G

C

in the oxidation of energy-rich molecules is used to make ATP from ADP and phosphate. These ATP molecules can then be used at any time and in any way that the needs of the cell dictate.

CHAPTER 34

INORGANIC QUALITATIVE ANALYSIS--CHEMICAL PRINCIPLES REVIEWED

Solutions to Odd-Numbered Questions and Problems

34.1 The goal of inorganic qualitative analysis is to identify the cations and the anions present in a substance of unknown composition. The two general steps in the analysis of anions include preliminary tests that indicate which anions or groups of anions may be present or absent. Distinctive reactions for each individual anion then allow verification of the presence or absence of specific anions. Cations are first separated into groups by precipitating the ions in one group, while leaving the ions of other groups in solution. The ions in each group are further separated by carefully chosen reactions under carefully controlled conditions. Finally, each ion is positively identified by a reaction that is characteristic of that ion.

34.3 (a) $SnS_2(s) + S^{2-} \rightarrow [SnS_3]^{2-}$

(b) $SnS_2(s) + 4H^+ + 6Cl^- \rightarrow [SnCl_6]^{2-} + 2H_2S(aq)$

(c) $Pb^{2+} + SO_4^{2-} \rightarrow PbSO_4(s)$

(d) $Fe^{2+} + 2NH_3(aq) + 2H_2O(l) \rightarrow Fe(OH)_2(s) + 2NH_4^+$

(e) $3NiS(s) + 8H^+ + 2NO_3^- \rightarrow 3Ni^{2+} + 2NO(g) + 4H_2O(l) + 3S(s)$

(f) $Co^{2+} + 2OH^- \rightarrow Co(OH)_2(s)$

34.5 According to the reaction types listed in Tables 17.2 and 17.7, reaction

(a) is nonredox--combination of compounds

(b) is nonredox--partner-exchange reaction

(c) is nonredox--partner-exchange reaction between ions in aqueous solution (formation of a precipitate)

(d) is nonredox--partner-exchange reaction between ions in aqueous solution (formation of a precipitate)

(e) is redox--does not fit into any of the categories in Table 17.7

(f) is nonredox--partner-exchange reaction between ions in aqueous solution (formation of a precipitate)

34.7 (a) $K = \dfrac{1}{[Hg_2^{2+}][Cl^-]^2}$ (b) $K = \dfrac{[NH_4^+][Cl^-]}{[NH_3]^2}$ (c) $K = \dfrac{[Ag(NH_3)_2^+][Cl^-]}{[NH_3]^2}$

34.9 (a) $2H_2O(l) \rightleftharpoons H_3O^+ + OH^-$
(b) $F^- + H_2O(l) \rightleftharpoons HF(aq) + OH^-$
(c) $NH_4^+ + H_2O(l) \rightleftharpoons H_3O^+ + NH_3(aq)$

34.11 $CaC_2O_4(s) \rightleftharpoons Ca^{2+} + C_2O_4^{2-}$
The effect of adding
(a) $Na_2C_2O_4$ is to provide additional $[C_2O_4^{2-}]$ that shifts the equilibrium to reactants ($CaC_2O_4(s)$) side
(b) $H_2C_2O_4$ is to provide additional $[C_2O_4^{2-}]$ that shifts the equilibrium to reactants ($CaC_2O_4(s)$) side
(c) NaCl produces no change in the equilibrium
(d) $CaCl_2$ is to provide additional $[Ca^{2+}]$ that shifts the equilibrium to reactants ($CaC_2O_4(s)$) side
(e) additional CaC_2O_4 is no effect on equilibrium--only more solid CaC_2O_4 is present
(f) HCl is to provide H^+ that reacts with $C_2O_4^{2-}$, giving a reduced $[C_2O_4^{2-}]$ that shifts the equilibrium to the products side

34.13 Assume that all NO_2^- initially forms HNO_2, which dissociates; let $x = [NO_2^-]$ at equilibrium.

	$HNO_2(aq)$ + $H_2O(l)$ ⇌	H_3O^+ +	NO_2^-
initial	0.1	0	0
change	$-x$	0	$+x$
equilibrium	$0.1 - x$	1	x

$$K_a = \frac{[H^+][NO_2^-]}{[HNO_2]} = \frac{(1)(x)}{(0.1 - x)} = 7.2 \times 10^{-4}$$

$x = 7 \times 10^{-5} - (7.2 \times 10^{-4})x$
$x = [NO_2^-] = 7 \times 10^{-5}$ mol/L

34.15 $CH_3COOH(aq) + H_2O(l) \rightleftharpoons CH_3COO^- + H_3O^+$
for pH = 5.00, $[H^+] = 1.0 \times 10^{-5}$ mol/L

$$K_a = \frac{[H^+][CH_3COO^-]}{[CH_3COOH]} = \frac{(1.0 \times 10^{-5})[CH_3COO^-]}{[CH_3COOH]} = 1.754 \times 10^{-5}$$

$$\frac{[CH_3COO^-]}{[CH_3COOH]} = (1.754 \times 10^{-5})/(1.0 \times 10^{-5}) = 1.8$$

for pH = 5.50, $[H^+] = 3.2 \times 10^{-6}$ mol/L

$$\frac{(3.2 \times 10^{-6})[CH_3COO^-]}{[CH_3COOH]} = 1.754 \times 10^{-5}$$

$$\frac{[CH_3COO^-]}{[CH_3COOH]} = (1.754 \times 10^{-5})/(3.2 \times 10^{-6}) = 5.5$$

34.17 Let $x = [H^+] = [HCO_3^-]$ at equilibrium.

	$H_2CO_3(aq) + H_2O(l)$	$\rightleftarrows$ H_3O^+	+ HCO_3^-
initial	0.034	0	0
change	$-x$	$+x$	$+x$
equilibrium	$0.034 - x$	x	x

$$K_{a_1} = \frac{[H^+][HCO_3^-]}{[H_2CO_3]} = \frac{(x)(x)}{(0.034 - x)} = 4.5 \times 10^{-7}$$

Assume that $0.034 - x \approx 0.034$.

$x^2 = (4.5 \times 10^{-7})(0.034) = 1.5 \times 10^{-8}$

$x = 1.2 \times 10^{-4}$ mol/L

The approximation is valid because $0.034 - 0.00012 \approx 0.034$.

Let $x = [CO_3]^{2-}$ at equilibrium.

	$HCO_3^- + H_2O(l)$	$\rightleftarrows$ H_3O^+	+ CO_3^{2-}
initial	1.2×10^{-4}	1.2×10^{-4}	0
change	$-x$	$+x$	$+x$
equilibrium	$1.2 \times 10^{-4} - x$	$1.2 \times 10^{-4} + x$	x

$$K_{a_2} = \frac{[H^+][CO_3^{2-}]}{[HCO_3^-]} = \frac{(1.2 \times 10^{-4} + x)(x)}{(1.2 \times 10^{-4} - x)} = 4.8 \times 10^{-11}$$

Assume that $(1.2 \times 10^{-4}) \pm x \approx 1.2 \times 10^{-4}$.

$x = (4.8 \times 10^{-11})(1.2 \times 10^{-4})/(1.2 \times 10^{-4}) = 4.8 \times 10^{-11}$ mol/L

The approximation is valid because $(1.2 \times 10^{-4}) \pm 4.8 \times 10^{-11} \approx 1.2 \times 10^{-4}$.

$[CO_3^{2-}] = 4.8 \times 10^{-11}$ mol/L

$PbCO_3(s) \rightleftarrows Pb^{2+} + CO_3^{2-}$

$Q_i = [Pb^{2+}][CO_3^{2-}] = (0.1)(4.8 \times 10^{-11}) = 5 \times 10^{-12}$

$Q_i > K_{sp}$, so an appreciable amount of $PbCO_3$ will form.

34.19 Selective oxidation (and reduction) means that conditions are adjusted so that one substance is oxidized (or reduced) while others are not. Separation is usually possible because of the formation of an aqueous solution and precipitate. For example, 2 M HNO_3 oxidizes PbS(s) and CuS(s) but not HgS(s) for separation.

34.21

$$
\begin{array}{rcl|l}
(2)[Fe^{3+} + e^- & \rightarrow & Fe^{2+}] & E° = 0.771 \text{ V} \\
2I^- & \rightarrow & I_2 + 2e^- & E° = -0.536 \text{ V} \\
\hline
2Fe^{3+} + 2I^- & \rightarrow & 2Fe^{2+} + I_2 & E° = 0.235 \text{ V}
\end{array}
$$

The formula of the compound must be FeI_2 because FeI_3 decomposes spontaneously under standard state conditions.

34.23

$$
\begin{array}{rcl|l}
(2)[Al(s) & \rightarrow & Al^{3+} + 3e^-] & E° = 1.662 \text{ V} \\
(3)[2H^+ + 2e^- & \rightarrow & H_2(g)] & E° = 0.000 \text{ V} \\
\hline
2Al(s) + 6H^+ & \rightarrow & 3H_2(g) + 2Al^{3+} & E° = 1.622 \text{ V}
\end{array}
$$

$$
\begin{array}{rcl|l}
Sn(s) & \rightarrow & Sn^{2+} + 2e^- & E° = 0.136 \text{ V} \\
2H^+ + 2e^- & \rightarrow & H_2(g) & E° = 0.000 \text{ V} \\
\hline
Sn(s) + 2H^+ & \rightarrow & Sn^{2+} + H_2(g) & E° = 0.136 \text{ V}
\end{array}
$$

$$
\begin{array}{rcl|l}
(2)[Al(s) & \rightarrow & Al^{3+} + 3e^-] & E° = 1.662 \text{ V} \\
(3)[Sn^{2+} + 2e^- & \rightarrow & Sn(s)] & E° = -0.136 \text{ V} \\
\hline
2Al(s) + 3Sn^{2+} & \rightarrow & 3Sn(s) + 2Al^{3+} & E° = 1.526 \text{ V}
\end{array}
$$

Thus oxidation of Al by H^+ occurs first and tin oxidation by H^+ is less favorable; any Sn^{2+} formed will be reduced by Al.

34.25 $[AgCl_4]^{3-} \rightleftharpoons Ag^+ + 4Cl^-$

$$K_d = \frac{[Ag^+][Cl^-]^4}{[AgCl_4^{3-}]} = \frac{[Ag^+](6)^4}{(0.01)} = 5 \times 10^{-6}$$

$[Ag^+] = (5 \times 10^{-6})(0.01)/(6)^4 = 4 \times 10^{-11}$ mol/L

34.27 The structure of thioacetamide is $CH_3\overset{\overset{S}{\|}}{C}NH_2$.

$$CH_3\overset{\overset{S}{\|}}{C}NH_2(aq) + H_3O^+ + H_2O \xrightarrow{\Delta} CH_3\overset{\overset{O}{\|}}{C}OH(aq) + NH_4^+ + H_2S(aq)$$

$$CH_3\overset{\overset{S}{\|}}{C}NH_2(aq) + 3OH^- \xrightarrow{\Delta} CH_3\overset{\overset{O}{\|}}{C}O^- + NH_3(aq) + H_2O(l) + S^{2-}$$

34.29 The common ion systems used to control the precipitation of (a) hydroxides is an NH_3/NH_4^+ buffer pair, (b) carbonates is NH_4^+/CO_3^{2-}, and (c) sulfides is H^+/S^{2-}.

34.31 $AgCl(s) \rightleftharpoons Ag^+ + Cl^-$

$$\quad\quad\quad\quad x \quad\; x$$

$K_{sp} = [Ag^+][Cl^-] = (x)(x) = 1.8 \times 10^{-10}$

$x = [Ag^+] = [Cl^-] = 1.3 \times 10^{-5}$ mol/L

$$\left[\frac{1.3 \times 10^{-5} \text{ mol AgCl}}{1 \text{ L}}\right]\left[\frac{143.32 \text{ g AgCl}}{1 \text{ mol AgCl}}\right] = 1.9 \times 10^{-3} \text{ g AgCl/L}$$

34.33 The relationship between the ion product and the solubility product (a) at equilibrium is $Q_i = K_{sp}$, (b) in order for precipitation not to occur is $Q_i < K_{sp}$, and (c) in order for precipitation to occur is $Q_i > K_{sp}$.

34.35 $BaCrO_4(s) \rightleftharpoons Ba^{2+} + CrO_4^{2-}$

$K_{sp} = [Ba^{2+}][CrO_4^{2-}] = (0.1)[CrO_4^{2-}] = 1.2 \times 10^{-10}$

$[CrO_4^{2-}] = 1 \times 10^{-9}$ mol/L

$Ag_2CrO_4(s) \rightleftharpoons 2Ag^+ + CrO_4^{2-}$

$K_{sp} = [Ag^+]^2[CrO_4^{2-}] = (0.1)^2[CrO_4^{2-}] = 2.5 \times 10^{-12}$

$[CrO_4^{2-}] = 3 \times 10^{-10}$ mol/L

Thus Ag^+ will begin to precipitate first for the 0.1 M solution of the cations.

For 0.01 M in Ba^{2+} and 0.01 M Ag^+

$K_{sp} = (0.01)[CrO_4^{2-}] = 1.2 \times 10^{-10}$

$[CrO_4^{2-}] = 1 \times 10^{-8}$ mol/L

$K_{sp} = (0.01)^2[CrO_4^{2-}] = 2.5 \times 10^{-12}$

$[CrO_4^{2-}] = 3 \times 10^{-8}$ mol/L

Thus Ba^{2+} will begin to precipitate first for the 0.01 M solution of the cations.

34.37 $PbCl_2(s) \rightleftharpoons Pb^{2+} + 2Cl^-$

$K_{sp} = [Pb^{2+}][Cl^-]^2 = (0.01)[Cl^-]^2 = 2 \times 10^{-5}$

$[Cl^-] = 4 \times 10^{-2}$ mol/L for $PbCl_2$ to precipitate.

$AgCl(s) \rightleftharpoons Ag^+ + Cl^-$

$K_{sp} = [Ag^+][Cl^-] = (0.01)[Cl^-] = 1.8 \times 10^{-10}$

$[Cl^-] = 2 \times 10^{-8}$ mol/L for AgCl to precipitate

$Hg_2Cl_2(s) \rightleftharpoons Hg_2^{2+} + 2Cl^-$

$K_{sp} = [Hg_2^{2+}][Cl^-]^2 = (0.01)[Cl^-]^2 = 1.3 \times 10^{-18}$

$[Cl^-] = 1 \times 10^{-8}$ mol/L for Hg_2Cl_2 to precipitate

The Hg_2^{2+} cation will precipitate first at $[Cl^-] = 1 \times 10^{-8}$ mol/L.

34.39

	$NH_3(aq) + H_2O(l)$	$\rightleftharpoons$	NH_4^+	$+ OH^-$
initial	6		0	0
change	$-x$		$+x$	$+x$
equilibrium	$6 - x$		x	x

where $x = [OH^-]$ at equilibrium

$$K_b = \frac{[NH_4^+][OH^-]}{[NH_3]} = \frac{(x)(x)}{6 - x} = 1.6 \times 10^{-5}$$

Assume that $6 - x \approx 6$.

$x^2 = (6)(1.6 \times 10^{-5}) = 1 \times 10^{-4}$

$x = 0.01$ mol/L

The approximation is valid and so $[OH^-] = 0.01$ mol/L.

$Ni(OH)_2(s) \rightleftharpoons Ni^{2+} + 2OH^-$

$K_{sp} = [Ni^{2+}][OH^-]^2 = [Ni^{2+}](0.01)^2 = 3 \times 10^{-16}$

$[Ni^{2+}] = 3 \times 10^{-12}$ mol/L

$[Ni(NH_3)_6]^{2+} \rightleftharpoons Ni^{2+} + 6NH_3(aq)$

$$K_d = \frac{[Ni^{2+}][NH_3]^6}{[Ni(NH_3)_6{}^{2+}]} = \frac{[Ni^{2+}](6)^6}{(0.1)} = 1 \times 10^{-9}$$

$[Ni^{2+}] = (1 \times 10^{-9})(0.1)/(6)^6 = 2 \times 10^{-15}$ mol/L

In 6 M NH_3, the majority of the nickel would be found in the form of $[Ni(NH_3)_6]^{2+}$.

34.41 $Ni(OH)_2(s) \rightleftharpoons Ni^{2+} + 2OH^-$

$K_{sp} = [Ni^{2+}][OH^-]^2 = (0.1)[OH^-]^2 = 3 \times 10^{-16}$

$[OH^-] = 5 \times 10^{-8}$ mol/L

$pOH = -\log [OH^-] = -\log(5 \times 10^{-8}) = 7.3$

$pH = 14.0 - pOH = 14.0 - 7.3 = 6.7$

34.43 $MS(s) \rightleftharpoons M^{2+} + S^{2-}$

$$K_{sp} = [M^{2+}][S^{2-}] = [M^{2+}]\left[\frac{3 \times 10^{-21}}{[H^+]^2}\right]$$

$$[H^+] = \left[\frac{[M^{2+}](3 \times 10^{-21})}{K_{sp}}\right]^{1/2}$$

for Zn^{2+}: $[H^+] = \left[\frac{(0.1)(3 \times 10^{-21})}{2 \times 10^{-24}}\right]^{1/2} = 10 \text{ mol/L}$

for Ni^{2+}: $[H^+] = \left[\frac{(0.1)(3 \times 10^{-21})}{3 \times 10^{-19}}\right]^{1/2} = 0.03 \text{ mol/L}$

$0.03 \text{ mol/L} \le [H^+] \le 10 \text{ mol/L}$

34.45 $[Ba^{2+}] = \frac{1.7 \times 10^{-10}}{[SO_4^{2-}]} = \frac{2.0 \times 10^{-9}}{[CO_3^{2-}]}$

$$\frac{[SO_4^{2-}]}{[CO_3^{2-}]} = \frac{1.7 \times 10^{-10}}{2.0 \times 10^{-9}} = 0.085$$

$[SO_4^{2-}] = (0.085)[CO_3^{2-}] = (0.085)(1.5) = 0.13 \text{ mol/L}$

34.47 Each successive reaction occurred because the added reagent reduced the silver ion concentration at equilibrium with the precipitate or complex ion that formed.

$2Ag^+ + 2OH^- \rightleftharpoons Ag_2O(s) + H_2O(l)$

$Ag_2O(s) + 4S_2O_3^{2-} + 2H^+ \rightleftharpoons 2[Ag(S_2O_3)_2]^{3-} + H_2O(l)$

$[Ag(S_2O_3)_2]^{3-} + Br^- \rightleftharpoons AgBr(s) + 2S_2O_3^{2-}$

$AgBr(s) + 2NH_3(aq) \rightleftharpoons [Ag(NH_3)_2]^+ + Br^-$

$[Ag(NH_3)_2]^+ + I^- \rightleftharpoons AgI(s) + 2NH_3(aq)$

34.49 $HCl(aq) + NH_3(aq) \rightarrow NH_4^+ + Cl^-$

$$(5 \text{ mL HCl soln})\left[\frac{1 \text{ L}}{1000 \text{ mL}}\right]\left[\frac{0.1 \text{ mol HCl}}{1 \text{ L HCl soln}}\right]\left[\frac{1 \text{ mol } NH_3}{1 \text{ mol HCl}}\right]\left[\frac{1 \text{ L } NH_3 \text{ soln}}{1 \text{ mol } NH_3}\right]$$

$$\times \left[\frac{1000 \text{ mL}}{1 \text{ L}}\right]\left[\frac{20 \text{ drops}}{1 \text{ mL}}\right] = 10 \text{ drops } NH_3 \text{ solution}$$

34.51 $2Br^- + S_2O_8^{2-} \rightarrow Br_2(l) + 2SO_4^{2-}$

$$(2 \text{ mL soln})\left[\frac{1 \text{ L}}{1000 \text{ mL}}\right]\left[\frac{0.03 \text{ mol } Br^-}{1 \text{ L soln}}\right]\left[\frac{1 \text{ mol } S_2O_8^{2-}}{2 \text{ mol } Br^-}\right] = 3 \times 10^{-5} \text{ mol } S_2O_8^{2-}$$

$2I^- + S_2O_8^{2-} \rightarrow I_2(aq) + 2SO_4^{2-}$

$$(2 \text{ mL soln})\left[\frac{1 \text{ L}}{1000 \text{ mL}}\right]\left[\frac{0.01 \text{ mol } I^-}{1 \text{ L soln}}\right]\left[\frac{1 \text{ mol } S_2O_8^{2-}}{2 \text{ mol } I^-}\right] = 1 \times 10^{-5} \text{ mol } S_2O_8^{2-}$$

total $S_2O_8^{2-} = 4 \times 10^{-5}$ mol

$$(4 \times 10^{-5} \text{ mol } S_2O_8^{2-})\left[\frac{1 \text{ mol } K_2S_2O_8}{1 \text{ mol } S_2O_8^{2-}}\right]\left[\frac{270.32 \text{ g } K_2S_2O_8}{1 \text{ mol } K_2S_2O_8}\right] = 0.01 \text{ g } K_2S_2O_8$$

CHAPTER 35

INORGANIC QUALITATIVE ANALYSIS--ANIONS, CATIONS, AND THE SCHEME

Solutions to Odd-Numbered Questions and Problems

35.1 The development of a brown to black color when a few drops of an unknown solution are added to a solution of manganese(II) chloride in concentrated hydrochloric acid indicates the presence of one or more of the oxidizing anions (NO_2^-, CrO_4^{2-}, and NO_3^-). The appearance of a dark blue suspension or precipitate when the unknown solution is added to a solution containing $FeCl_3$, $K_3[Fe(CN)_6]$, and dilute HCl indicates the presence of one or more of the reducing anions (S^{2-}, SO_3^{2-}, I^-, and NO_2^-). The release of gas when 6 M $HClO_4$ is added and the solution is warmed indicates the presence of one or more of the following anions: S^{2-}, SO_3^{2-}, CO_3^{2-}, and NO_2^-. After cooling the $HClO_4$ solution, addition of $AgNO_3$ will precipitate any of the following anions: I^-, Br^-, and Cl^-. After removal of any silver salt precipitate, making the solution just alkaline with 6 M NH_3 will precipitate any of the following anions: PO_4^{3-}, CrO_4^{2-}, and SO_4^{2-}. Anions that may remain in solution include NO_3^- and SO_4^{2-}. A solid can be tested with concentrated sulfuric acid: a color change indicates CrO_4^{2-}; a colorless, odorless gas indicates CO_3^{2-}; a colorless, odoriferous gas indicates the presence of one or more of the following anions: S^{2-}, SO_3^{2-}, and Cl^-; and a colored gas indicates the presence of one or more of the following anions: NO_2^-, I^-, and Br^-.

35.3 The anions in the schemes that are
(a) strong Brønsted-Lowry bases include S^{2-}, PO_4^{3-}, and CO_3^{2-}
(b) weak Brønsted-Lowry bases include SO_3^{2-}, CrO_4^{2-}, NO_2^-, and SO_4^{2-}
(c) strong oxidizing agents include NO_3^-, CrO_4^{2-}, and sometimes NO_2^-
(d) reducing agents include I^-, S^{2-}, SO_3^{2-}, and sometimes NO_2^-

35.5 The solution containing one or more of the halide ions (I^-, Br^-, and Cl^-) is made slightly acidic with dilute HCl and a small amount of CCl_4 added.

Addition, with shaking, of a few drops of chlorine water will give a violet color in the CCl_4 layer, indicating the presence of iodide ion. Addition of more chlorine water oxidizes the iodine to colorless IO_3^- and then any Br^- present to Br_2, which then will appear as a yellow to brown color in the CCl_4. If either Br^- or I^- is present, they are oxidized in H_2SO_4 solution at elevated temperature by $S_2O_8^{2-}$. The free halogens are extracted into CCl_4, and $AgNO_3$ is added to the acidified water layer to precipitate Cl^- as AgCl. Dissolution of the precipitate in aqueous ammonia unequivocally shows the presence of chloride ion. If I^- and Br^- were absent, the oxidation step would be omitted.

$2I^- + Cl_2(aq) \xrightarrow{CCl_4} I_2(CCl_4) + 2Cl^-$

$I_2(CCl_4) + 5Cl_2(aq) + 6H_2O(l) \rightarrow 2IO_3^- + 10Cl^- + 12H^+$

$2Br^- + Cl_2(aq) \xrightarrow{CCl_4} Br_2(CCl_4) + 2Cl^-$

$2X^- + S_2O_8^{2-} \xrightarrow{H_2SO_4} X_2 + 2SO_4^{2-}$ X = Br or I

$Cl^- + Ag^+ \rightarrow AgCl(s)$

$AgCl(s) + 2NH_3(aq) \rightarrow [Ag(NH_3)_2]^+ + Cl^-$

35.7 Anion unknowns that give strong positive tests for NO_2^- often give faint tests for NO_3^- because of air oxidation of the NO_2^- to NO_3^-. Likewise, those containing S^{2-} or SO_3^{2-} can give faint tests for SO_4^{2-} for the same reason.

35.9 The addition of $MnCl_2(aq)$ and HCl(aq) would give a black-brown color if NO_3^- were present. A separate test with a solution of $FeCl_3$, $K_3[Fe(CN)_6]$, and dilute HCl would give a blue color if S^{2-} and/or SO_3^{2-} were present. Treatment of the sample with 6 M $HClO_4$ and heating would give a colorless, odorless gas from CO_3^{2-} and colorless, odoriferous gases from SO_3^{2-} and/or S^{2-}. Adding $AgNO_3$ and making the solution just alkaline with 6 M NH_3 will give a yellow precipitate for PO_4^{3-}.

for S^{2-}: $S^{2-} + 2H^+ \rightarrow H_2S(aq)$

$Pb^{2+} + H_2S(aq) \rightarrow PbS(s) + 2H^+$

for SO_3^{2-}: $SO_3^{2-} + 2H^+ \rightarrow SO_2(g) + H_2O(l)$

$5SO_2(g) + 2MnO_4^- + 2H_2O(l) \rightarrow 5SO_4^{2-} + 2Mn^{2+} + 4H^+$

$SO_4^{2-} + Ba^{2+} \rightarrow BaSO_4(s)$

for CO_3^{2-}: $CO_3^{2-} + 2H^+ \rightarrow CO_2(g) + H_2O(l)$

$CO_2(g) + Ba^{2+} + 2OH^- \rightarrow BaCO_3(s) + H_2O(l)$

for PO_4^{3-}: $PO_4^{3-} + 12MoO_4^{2-} + 3NH_4^+ + 24H^+ \rightarrow (NH_4)_3P(Mo_3O_{10})_4(s) + 12H_2O(l)$

for NO_3^-: $NO_3^- + 3Fe^{2+} + 4H^+ \rightarrow NO(aq) + 3Fe^{3+} + 2H_2O(l)$

$Fe^{2+} + NO(aq) \rightarrow [Fe(NO)]^{2+}$

35.11 The possible anions include S^{2-}, SO_3^{2-}, CO_3^{2-}, NO_2^-, I^-, Br^-, Cl^-, PO_4^{3-}, CrO_4^{2-} or $Cr_2O_7^{2-}$, NO_3^-, and SO_4^{2-}.

(a) The dark color indicates NO_2^-, CrO_4^{2-}, and NO_3^- may be present.
(b) The negative results indicate S^{2-}, SO_3^{2-}, I^-, and NO_2^- are absent.
(c) No gases indicates CO_3^{2-} is absent.
(d) The white precipitate indicates Cl^- is present; Br^- is absent.
(e) The second precipitate indicates PO_4^{3-}, $Cr_2O_7^{2-}$, and SO_4^{2-} may be present.
(f) The gas again indicates Cl^- is present.
(g) The lack of color confirms both I^- and Br^- are absent.
(h) Dissolution of the precipitate confirms Cl^-.
(i) The negative result indicates PO_4^{3-} is absent.
(j) The brown ring confirms NO_3^-.
(k) The white barium precipitate confirms SO_4^{2-} and indicates CrO_4^{2-} and $Cr_2O_7^{2-}$ are absent.

The anions present are SO_4^{2-}, NO_3^-, and Cl^-.

35.13 Too large an excess of Cl^- during the precipitation of Group I cations dissolves the silver and lead chlorides by complex formation.

(a) $Pb^{2+} + 2Cl^- \rightleftharpoons PbCl_2(s)$
$Hg_2^{2+} + 2Cl^- \rightleftharpoons Hg_2Cl_2(s)$
$Ag^+ + Cl^- \rightleftharpoons AgCl(s)$

(b) $PbCl_2(s) + 2Cl^- \rightleftharpoons [PbCl_4]^{2-}$
$AgCl(s) + Cl^- \rightleftharpoons [AgCl_2]^-$

35.15 The Pb^{2+} ion is present.

35.17 The Pb^{2+} ion appears in both Cation Groups I and II because the $PbCl_2$ is soluble enough that it is impossible to avoid carrying some Pb^{2+} over into the filtrate. The reason mercury appears in both of these groups is different--it has a +1 oxidation state as Hg_2^{2+} in Group I and a +2 oxidation state as Hg^{2+} in Group II.

35.19 $Sn(OH)_2(s) + 2HCl(aq) \rightleftharpoons Sn^{2+} + 2Cl^- + 2H_2O(l)$

$Sn(OH)_2(s) + 2OH^- \rightleftharpoons [Sn(OH)_4]^{2-}$

35.21 Yes, Pb^{2+} and Cu^{2+} were in the filtrate; $[Cu(NH_3)_4]^{2+}$ is blue and $Pb(OH)_2$ is a white solid.

35.23 The Cu^{2+} ion is present.

35.25 The procedures in which oxidation-reduction reactions occur are 2, 3, 4, 5, 6, and 15.

$Fe_2S_3(s) + 4H^+ \rightarrow 2Fe^{2+} + 2H_2S(g) + S(s)$

$3Fe^{2+} + NO_3^- + 4H^+ \rightarrow NO(g) + 3Fe^{3+} + 2H_2O(l)$

$Mn^{2+} + H_2O_2(aq) + 2OH^- \rightarrow MnO_2(s) + 2H_2O(l)$

$2[Cr(OH)_4]^- + 2OH^- + 3H_2O_2(aq) \rightarrow 2CrO_4^{2-} + 8H_2O(l)$

$MnO_2(s) + 2H^+ + H_2O_2(aq) \rightarrow Mn^{2+} + 2H_2O(l) + O_2(g)$

$3Mn^{2+} + ClO_3^- + 3H_2O(l) \rightarrow 3MnO_2(s) + 6H^+ + Cl^-$

$MnO_2(s) + HNO_2(aq) + H^+ \rightarrow Mn^{2+} + H_2O(l) + NO_3^-$

$2Mn^{2+} + 14H^+ + 5NaBiO_3(s) \rightarrow 2MnO_4^- + 5Bi^{3+} + 7H_2O(l) + 5Na^+$

$Cr_2O_7^{2-} + 4H_2O_2(aq) + 2H^+ \rightleftharpoons 2CrO_5(\text{amyl alcohol}) + 5H_2O(l)$

35.27 The reagent(s) to use to distinguish between the following cations:

(a) NaOH gives pink $Co(OH)_2$ and green $Ni(OH)_2$ or NH_3 gives tan $[Co(NH_3)_6]^{2+}$ and blue $[Ni(NH_3)_6]^{2+}$.

(b) NaOH gives colorless $[Zn(OH)_4]^{2-}$ and green $Fe(OH)_2$, NH_3 gives colorless $[Zn(NH_3)_4]^{2+}$ and green $Fe(OH)_2$, or S^{2-} gives white ZnS and black FeS.

(c) NH_3 gives green $Fe(OH)_2$ and blue $[Ni(NH_3)_6]^{2+}$.

(d) NaOH gives pink $Mn(OH)_2$ and red-brown $Fe(OH)_3$, NH_3 gives red-brown $Fe(OH)_3$ and no reaction with Mn^{2+} (in the presence of sufficient $[NH_4^+]$), or S^{2-} gives pink MnS and black Fe_2S_3.

35.29 The ions present are Co^{2+} and Ni^{2+}.

35.31 The reagent K_2CrO_4 can be used to distinguish Ca^{2+} from Ba^{2+}.

35.33 Any NH_4^+ must be removed before performing the confirmatory test for K^+ because NH_4^+ forms a tetraphenylborate precipitate like K^+. This removal is done by adding HNO_3 to form NH_4NO_3, evaporating, and heating to decompose the NH_3NO_3 into $N_2O(g)$ and $H_2O(g)$. The test for NH_4^+ should be done on the original sample because NH_3, which is a source of NH_4^+, is added several times to effect cation group separations.

35.35 The anions that cannot be present in a water-soluble sample known to contain Cu^{2+} and Zn^{2+} are S^{2-}, CO_3^{2-}, and PO_4^{3-}.

35.37 Only certain metallic ions give flame tests because not all of them have electronic transitions that fall in the visible spectral range.

35.39 Possible cations include Ag^+, Pb^{2+}, Hg_2^{2+}, Hg^{2+}, Cu^{2+}, SbO^+, Sn^{2+}, Sn^{4+}, Mn^{2+}, Fe^{2+}, Fe^{3+}, Co^{2+}, Ni^{2+}, Al^{3+}, Cr^{3+}, Zn^{2+}, Ba^{2+}, Ca^{2+}, Mg^{2+}, K^+, and Na^+.

(a) No precipitate indicates Hg_2^{2+}, Pb^{2+}, and Ag^+ are absent.

(b) Dark precipitate A indicates HgS, CuS, Sb_2S_3, and SnS_2 could be present.

(c) Dark precipitate C could be MnS, FeS, Fe_2S_3, CoS, NiS, $Al(OH)_3$, $Cr(OH)_3$, and ZnS.

(d) Subsequent treatment indicates Ba^{2+}, Ca^{2+}, Mg^{2+}, K^+, and Na^+ are absent.

(e) No change indicates Sn^{4+} and Sb^{3+} are absent.

(f) Dissolution of precipitate A indicates Hg^{2+} is absent and that Cu^{2+} is probably present.

(g) The blue color of E indicates Cu^{2+} is present.

(h) Both results confirm Cu^{2+}.

(i) That part of C that did not dissolve in HCl but did in HNO_3 indicates Co^{2+} and/or Ni^{2+} is present as well as Mn^{2+}, Fe^{2+}, Al^{3+}, Cr^{3+}, and/or Zn^{2+}.

(j) The green precipitate G indicates the presence of $Ni(OH)_2$; the colorless H indicates the absence of Cr^{3+}.

(k) The tests confirm Ni^{2+}, and indicate the absence of Mn^{2+}, Fe^{2+}, and Co^{2+}.

(l) No effect with NH_3 indicates the absence of Al^{3+}; the white precipitate that dissolved in HCl confirms Zn^{2+}.

The metals present in German silver include copper, nickel, and zinc.

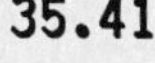

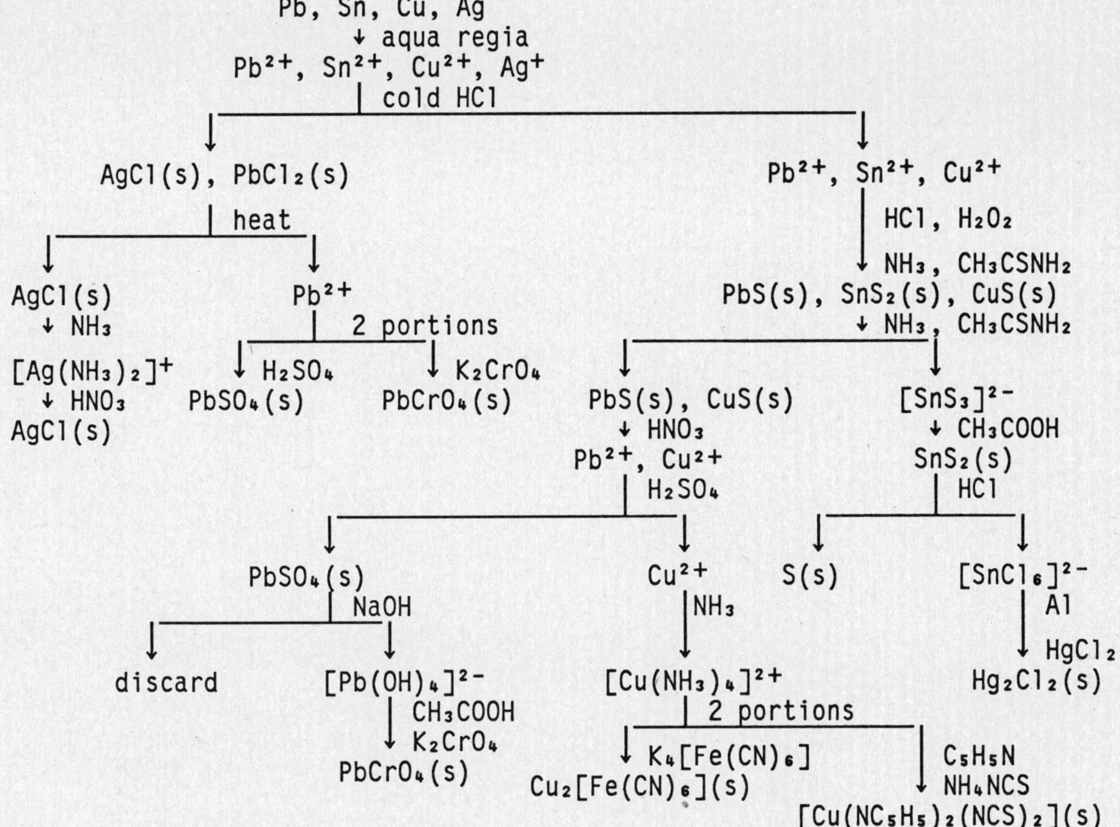

A 9
B 0
C 1
D 2
E 3
F 4
G 5
H 6
I 7
J 8